MODERN DIESEL TECHNOLOGY: PREVENTIVE MAINTENANCE AND INSPECTION

John Dixon

Centennial College, Toronto, Ontario, Canada

DELMAR
CENGAGE Learning™

Australia • Brazil • Japan • Korea • Mexico • Singapore • Spain • United Kingdom • United States

Modern Diesel Technology: Preventive Maintenance and Inspection
John Dixon

Vice President, Editorial: Dave Garza

Director of Learning Solutions: Sandy Clark

Executive Editor: David Boelio

Managing Editor: Larry Main

Senior Product Manager: Sharon Chambliss

Editorial Assistant: Lauren Stone

Vice President, Marketing: Jennifer McAvey

Executive Marketing Manger:
    Deborah S. Yarnell

Marketing Manager: Jimmy Stephens

Marketing Specialist: Mark Pierro

Production Director: Wendy Troeger

Production Manager: Mark Bernard

Content Project Manager: Cheri Plasse

Art Director: Benj Gleeksman

Technology Project Manager:
    Chrstopher Catalina

Production Technology Analyst: Thomas Stover

Library of Congress Control Number: 2007922077

ISBN-13: 978-1-4180-5391-8

ISBN-10: 1-4180-5391-0

Delmar
Executive Woods
5 Maxwell Drive
Clifton Park, NY 12065
USA

Cengage Learning is a leading provider of customized learning solutions with office locations around the globe, including Singapore, the United Kingdom, Australia, Mexico, Brazil, and Japan. Locate your local office at **www.cengage.com/global**

To learn more about Delmar, visit **www.cengage.com/delmar**

Purchase any of our products at your local bookstore or at our preferred online store **www.cengagebrain.com**

Notice to the Reader

Publisher does not warrant or guarantee any of the products described herein or perform any independent analysis in connection with any of the product information contained herein. Publisher does not assume, and expressly disclaims, any obligation to obtain and include information other than that provided to it by the manufacturer. The reader is expressly warned to consider and adopt all safety precautions that might be indicated by the activities described herein and to avoid all potential hazards. By following the instructions contained herein, the reader willingly assumes all risks in connection with such instructions. The publisher makes no representations or warranties of any kind, including but not limited to, the warranties of fitness for particular purpose or merchantability, nor are any such representations implied with respect to the material set forth herein, and the publisher takes no responsibility with respect to such material. The publisher shall not be liable for any special, consequential, or exemplary damages resulting, in whole or part, from the readers' use of, or reliance upon, this material.

Printed in the United States of America
3 4 5 6 7   26 25 24 23 22

Brief Contents

Contents

Preface

The goal for writing this textbook is to give entry-level truck technicians a solid foundation in the area of preventive maintenance. The book starts with general shop safety, hand tools, and fasteners. This is basic knowledge that must be mastered before a technician ever gets to the shop floor. The remaining chapters provide detailed preventive maintenance procedures for specific areas of the vehicle. These chapters contained brief explanations of how the component operates. My beliefs are that if a technician understands how something is supposed to function, he or she will have a greater ability to identify potential problems and have the opportunity.

The text is written in a step-by-step format for the entry-level technician in appropriate language and identifies some of the jargon that is used in the trucking industry.

The second part of the text deals with trailer preventive maintenance including trailer refrigeration units, vans, flatbeds, dry bulk, and bulk liquid tank trailers.

The secondary objective of this book it to cover some of the ASE task objectives. The learning outcome objectives are designed to meet or exceed ASE task objectives, specifically the T8 tasks. This text also covers many of the T1, T2, T3, T4, T5, T6, and T7 ASE tasks.

The text starts with general shop safety. This is one of those important subjects that just can't be stressed enough. It then progresses to hand tools and fasteners that all technicians must master and use every day to perform their job.

The fourth chapter on preventive maintenance explains why this is such an important task in the trucking industry and why it must be strictly adhered to. Preventative maintenance is usually the first job that will be assigned to new technicians in most truck shops. As a technician performing service work will evolve and as experience is gained. Technicians will be able to pick out problems and detect future problems so that the vehicle will hopefully never suffer the dreaded road-side breakdown.

The remaining chapters break the vehicle down in subsystems, explaining the purpose of that system and then the required preventive maintenance that system will require.

This book should be a very usable study resource for entry-level truck technicians and should be a good basis for a basic preventive maintenance program. There is no one preventive program that will be a perfect fit for all trucking companies because all fleets are different and will require a specifically tailored program. This is the job of the service manager.

Although there are many people whom I am deeply indebted to for sharing their insight and experience, it was one of my former students, Ben Wurthman, who really shed some light on the subject of tank trailer maintenance. The trucking industry is very diverse, and in your career as a technician, there are some areas of maintenance that you will never work on. For me, tank trailers were one of them. Ben's general manager, Mr. John Corrigan, graciously allowed me into the PRO-KLEEN tank trailer facility to photograph Ben performing a service on a bulk liquid tank trailer all the while explaining all the procedures in great detail. Thanks again, Ben.

Don Coldwell of Volvo Trucks Canada was also a great resource, providing me with detailed preventive maintenance literature. This is a subject that Don feels is vital to the operation of a fleet or vehicle owner.

I would also like to make special thanks to Al Thompson of Centennial College for taking time out for me to share his ultimate wisdom of photography. Al also photographed a few of the more difficult shots.

I must also thank a very good friend and colleague, David Weatherhead, for all of his technical assistance and computer skills. David was able to save my work when I suffered two computer crashes, in a six-month period. Thanks, Dave.

John Dixon, June 2008

ABOUT THE SERIES

The Modern Diesel Technology (MDT) series has been developed to address a need for *modern,* system-specific text-books in the field of truck and heavy equipment technology. This focused approach gives schools more flexibility in designing programs that target specific ASE certifications. Because each text-book in the series focuses exclusively on the competencies identified by its title, the series is an ideal review and study vehicle for technicians prepping for certification examinations.

Titles in the Modern Diesel Technology Series include:

MDT: Electricity and Electronics, by Joe Bell; ISBN: 1401880134

MDT: Heating, Ventilation, Air Conditioning and Refrigeration, by John Dixon; ISBN: 1401878490

MDT: Electronic Diesel Engine Diagnosis, by Sean Bennett; ISBN: 1401870791

MDT: Brakes, Suspension, and Steering Systems, by Sean Bennett; ISBN: 1418013722

MDT: Heavy Equipment Systems, by Robert Huzij, Angelo Spano, Sean Bennett, and George Parsons; ISBN: 1418009504

MDT: Preventive Maintenance and Inspection, by John Dixon; ISBN: 1418053910

ACKNOWLEDGMENTS

I feel it is important to thank my apprenticeship students for their feedback over the years. While developing this text, I was able to teach from a couple of the sections that were appropriate for the class, allowing me to gather some feedback from the students who are my greatest critics. My rational is that if my students didn't understand a concept, I could try another explanation until they did. Many of my students have been working in the trade for 5 years or more on the front line of new technology. Their feedback was and is paramount to me.

I must thank my wife Connie and our three daughters, Alyzza, Jaymee, and Olyvia, for affording me the time to work on this text at home. I can usually be found in my basement with a laptop surrounded by resource materials. I now have many projects backed up that I must attend to in the near future.

Finally, I must thank Sean Bennett for putting this book series, Modern Diesel Technology, together and inviting me to participate. As always, Sean is a wealth of information, a great mentor, colleague, and friend. Without his encouragement, expertise, and patience, this book would not have been possible.

Individuals

Charles Arsenault, Arsenault Associates

Jim Bardeau, Volvo Training, Centennial College

Sean Bennett, Centennial College

Susan Bloom, Centennial College

Dan Bloomer, Centennial College

Ray Camball, Trailmobile

Mike Cerato, Centennial College

Dan Cushing, Ryder Canada

Don Coldwell, Volvo Trucks Canada, Inc.

Owen Duffy, Centennial College

Boyce H. Dwiggins, Cengage Delmar Author

Helmut Hryciuk, Centennial College

Serge Joncas, Volvo Training

John Kramar, Centennial College

George Liidermann, Freightliner Training

Bob Marshal, University of Northwestern Ohio

Dale McPherson, Cengage Delmar Author

Alan McClelland, Centennial College

John Montgomery, Volvo Trucks Canada, Inc.

David Morgan, Freightliner Training

John Murphy, Centennial College

Daniela Perriccioli, Centennial College

Steve Plaskos, City of Toronto

Ed Roeder, Anderson Haulage LTD

Darren Smith, Centennial College

Angelo Spano, Centennial College

Gino Tamburro, Centennial College

Al Thompson, Centennial College

David Weatherhead, Centennial College

Gus Wright, Centennial College

Contributing Companies

We would like to thank the following companies who provided technical information and art for this book:

Arsenault Associates
ASE
Battery Council International
Freightliner LLC
SAF Holland
Snap-On Tools Company

Technical and Maintenance Council (TMC)
Thermo King Corporation.
Volvo Trucks North America, Inc

.

SUPPLEMENT

An Instructor Resources CD is available with the textbook. Components of the CD include an electronic copy of the Instructor's Guide and an Image Gallery that includes an electronic copy of the images in the book.

General Shop Safety

Objectives

Upon completion and review of this chapter, the student should be able to:

- Use personal safety equipment.
- Demonstrate safe lifting and carrying of heavy objects.
- Explain personal safety warnings as they apply in a truck shop.
- Identify the different fire classifications and how to properly extinguish that fire.
- List the four different classifications of hazardous waste and their respective hazards to your health and the environment.
- Explain the laws regarding hazardous materials, including both the Right-to-Know Law and employee/employer obligations.
- Identify the type of records that must be kept at a maintenance facility that are required by law for commercial vehicles involved in interstate shipping.

Key Terms

corrosive

flammable liquids

hazardous materials

Material Safety Data Sheet (MSDS)

parts requisition

reactive

Resource Conservation and Recovery Act (RCRA)

Right-to-Know Law

solvents

spontaneous combustion

toxic

Workplace Hazardous Material Information Systems (WHMIS)

INTRODUCTION

Not enough can be said about the importance of personal safety. Far too often, injuries that occur in the shop or around heavy trucks are completely avoidable. Applying a few simple rules and common sense can make all the difference ensuring a technician remains accident free through entire career. It is not only your employer's responsibility to make sure you are aware of the dangers that you are working around and ensure proper training, but also your responsibility to make sure you understand them and ask questions whenever in doubt.

SYSTEM OVERVIEW

This chapter deals with issues that a truck technician must understand to keep themselves and their co-workers safe in a shop environment. The proper use of personal safety equipment, shop behavior, fire safety, and hazardous materials are discussed, as well as the importance of keeping accurate shop records.

PERSONAL SAFETY EQUIPMENT

It is important for technicians to have well-maintained safety equipment that is required to perform the intended tasks. It is more important that the technician get into the habit of wearing such equipment and using it properly.

Eye Protection

The human eye is vulnerable to injury from many different things in a shop like dust, metal shavings, and liquids. Grinding or cutting metal creates tiny particles that are thrown off at high speed. Gasses and liquids escaping a broken hose or fuel line fitting can be sprayed great distances under great force. Dirt and corrosion can easily find its way into your eyes while working underneath a vehicle.

Many shops have already adopted the rule of mandatory safety glasses at all times. Although many technicians already have some type of safety glasses, they are often not in the habit of wearing them. Many times the safety glasses are worn on top of a ball cap for show and will provide absolutely no protection for the eyes when worn in this manner. For a person who does not regularly use eyeglasses, wearing safety glasses is only an irritation for the first few days, and it will gradually become second nature to wear them. Technicians should wear their safety glasses routinely, in the same way as wearing a seat belt while in a vehicle.

One problem encountered by technicians is that they try to wear the cheap, ill-fitting, vision-impairing glasses that are often provided to visitors on entry to a shop. The investment in good, well-fitting safety glasses that feel comfortable to you should be considered a tool necessary to perform your job. If the safety glasses protect your eyes even once from an injury over the length of your career, you will consider them a bargain. Just ask anyone who has ever experienced an eye injury (see **Figure 1-1**).

There are many types of eye protection available. The lenses must be made of safely glass or plastic and offer some type of side protection. Regular prescription glasses are not sufficient.

If chemicals such as battery acid, fuel, or solvents enter your eyes, flush them continuously with clean water until you can get medical help.

CAUTION *Persons not wearing safety glasses should never be permitted into the shop area.*

A

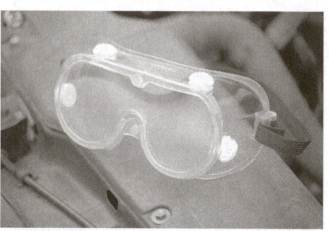

B

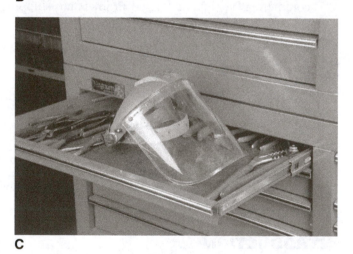

C

Figure 1-1 (A) Safety glasses; (B) splash goggles; (C) face shield. *(Courtesy of Goodson Tools & Supplies for Engine Builders)*

Footwear

Foot protection must be worn in any shop situation. Service work includes the handling of heavy components that, if accidentally dropped, could severely damage your foot or, if sharp objects are

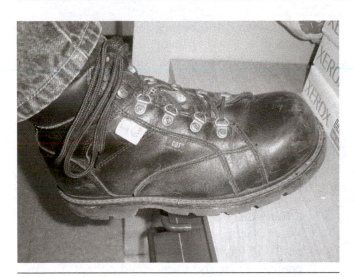

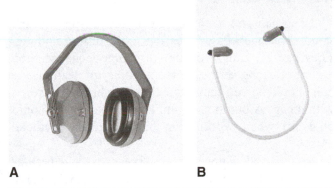

Figure 1-3 Typical (A) ear muffs and (B) ear plugs. *(Courtesy of Dalloz Safety)*

Figure 1-2 Safety boots identified by a green triangle. *(Courtesy of John Dixon)*

accidentally stepped on, may puncture the bottom of your foot. Always wear steel-toe, steel-shank, safety boots or shoes with nonslip soles. Approved footwear can be identified by a green triangle placed somewhere on the boot or shoe (see **Figure 1-2**).

CAUTION *Persons wearing athletic shoes, street shoes, or sandals should never be permitted into the shop area.*

Hand Protection

Technicians should wear heavy, properly fitting work gloves to protect the hands from cuts and abrasions as well as burns from chemicals or high temperature components. Injuries to the hands can seriously compromise your ability to work for many days. Work gloves should be worn during operations such as cutting and welding or handling caustic chemicals such as batteries.

CAUTION *Never wear gloves around moving equipment such as stationary grinders or drill presses that could accidentally snag a glove, resulting in severe hand injuries.*

Hearing Protection

Hearing damage in the workplace is completely avoidable. Wearing simple ear plugs or earphone-type protectors is usually all that is required. Exposure to high noise levels for extended periods of time can lead to permanent ear damage and hearing loss. Always wear hearing protection when working around engine

or chassis dynamometers or operating tools such as impact guns of air hammers (see **Figure 1-3**).

Clothing

It is important for technicians to look professional while on the job. It is even more important that the clothing they wear be well-fitting, comfortable, and durable. Loose-fitting clothing can get caught on moving components or shop tooling, causing serious personal injury or death. Neckties should never be worn while working in a shop environment. Many service technicians prefer to wear coveralls or shop coats to protect their personal clothing. Cut-offs and short pants do not offer the protection required to work in a shop environment (see **Figure 1-4**).

Figure 1-4 Shop coats can be worn to protect clothing.

Hair and Jewelry

Long hair and dangling jewelry can result with the same consequences as wearing loose-fitting clothing. They can get caught on moving components or shop tooling, causing serious personal injury or death. Long hair should be tied securely behind the head or covered with a cap. A bump cap (similar to a hard hat) should be worn when working in pits or under overhead hoists.

Remove all watches, rings, bracelets, and necklaces. These items can also be caught in moving parts, causing serious injury or electrically arc on live circuits.

SAFE LIFTING AND CARRYING

To avoid personal injury, technicians should know the proper technique for lifting and carrying heavy objects. You should never exceed your own physical abilities and seek help from others when you are unsure if you can handle the size or weight of the material or object. Even small, compact auto parts can be surprisingly heavy or unbalanced. Always size up the task before lifting an object. When lifting any object, follow these steps (see **Figure 1-5**).

1. Place your feet close to the load, making sure that you are balanced properly.
2. Keep your back and elbows as straight as possible. Bend your knees until your hands reach the best place for getting a strong grip on the ends of the load.
3. When the component is shipped in a cardboard box, ensure that the box is in good condition. If the box is damp or inadequately sealed, it could split open during the lift, allowing a heavy component to fall through the side or bottom of the box, causing injury or damage.
4. Grasp the component, keeping it close to your body, and lift by straightening your legs. Using your leg muscles instead of your back muscles can prevent back injury.
5. If you intend to change direction while carrying a load, do not twist at the hips; instead do so by turning the whole body, including the feet.
6. When placing an object on a counter or shelf, do not bend forward. Instead place the edge of the component on the surface and slide it forward, paying attention not to pinch your fingers.
7. When lowering a load back down, do not bend forward at the hips, because this will strain the back muscles. Instead, keep your back straight and bend at the knees.
8. Use wood blocks to protect your fingers when picking up heavy objects or lowering them to the floor.

PERSONAL SAFETY WARNINGS

Technicians must act professionally while working in a truck shop environment. Horseplay in a shop can be extremely dangerous and could cause someone to make an unplanned trip to the hospital. Such things as air nozzle or grease gun fights, creeper races, and practical jokes have no place in a truck shop.

Smoking is also an activity that should not be done while working on any vehicle or machinery in a shop. Fumes from fuel, lubricants, or batteries can easily be ignited by a cigarette, causing serious personal injury or death.

When tilting the cab of a truck, make sure the area in front of the truck is clear of any obstructions and people. Make sure the truck engine is not running while tilting the cab because the action of tilting the cab could cause the transmission to shift into gear, allowing the vehicle to move, causing an accident that could result in serious personal injury, death, or property damage (see **Figure 1-6**).

Cutting and welding require the use of a welding helmet or goggles equipped with the properly shaded lens for the task being performed. These devices protect the eyes and face from airborne molten steel and harmful light rays. Never use any welding equipment that you have not been specifically trained on (see **Figure 1-7**).

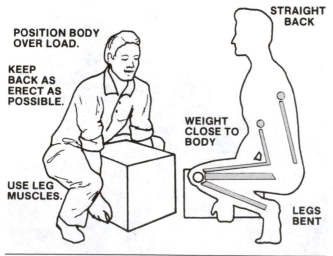

Figure 1-5 Use your leg muscles, never your back, when lifting any heavy load.

Figure 1-6 The tilting of a hood must be undertaken with care. Make sure there is adequate clearance in front of the vehicle and that the area is free of people and all objects. Do not tilt a cab with the engine running because it could cause the transmission to go into gear. With the engine running, the vehicle could move causing an accident that could result in personal injury, death, and property damage.

A

B

Figure 1-7 (A) Welding helmet and (B) oxyacetylene goggles.

Trucks that have just come in from the road or have been running can cause serious burns. Avoid contact with components such as exhaust manifolds, turbochargers, mufflers, tailpipes, and radiators.

Whenever you work with a hydraulic press, make sure you work safely, ensuring the component is properly supported to prevent slippage or breakage. If possible, stand to one side of the press and always wear safety glasses or goggles.

Tools and parts should always be properly stored after every use. This not only keeps the workplace neat and tidy but also reduces the tripping hazards involved with leaving tools or components on the shop floor. There is also the reduction of time wasted by technicians looking all over the shop for misplaced tools or parts.

SHOP TOOLS

Understanding the proper use of shop tools can eliminate accidents right from the start. These tools include hand tools and handheld electric and pneumatic power tools as well as stationary equipment. The following is a list of rules that should be followed whenever you work with shop tools:

- Use tools only for the task they are designed for.
- Chose the correct size of tool for the job.
- Always keep tools in safe working condition.
- Store tools properly when not in use.
- Report any breakage or malfunctions to your instructor or service manager.
- Never use tools with broken, loose, or missing handles.
- Ensure cutting tools are sharpened properly and are kept in good condition.
- Never use a tool that you don't know how to operate. Seek proper training first.

Only the people using tools can prevent accidents that injure themselves or personnel in the vicinity. Knowing what a tool is designed to do and how to use it safely is the key.

WORK AREA SAFETY

The work area should always be kept safe. All surfaces should be kept clean, dry, and orderly. Spilled liquids such as oil, coolant, or grease on the shop floor can cause slips, resulting in serious personal injury. Be sure to use a commercially available oil absorbent when cleaning up oil spills. Water should never be allowed to lie on the shop floor because it can conduct electricity and can also be slippery. Walking paths throughout the shop should be kept clean and wide enough for safe clearance. Ensure there is enough working space around any machines. Also, keep workbenches clean and orderly.

Adequate ventilation is another safety concern that must be addressed in all shops. While diesel engines produce less carbon monoxide (CO) fumes than gasoline engines, their exhaust pipes must be connected to an operating shop exhaust extraction system because of the harmful particulates they produce. Deadly CO can also be produced by space heaters used in some shops. These devices must be periodically inspected to make sure they are adequately vented and have not become blocked. Proper ventilation is also important in any area where flammable chemicals are used or where batteries are being charged.

An up-to-date list of emergency phone numbers should be posted by the telephone. These numbers should include a doctor, hospital, ambulance, and fire and police departments. All work areas should have a first-aid kit for treating minor injuries. An eyewash station should also be accessible to the shop area to flush the eyes when required.

Diesel fuel and solvents are **flammable liquids.** Flammable liquids area easily ignited. For this reason, solvents or diesel fuel should be stored in an approved container and never used to wash hands or tools.

All **solvents** (chemicals that dissolve other substances) should be handled with care to avoid spillage. Containers should be kept tightly sealed except when dispensing the product. Exercise extreme caution when dispensing or moving solvents in bulk containers (see **Figure 1-8**). Static electricity can build up to a level that could create a spark, resulting in an explosion. Discard or clean any empty solvent containers; fumes inside an empty container can start a fire or explosion. Do not allow any open flame or spark to occur anywhere near flammable solvents or chemicals, including battery acids. Solvents and other combustible materials must be stored in approved storage cabinets or rooms that have adequate ventilation (see **Figure 1-9**). Combustible materials should never be stored around exits or stairways.

Oily rags should also be stored in an approved metal container. When oily, greasy, or paint-soaked rags are left lying about or are stored improperly, they are possible sources of **spontaneous combustion**, a fire that starts by itself.

Drain covers on the shop floor should be secured in place flush to the shop floor to eliminate tripping hazards.

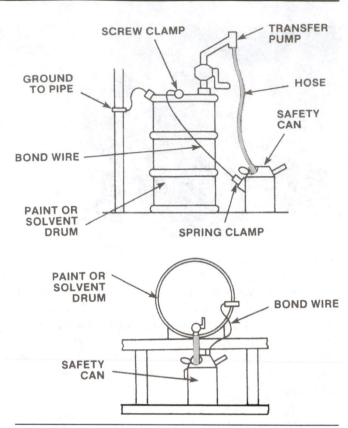

Figure 1-8 Safe methods of transferring flammable materials from bulk storage. (*Courtesy of DuPont Automotive Finishes*)

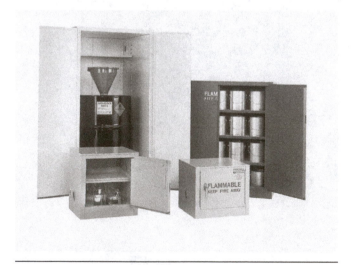

Figure 1-9 Store combustible materials in approved safety cabinets. (*Courtesy of the Protectoseal Company*)

FIRE SAFETY

Ensure that you know where to locate and how to properly operate firefighting equipment in the work area (see **Figure 1-10**). All fires are classified in one or more of the following categories.

Figure 1-10 Typical fire extinguisher that can be used on A, B, or C type fires. *(Courtesy of John Dixon)*

Class A

Fires in which the burning materials are ordinary combustibles, such as paper, wood, cloth, or trash. Putting out this type of fire requires drowning with water or foam solutions containing a high percentage of water, or a multipurpose dry chemical extinguisher.

Class B

Fires in which the burning material is a liquid, such as gasoline, diesel fuel, oil, grease, or solvents. Extinguishing this type of fire requires a smothering action.

Class C

Fires in which the burning material is "live" electrical equipment: motors, switches, generators, transformers, or general wiring. Extinguishing this type of fire requires a nonconductive smothering action, such as carbon dioxide or dry chemical extinguisher. Do not use water on this type of fire.

Class D

Fires in which the burning materials are combustible metals. Special extinguishing agents are required to put out this type of fire (see **Figure 1-10**).

The following are some general tips for the proper operation of various fire extinguishers based on the type of extinguishing agent they use:

- **Foam:** Do not spray the jet into the burning liquid. Allow the foam to fall lightly on the fire.
- **Carbon dioxide:** Direct discharge as close to the fire as possible, first at the edge of the flames and gradually forward and upward.
- **Soda-acid, gas cartridge:** Direct stream at the base of the flame.
- **Pump tank:** Place foot on footrest and direct the stream at the base of the flame. Hand pump tanks are also available.
- **Dry chemical:** Direct at the base of the flame. In the case of Class A fires, follow up by directing the dry chemicals at the remaining material that is burning.

CAUTION *Once a fire extinguisher has been used, it must be reported immediately so that it may be replaced.*

HAZARDOUS MATERIALS

While working in a shop environment, technicians can come into contact with many **hazardous materials.** Hazardous materials are products that can cause harm to a person's well-being.

Hazardous materials can also damage and pollute land, air, or water. There are four types of hazardous waste:

- **Flammable:** These are materials that can easily catch fire.
- **Corrosive:** These materials are so caustic that they can dissolve metals and burn skin and eyes.
- **Reactive:** These materials will become unstable (burn, explode, or give off toxic fumes) if mixed with air, water, heat, or other materials.
- **Toxic:** These materials can cause illness or death after being inhaled or contacting the skin.

Laws Regulating Hazardous Materials

The Occupational Safety and Health Administration (OSHA) administers a law called the

	Class of Fire	Typical Fuel Involved	Type of Extinguisher
Class △A△ **Fires** (green)	**For Ordinary Combustibles** Put out a Class A fire by lowering its temperature or by coating the burning combustibles.	Wood Paper Cloth Rubber Plastics Rubbish Upholstery	Water*[1] Foam* Multipurpose dry chemical[4]
Class ☐B☐ **Fires** (red)	**For Flammable Liquids** Put out a Class B fire by smothering it. Use an extinguisher that gives a blanketing, flame-interrupting effect; cover whole flaming liquid surface.	Gasoline Oil Grease Paint Lighter fluid	Foam* Carbon dioxide[5] Halogenated agent[6] Standard dry chemical[2] Purple K dry chemical[3] Multipurpose dry chemical[4]
Class ◯C◯ **Fires** (blue)	**For Electrical Equipment** Put out a Class C fire by shutting off power as quickly as possible and by always using a nonconducting extinguishing agent to prevent electric shock.	Motors Appliances Wiring Fuse boxes Switchboards	Carbon dioxide[5] Halogenated agent[6] Standard dry chemical[2] Purple K dry chemical[3] Multipurpose dry chemical[4]
Class ☆D☆ **Fires** (yellow)	**For Combustible Metals** Put out a Class D fire of metal chips, turnings, or shavings by smothering or coating with a specially designed extinguishing agent.	Aluminum Magnesium Potassium Sodium Titanium Zirconium	Dry powder extinguishers and agents only

*Cartridge-operated water, foam, and soda-acid types of extinguishers are no longer manufactured. These extinguishers should be removed from service when they become due for their next hydrostatic pressure test.

Notes:

(1) Freezes in low temperatures unless treated with antifreeze solution, usually weighs over 20 pounds (9 kg), and is heavier than any other extinguisher mentioned.

(2) Also called ordinary or regular dry chemical (sodium bicarbonate).

(3) Has the greatest initial fire-stopping power of the extinguishers mentioned for class B fires. Be sure to clean residue immediately after using the extinguisher so sprayed surfaces will not be damaged (potassium bicarbonate).

(4) The only extinguishers that fight A, B, and C classes of fires. However, they should not be used on fires in liquefied fat or oil of appreciable depth. Be sure to clean residue immediately after using the extinguisher so sprayed surfaces will not be damaged (ammonium phosphates).

(5) Use with caution in unventilated, confined spaces.

(6) May cause injury to the operator if the extinguishing agent (a gas) or the gases produced when the agent is applied to a fire is inhaled.

Figure 1-11 Guide to fire extinguisher selection.

Right-to-Know Law. The law mandates that any company that uses or produces hazardous material must inform their employees, customers, and vendors of any potential hazards that could result by using the product.

It is most important that you keep yourself informed about the safe handling of the materials that you come in contact with in your workplace. You are the only one who can protect yourself and keep your co-workers safe from the harms of hazardous materials.

The following are some of the major topics of the Right-to-Know Law:

- You have the right to know what hazards you may face on the job.
- Your have a right to learn about these materials and how to protect yourself from them.
- You cannot be fired or discriminated against for requesting information and training on how to handle hazardous materials.

- You have the right for your doctor to receive the same hazardous material information that you receive.

Employee/Employer Obligations

It is your responsibility to familiarize yourself with the **Material Safety Data Sheet (MSDS) (see Figure 1-12)**. By familiarizing yourself with the MSDS, you are much more likely to avoid personal injury. Your employer or school facility that uses hazardous materials has the responsibility to:

- Provide a safe workplace.
- Educate employees about the hazardous materials they will encounter while on the job.
- Recognize, understand, and use warning labels and MSDS and **Workplace Hazardous Material Information Systems (WHMIS)** training.
- Provide personal protective clothing and equipment and train employees to use them properly.

As an employee or student, it is your responsibility to:

- Read the warning labels on the materials.
- Follow the instruction and warnings on the MSDS or WHMIS.
- Take the time to learn to use protective equipment and clothing.
- Use common sense when working with hazardous materials.
- Ask the service manager if you have any question about a hazardous material.

Personal Protection from Hazardous Materials

The long-term effects of exposure to hazardous materials should be of great concern for all personnel working in a truck shop. These products may include solvents, cleaning agents, and paint products. These concerns have become evident over the concerns with the use of asbestos for various truck parts. When first introduced on the market, asbestos was widely used in the manufacture of brake pads, shoes, and clutches. It is now widely known that asbestos fibers pose a health risk and that long-term exposure to even small amounts can cause lethal heath problems. Consequently, asbestos has been gradually phased out of both the automotive and truck parts markets.

In order for you as a technician to handle hazardous materials safely, it is important that you:

- Know what the material is.
- Know the material is dangerous.
- Know the correct safety equipment needed for working with the material.
- Know how to use the safety equipment properly (see **Figure 1-13**).
- Make sure that the safety equipment fits properly and is in good working condition.

A person's own personal hygiene can also have a big effect on minimizing the risk of exposure to hazardous materials. Some key ways of minimization:

- Do not smoke while working around hazardous materials.
- Wash hands before eating.
- Shower after working.
- Change into work clothing when you get to work and back into your street clothing after work. Never bring your work clothes home with you.

Hazardous Waste Handling and Disposal

There are specific laws governing the disposal of hazardous wastes. It is your responsibility and your shop's responsibility to be aware of how these laws apply to your situation.

These laws include the **Resource Conservation and Recovery Act (RCRA)**. This law states that after you have used hazardous materials, they must be properly stored until an approved hazardous waste hauler arrives to take them to the disposal site (see **Figure 1-14**). In addition, your responsibility continues until the materials arrive at an approved disposal site and are processed in accordance with the law.

When dealing with hazardous wastes:

- Consult the MSDS or WHMIS under the "Waste Disposal Method" category.
- Check with your instructor or service manager for the exact method for correct storage and disposal.
- Follow their recommendations exactly.
- Absolutely never dispose of hazardous materials into a dumpster.
- Absolutely never dump waste anywhere but into a collection site of a licensed facility.
- Absolutely never pour waste down drains, toilets, sinks, or floor drains.
- Absolutely never use hazardous waste to kill weeds or to control dust on gravel roads.

MATERIAL SAFETY DATA SHEET

24 Hour Emergency Phone (316) 524-5751

Division of Vulcan Materials Company / P. O. Box 530390 • Birmingham, AL 35253-0390

I – IDENTIFICATION

CHEMICAL NAME	CHEMICAL FORMULA	MOLECULAR WEIGHT
Sodium Hydroxide Solution	NaOH	40.00

TRADE NAME
Caustic Soda, 73%, 50% and Weaker Solutions

SYNONYMS	DOT IDENTIFICATION NO.
Liquid Caustic, Lye Solution, Caustic, Lye, Soda Lye	UN 1824

II– PRODUCT AND COMPONENT DATA

COMPONENT(S) CHEMICAL NAME	CAS REGISTRY NO.	% (wt.) Approx.	OSHA PEL
Sodium Hydroxide	1310-73-2	73, 50 and less	2 mg/m³ Ceiling

Note: This Material Safety Data Sheet is also valid for caustic soda solutions weaker than 50%. The boiling point, vapor pressure, and specific gravity will be different from those listed.

* Denotes chemical subject to reporting requirements of Section 313 of Title III of the 1986 Superfund Amendments and Reauthorization Act (SARA) and 40 CFR Part 372

III – PHYSICAL DATA

APPEARANCE AND ODOR	SPECIFIC GRAVITY
Colorless or slightly colored, clear or opaque; odorless	50% Solution: 1.53 @ 60°F/60°F 73% Solution: 1.72 @ 140°F/4°F

BOILING POINT	VAPOR DENSITY IN AIR (Air = 1)
50% Solution: 293°F (145°C) 73% Solution: 379°F (192.8°C)	N/A

VAPOR PRESSURE	% VOLATILE, BY VOLUME
50% = 6.3 mm Hg @ 104°F 73% = 6.0 mm Hg @ 158°F	0

EVAPORATION RATE	SOLUBILITY IN WATER
0	100%

IV – REACTIVITY DATA

STABILITY	CONDITIONS TO AVOID Mixture with water, acid or incompatible materials can cause splattering and release of large amounts of heat (Refer to Section VIII). Will react with some metals forming flammable hydrogen gas.
Stable	

INCOMPATIBILITY (Materials to avoid)
Chlorinated and fluorinated hydrocarbons (i.e. chloroform, difluoroethane), acetaldehyde, acrolein, aluminum, chlorine trifluoride, hydroquinone, maleic anhydride, phosphorous pentoxide and tetrahydrofuran.

HAZARDOUS DECOMPOSITION PRODUCTS

Will not decompose

HAZARDOUS POLYMERIZATION

Will not occur

Figure 1-12 Typical Material Safety Data Sheet (MSDS). *(Courtesy of Vulcan Materials Company)*

V – FIRE AND EXPLOSION HAZARD DATA

FLASHPOINT (Method used)	FLAMMABLE LIMITS IN AIR
None	None

EXTINGUISHING AGENTS

N/A   NFPA Hazard Ratings: Health 3; Flammability 0; Reactivity 1

UNUSUAL FIRE AND EXPLOSION HAZARDS

Firefighters should wear self-contained positive pressure breathing apparatus, and avoid skin contact. Refer to Reactivity Data, Section IV.

VI – TOXICITY AND FIRST AID

EXPOSURE LIMITS (When exposure to this product and other chemicals is concurrent, the exposure limit must be defined in the workplace.)

ACGIH: 2 mg/m^3 Ceiling

OSHA 2 mg/m^3 Ceiling

IDLH: 250 mg/m^3

Effects described in this section are believed not to occur if exposures are maintained at or below appropriate TLVs.
Because of the wide variation in individual susceptibility, these exposure limits may not be applicable to all persons and those with medical conditions listed below.

MEDICAL CONDITIONS AGGRAVATED BY EXPOSURE

May aggravate existing skin and/or eye conditions on contact.

ACUTE TOXICITY   Primary route(s) of exposure: ☒ Inhalation   ☒ Skin Absorption   ☐ Ingestion

Inhalation: Inhalation of solution mist can cause mild irritation at 2 mg/m^3. More severe burns and tissue damage at the upper respiratory tract, can occur at higher concentrations. Pneumonitis can result from severe exposures.

Skin: Major potential hazard - contact with the skin can cause severe burns with deep ulcerations. Contact with solution or mist can cause multiple burns with temporary loss of hair at burn site. Solutions of 4% may not cause irritation and burning for several hours, while 25 to 50% solutions can cause these effects in less than 3 minutes.

Eyes: Major potential hazard - Liquid in the eye can cause severe destruction and blindness. These effects can occur rapidly effecting all parts of the eye. Mist or dust can cause irritation with high concentrations causing destructive burns.

Ingestion: Ingestion of sodium hydroxide can cause severe burning and pain in lips, mouth, tongue, throat and stomach. Severe scarring of the throat can occur after swallowing. Death can result from ingestion.

FIRST AID

Inhalation: Move person to fresh air. If breathing stops, administer artificial respiration. Get medical attention immediately.

Skin: Remove contaminated clothing immediately and wash skin thoroughly for a minimum of 15 minutes with large quantities of water (preferably a safety shower). Get medical attention immediately.

Eyes: Wash eyes immediately with large amounts of water (preferably eye wash fountain), lifting the upper and lower eyelids and rotating eyeball. Continue washing for a minimum of 15 minutes. Get medical attention immediately.

Ingestion: If person is conscious, give large quantities of water to dilute caustic. Do not induce vomiting. Get medical attention immediately. Do not give anything by mouth to an unconscious person.

Figure 1-12 (*continued*)

CHRONIC TOXICITY

No known chronic effects

Carcinogenicity: No studies were identified relative to sodium hydroxide and carcinogenicity. Sodium hydroxide is not listed on the IARC, NTP or OSHA carcinogen lists.

Reproductive Toxicity: No studies were identified relative to sodium hydroxide and reproductive toxicity.

VII – PERSONAL PROTECTION AND CONTROLS

RESPIRATORY PROTECTION

Where concentrations exceed or are likely to exceed 2mg/m^3 use a NIOSH/MSHA approved high-efficiency particulate filter with full facepiece or self-contained breathing apparatus. Follow any applicable respirator use standards and regulations.

VENTILATION

As necessary to maintain concentration in air below 2 mg/m^3 at all times.

SKIN PROTECTION

Wear neoprene, PVC, or rubber gloves; PVC rain suit; rubber boots with pant legs over boots.

EYE PROTECTION

Chemical goggles which are splashproof and faceshield.

HYGIENE

Avoid contact with skin and avoid breathing mist. Do not eat, drink, or smoke in work area. Wash hands prior to eating, drinking, or using restroom. Any protective clothing or shoes which become contaminated with caustic should be removed immediately and thoroughly laundered before wearing again.

OTHER CONTROL MEASURES

Safety shower and eyewash station must be located in immediate work area. To determine the exposure level(s), monitoring should be performed regularly.
NOTE: Protective equipment and clothing should be selected, used, and maintained according to applicable standards and regulations. For further information, contact the clothing or equipment manufacturer or the Vulcan Chemicals Technical Service Department.

Figure 1-12 (continued)

VIII – STORAGE AND HANDLING PRECAUTIONS

Follow protective controls set forth in Section VII when handling this product.
Store in closed, properly labeled tanks or containers. Do not remove or deface labels or tags.
When diluting with water, slowly add caustic solution to the water. Heat will be produced during dilution.
Full protective clothing, goggles and faceshield should be worn. Do not add water to caustic because
excessive heat formation will cause boiling and spattering.
Contact of caustic soda cleaning solutions with food and beverage products (in enclosed vessels or
spaces) can produce lethal concentrations of carbon monoxide gas. Do not enter confined spaces such as tanks
or pits without following proper entry procedures as required by 29 CFR 1910.146.

SARA Title III Hazard Categories: Immediate Health.

IX – SPILL, LEAK AND DISPOSAL PRACTICES

STEPS TO BE TAKEN IN CASE MATERIAL IS RELEASED OR SPILLED

Cleanup personnel must wear proper protective equipment (refer to Section VII). Completely contain
spilled material with dikes, sandbags, etc., and prevent run-off into ground or surface waters or sewers.
Recover as much material as possible into containers for disposal. Remaining material may be diluted with
water and neutralized with dilute hydrochloric acid. Neutralization products, both liquid and solid, must be
recovered for disposal. Reportable Quantity (RQ) is 1000 lbs. Notify National Response Center (800/424-8802)
of uncontained releases to the environment in excess of the RQ.

WASTE DISPOSAL METHOD

Recovered solids or liquids may be sent to a licensed reclaimer or disposed of in a permitted waste management facility. Consult federal, state, or local disposal authorities for approved procedures.

X – TRANSPORTATION

DOT HAZARD CLASSIFICATION

Sodium Hydroxide Solution, 8, UN 1824, PG II, RQ

PLACARD REQUIRED

Corrosive, 1824, Class 8

LABEL REQUIRED

Corrosive, Class 8. Label as required by OSHA Hazard Communication Standard, and any applicable state
and local regulations.

Medical Emergencies

**Call collect 24 hours a day
for emergency toxicological
information 415/821-5338**

Other Emergency information

Call 316/524-5751 (24 hours)

DATE OF PREPARATION: November 1, 1993

For any other information contact:

**Vulcan Chemicals
Technical Service Department
P.O. Box 530390
Birmingham, AL 35253-0390
800/873-4898
8 AM to 5 PM Central Time
Monday Through Friday**

Figure 1-12 (*continued*)

Figure 1-13 Wear proper safety equipment when handling hazardous waste. *(Courtesy of DuPont Automotive Finishes)*

WARNING *A service facility is ultimately responsible for the safe disposal of hazardous wastes, even after the waste leaves the shop. In the event of an emergency hazardous waste spill, contact the National Response Center (1-800-424-8802) immediately. Failure to do so can result in a fine, a year in jail, or both.*

Figure 1-14 Many automotive shops hire full-service haulers for hazardous waste removal. *(Courtesy of DuPont Automotive Finishes)*

SHOP RECORDS

The law requires that certain records be kept by service facilities if the trucks being serviced are involved in interstate shipping. These records include the following:

- Identification of each vehicle (VIN) including the company unit number, model, serial number, year, and tire size.
- A schedule showing the nature and due date of the various inspections and maintenance to be performed.
- A record of the nature and date of inspections, maintenance, and repairs made.
- Lubrication record.

Service records are important even if the truck is not involved in interstate shipping. These records are important because:

- They are required by the Department of Transportation (DOT).
- Component failures can be tracked by manager, not only to alert them of problems but also to highlight components that have performed well.
- A service manager cannot develop a good preventative maintenance (PM) program without the use of these records.
- If a serious accident occurs, a vehicle's maintenance records are usually beneficial in the defense of a lawsuit.

Vehicle Identification Number (VIN)

The VIN is a series of letters and numbers unique to the vehicle for which it registered to. The Federal Motor Vehicle Safety Standard (FMVSS) specified that all vehicles in the United States be assigned a VIN. The VIN is located on the left frame rail over the front axle and on the Vehicle Specification Decal (verify location of decal with driver's manual). Heavy duty trucks are assigned a 17-character VIN. Using a combination of letters and numerals, the VIN codes the vehicle make, series or type, application, chassis, cab, axle configuration, gross vehicle weight rating (GVWR), engine type, model year, manufacturing plat location, and production serial number. A check digit (ninth position) is determined by assignment of weighted values to the other 16 characters. These weighted values are processed through a series of equations designed to check the validity of the VIN and to detect VIN alteration. The VIN can also be accessed off the chassis data bus in current trucks (see **Figure 1-15**).

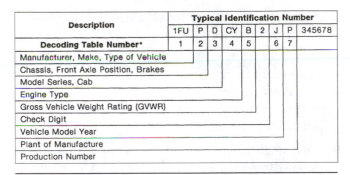

Description	Typical Identification Number							
	1FU	P	D	CY	B	2	J	P	345678
Decoding Table Number*	1	2	3	4	5		6	7
Manufacturer, Make, Type of Vehicle								
Chassis, Front Axle Position, Brakes								
Model Series, Cab								
Engine Type								
Gross Vehicle Weight Rating (GVWR)								
Check Digit								
Vehicle Model Year								
Plant of Manufacture								
Production Number								

Figure 1-15 Chart explaining a typical VIN number.

Work Orders

A work order or repair order is used in most shops and is also used by more than one person, usually the truck driver, service manager, and the technician. The driver lists complaints that are noted during a trip, and this information is relayed to the service manager, who then writes this information on the work order. The service manager should also write down any other scheduled maintenance that should also be performed at this time. The service manager then passes the work order over to the technician, who performs any scheduled services or repairs at this time. The technician notes everything that he has done to the vehicle.

Depending on the shop, the parts used by the technician may be noted on the work order or they may be listed on a **parts requisition.** A parts list also identifies the vehicle VIN or company identification folder.

Most fleets maintain a file on every vehicle, whether it be tractor, trailer, or dolly. This file includes all the vehicle maintenance and repair records, schedule, and PM inspection results. Work orders performed on the particular vehicle are generally kept in its history folder.

FMVSS regulations require that the mandatory records of a vehicle file be retained where the vehicle is housed or maintained for a period of one year. They also require these records to be kept for six months after the vehicle has left the carrier's control. The same regulations require that the driver's vehicle condition report be retained for a period of three months from the date of the report.

LIFT AND HOIST SAFETY

The safe lifting of a heavy duty truck or trailer by a lift or frame machine requires special care on the part of the technician. Specialized adapters and hoist plates must be positioned correctly on twin post and rail type lifts to prevent damage to components underneath the vehicle. Each vehicle has specified lifting points that must be used to evenly support the weight of the vehicle. These locations for lifting can be found in the vehicle service literature. Before operating a lift or alignment machine, carefully read the operator's manual supplied by the manufacturer of the equipment and understand all the operating and maintenance instructions.

Heavy truck components such as engines or transmissions should be lifted using chain hoists or cranes. These devices must be properly attached to the part being lifted in order to prevent serious personal injury. Use bolts and lift shackles with adequate strength for the load rating of the component you intend to lift. Secure the lifting chain or cable to the component that is to be lifted.

The following are some general rules for using jacks, lifts, frame machines, and hoists:

- Do not allow anyone to remain in a vehicle when it is being raised.
- Make sure you know how to operate the equipment and that you know its limitations.
- Never overload a lift, hoist, or jack.
- Chain hoists and cranes must be properly attached to the parts being lifted. Use bolts and lift shackles with adequate strength for the load rating of the component you intend to lift.
- Mechanical locks or stands must be engaged after lifting a truck on any kind of hoist.
- Do not use any lift, hoist, or jack that you believe to be defective or not operating properly. Tag it and report it to your instructor or service manager immediately.
- Make sure all personnel and obstructions are clear before raising or lowering and engine or vehicle.
- Avoid working, walking, or standing under suspended objects that are not mechanically supported.

Summary

- It is important that a technician get into the habit of wearing safety equipment and using it properly.
- Many shops have already adopted the rule of mandatory safety glasses at all times.
- Always wear steel-toe, steel-shank, safety boots or shoes with nonslip soles.

- Work gloves should be worn during operations such as cutting and welding, or handling caustic chemicals such as batteries.
- Always wear hearing protection when working around engine or chassis dynamometers or operating tools such as impact guns on air hammers.

- Loose-fitting clothing can get caught on moving components or shop tooling. Neckties should never be worn while working in a shop environment.

- Technicians should know the proper technique for lifting and carrying heavy objects and never exceed their own physical abilities.

- Horseplay in a shop can be extremely dangerous and could cause someone to make an unplanned trip to the hospital.

- Understanding the proper use of shop tools can eliminate accidents right from the start.

- The work area should always be kept safe. All surfaces should be kept clean, dry, and orderly.

- Emergency telephone numbers and a first-aid kit should be kept handy at all times.

- Ensure that you know where to locate and how to properly operate firefighting equipment in the work area. Use water or foam on ordinary combustibles; foam, carbon dioxide, or dry chemicals on burning liquids; carbon dioxide or dry chemicals on burning "live" electrical equipment; and special extinguishing agents on burning metals.

- Technicians can come into contact with many hazardous materials. Hazardous materials are products that can cause harm to a person's well-being.

- Your employer is obligated to inform you of potential hazards in your workplace, and you have a right to protect yourself from them.

- Specific laws govern the disposal of hazardous wastes, including oil, antifreeze/coolant, refrigerants, batteries, battery acids, acids and solvents used for cleaning, and paint and body repair products wastes. Hazardous wastes may be recycled in the shop or removed by a licensed disposal hauler.

- The law requires that certain records be kept by service facilities if the trucks being serviced are involved in interstate shipping.

Review Questions

1. To prevent eye injury, the best method is to:

   A. take care when performing tasks like grinding or using tools that throw off particles.

   B. always wear safety glasses.

   C. always wear a bump cap.

   D. ensure that there is a source of clean running water available to flush foreign particles out of the eyes.

2. Which of the following describes a safe lifting and carrying practice?

   A. Twist your body when changing your direction of travel while carrying a heavy object.

   B. Bend forward when placing heavy objects on a shelf or counter.

   C. Lift by bending and then straightening your legs, rather than by using your back.

   D. Position your feet as far as possible from the load when you begin to lift.

3. Oil spills on the shop floor should be cleaned up using a _____.

4. The truck's exhaust pipe should be connected to a shop exhaust system whenever the vehicle engine is running inside the shop to protect personnel from:

   A. carbon monoxide.

   B. carbon dioxide.

   C. fire.

   D. particulates.

5. Do **not** attempt to put out a Class B fire using:

   A. foam.

   B. carbon dioxide.

   C. a dry chemical type extinguisher.

   D. water.

6. A Class C fire involves:

    A. ordinary combustibles, such   C. live electrical equipment.
       as paper or cloth.
                                             D. combustible metals.
    B. a flammable liquid.

7. When using carbon dioxide as a extinguishing agent for a fire, direct the discharge:

    A. at the top of the flames.   C. at the base of the flames.

    B. first at the edge of the fire,   D. several feet over the top of the flames.
       then forward and upward.

8. List four types of hazardous wastes that are discussed in this chapter.

9. The Right-to-Know Law was passed by the government to:

    A. require any company that   C. require industries to compensate employees injured by contact
       disposes of hazardous materials   with hazardous materials.
       to inform its community.
                                             D. require chemical industries to reveal complete information about
    B. protect employees, customers,   the chemicals it produces.
       or vendors from hazards in
       the workplace caused by
       hazardous chemicals.

10. Which of the following is covered under the Resource Conservation and Recovery Act?

    A. Waste water   C. Cleaning solvents

    B. Waste oil   D. All of the above

11. Which of the following methods is approved for the disposal of hazardous waste?

    A. Washing them down the   C. Using them as a weed killer
       drain with plenty of water
                                             D. Placing them in leak-proof containers and disposing of them in
    B. Recycling them by reusing   an RCRA-approved method
       them in the shop

12. Which record must be kept by shops that work on trucks involved in interstate shipping?

    A. Out-of-service times   C. Names of all service personal

    B. Names of all the drivers   D. Nature and date of inspections

13. What information is provided by the first digit of the VIN of a heavy-duty truck?

    A. Model year   C. Manufacturer

    B. Axle configuration   D. Gross weight rating

14. Which of the following is **not** a safe practice when using truck lifts and hoists?

    A. Allowing the driver to remain   C. Using cables or chains to secure the object being lifted
       in the vehicle
                                             D. Checking that the attachments are secure
    B. Locating the recommended
       lift points

CHAPTER

2 Hand Tools

Objectives

Upon completion and review of this chapter, the student should be able to:

- List and describe the use of common hand tools used in heavy duty truck shops.
- Describe how to use common pneumatic, electrical, and hydraulic power tools used in heavy duty truck shops.
- Identify the mechanical and electronic measuring tools used in heavy-duty truck shops.
- Demonstrate an accurate measurement with a micrometer.

Key Terms

adjustable pliers

adjustable wrench

air ratchet

box-end wrench

chisels

combination wrench

deburring

diagonal cutting pliers

dial caliper

die stocks

feeler gauge

file

hacksaw

hand-threading dies

hex

hydraulic

impact sockets

impact wrench

interference fit

International Organization for Standardization

locking pliers

machinist's rule

micrometer

needlenose pliers

open-end wrench

Phillips screwdriver

pliers

pneumatic

Posi-Drive™ screwdriver

press

punches

ratchet

screw pitch gauge

shrink fit

sidecutters

socket wrench

tap

thickness gauge

torque wrenches

TORX® fastener

vernier caliper

Vise Grips™

wrench

INTRODUCTION

In most instances, truck technicians are required to obtain and maintain their own hand tools. There are many manufacturers of these tools, and it comes down to personal choice by the individual. The price range of these tools can vary considerably, but note that you usually get what you pay for. Before making your selection, you should also investigate any warranty policies.

When starting in the trade, a new technician can be overwhelmed by the quantity and cost of these tools. Not every tool needs to be purchased all at once. Start by buying the tools that are required by you most in your working day before buying the less common specialty tools. Over time, your toolbox will gradually fill up.

Tools give the technician the ability to put thoughts into action. Although it is necessary to know how to make a repair, not even the best technician could perform without tools.

Many of the tools used by truck technicians are general-purpose hand and handheld power tools. The sizes of these tools must fit the fasteners that are used by the truck manufacturers. Most truck components and shop equipment use common fasteners. Depending on the manufacturer of the vehicle, the fasteners can be standard SAE or metric-size fasteners. A well-equipped technician should have both standard and metric wrenches and sockets in a range of sizes and styles. A technician's knowledge of how to use the correct tool for the job is a key factor in performing high-quality service work.

SYSTEM OVERVIEW

This chapter deals with introducing a new technician to some of the many hand and handheld power tools that are used in the trade and explains how to use these tools correctly in a safe manner. The chapter also deals with the use of specialty measuring tools and how to make accurate measurement readings with them, and finally, the use of common fasteners, fastener identification, and thread repair.

SCREWDRIVERS

There are assortments of fasteners used in the trucking industry that are driven by a screwdriver. Each style of fastener requires a specific kind of screwdriver, and a well-equipped technician will have several sizes of each.

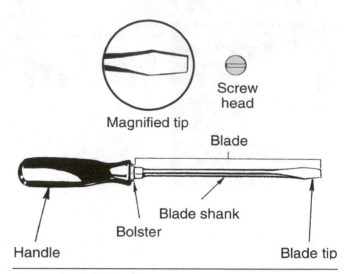

Figure 2-1 A common slotted screwdriver.

All screwdrivers, apart from the drive tip, have several things in common. Screwdriver size is determined by the length of the shank and the size of the tip. Handle size is also important. The larger diameter of the handle allows the technician to get a better grip on it and therefore apply more torque. A screwdriver handle should also be insulated from the blade and manufactured with a nonconductive material.

Standard Screwdrivers

Standard screwdrivers, often referred to as slotted screwdrivers, are probably the most common style of driver and are used for turning carriage bolts, machine screws, and sheet metal screws (see **Figure 2-1**).

Phillips Screwdrivers

The tip of a **Phillips screwdriver** had four prongs that fit to the four slots in a Phillips head screw. This is a very common fastener style in the truck and coach industry. It has a better appearance than a slot head screw and is easier to install and remove. The four recessed slots hold the screwdriver tip so that there is less likelihood of slippage or damage (see **Figure 2-2**).

Specialty Screwdrivers

A number of specialty fasteners can be used in place of slot or Phillips head screws. The objective is to improve transfer of torque from the screwdriver to the fastener, minimize slip, and enable installation on robotic assembly lines.

The **Posi-Drive™ screwdriver** (see **Figure 2-3**) is very much like a Phillips but has a tip that is flatter and blunter. The squared tip grips the screw head and slips less than an equivalent Phillips screwdriver.

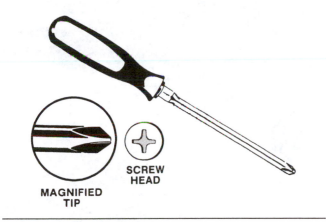

Figure 2-2 The Phillips screwdriver tip.

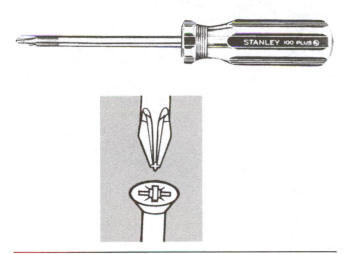

Figure 2-3 The Posi-Drive™ screwdriver is comparable to the Phillips design. (© 1998 Stanley Tools, a Product Group of the Stanley Works, New Britain, CT)

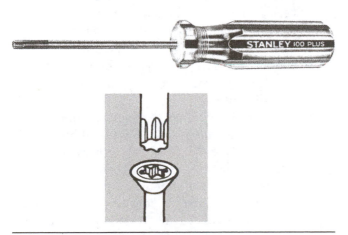

Figure 2-4 TORX® screwdrivers feature a six-prong tip ensuring a more positive grip than the slot of the Phillips style. (© 1998 Stanley Tools, a Product Group of the Stanley Works, New Britain, CT; TORX® is a registered trademark of Textron, Inc. Acument™ Global Technologies)

A **TORX® fastener** (see **Figure 2-4**) is often used to secure headlight assemblies, mirrors, and brackets. Not only does its six-prong tip sustain greater torque and less slippage, but it also provides a measure of tamper resistance.

Screwdriver Safety

- Use screwdrivers only for turning screws. Using them as punches or prybars breaks handles, bends shanks, and dulls and twists the tips. Abuse makes them unfit to tighten or loosen screws safely.
- A slotted screwdriver tip can be easily dressed back to it original shape. Always use a hand file: not only will you have much better control over the shape of the tip but you will not remove the metal's temper. A bench grinder will produce too much heat.
- A screwdriver blade that fits properly will allow you to get the greatest turning power while applying the least amount of pressure. A blade tip that does not fit properly will not only damage the screw slot but also possibly the tip itself.
- Always keep your free hand clear of the screwdriver tip when putting force on any type of screwdriver.
- Whether driving or removing a screw, make sure the blade is lined up properly with the screw. You will not get a good grip on the slot if the tip is held at an angle.
- Screwdrivers designed for use with wrenches have either a square shank or a special bolster at the handle to withstand the application of extra force.
- Don't try to hold parts in your hand. Instead, put the part in a vice or on a work bench to avoid the likelihood of stabbing your hand with the screwdriver tip.
- Whenever working on electrical devices, use a screwdriver with an insulated or molded plastic handle to avoid a shock (wooden handles with set screws are not acceptable).

HAMMERS

A hammer is generally classified by the weight and material of the head. The heads themselves are categorized as steel and soft face. The heads of steel-face hammers are made from high-grade alloy steel. The steel is deep forged and heat treated to a suitable degree of hardness. Soft-face hammers have a contact

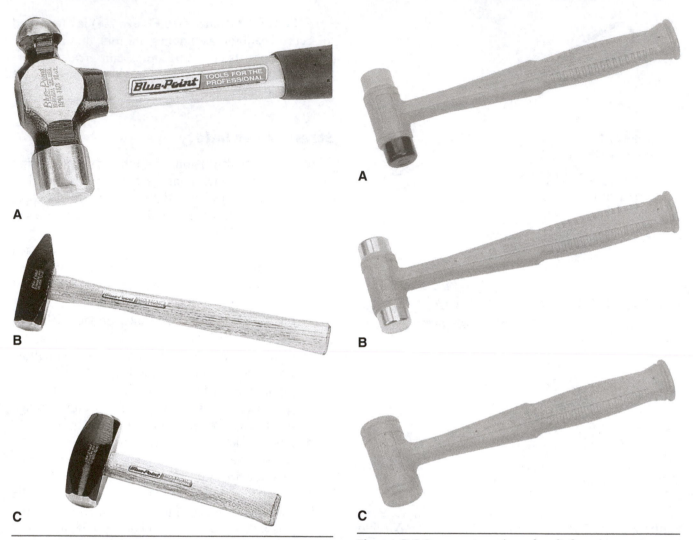

Figure 2-5 Some examples of steel-face hammers.

Figure 2-6 Some examples of soft-face hammers.

surface that yields when it comes into contact with a hard surface. Soft-face hammers are preferred when machined or precision surfaces are involved or when damage to a finish is undesirable. For example, a brass hammer can be used to drive in gears or shafts. **Figure 2-5** and **Figure 2-6** show both steel-face and soft-face hammers.

Hammer Safety

- Always wear eye protection when using a hardened tool against a hardened surface. This will protect your eyes from flying chips. Whenever possible, use soft-face hammers (plastic, wood, or rawhide) when striking hardened surfaces.
- Never strike two hammers together or against another hardened surface. A hammer can chip and cause serious bodily injury as well as damage to the tool itself.

- Make sure the handle is secured properly to the head of the hammer. It should be wedged tightly to prevent injury to yourself and others.
- Replace handles that are cracked or splintered and do not use the handle for prying or bumping. Handles can be easily damaged from this type of abuse.
- Use the correct size hammer for the job. A hammer that is too small (light) will bounce off the work. One that is too large (heavy) is more difficult to control and can damage the part.
- The hammer handle should be grasped near the end. This will increase your leverage as well as reduce physical fatigue. There will also be less chance of crushing your fingers between the handle and the projecting parts and edges of the workplace if you should miss.
- Prevent injuries to others. Keep the arc of your swing away form the path of other people in your work area in case you lose your grip on the

handle. Keep the handle dry and free of grease and oil.

- Keep the hammer face parallel with your work. Force is then spread over the full hammer face, reducing the tendency of the edges of the hammerhead to chip or slip off the object.

SAWS AND KNIVES

One very common tool used by the truck technician is the **hacksaw.** This tool can be used to cutting bolts, angle iron tubing, and so on.

Hacksaw blades are usually available with 14, 24, or 32 cutting teeth per inch. The blade with 14 teeth per inch is used for fairly thick metal, while the 18-tooth one is used for cutting medium thick material. The 24-tooth blade is usually used on heavy sheet metal, copper, brass, and medium tubing. For thin sheet metal and thinwall tubing, use a 32-tooth blade. Blades may be manufactured from a variety of materials; high-speed and tungsten steels are best for cutting alloy steels. **Figure 2-7** illustrates saw blade selection for various materials.

When using a hacksaw, be sure the blade is held with some tension within the frame and that the teeth of the blade angle away from the handle. The work piece should be secured firmly to prevent it from slipping. Hold the top of the frame with one hand

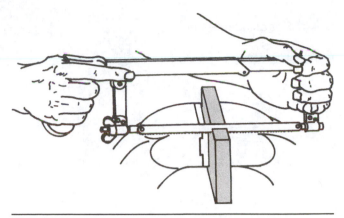

Figure 2-8 A demonstration of the proper way to use a hacksaw. Notice the hacksaw is level.

while holding the handle with the other one. Apply a slight pressure on the forward stroke to ensure cutting, and then release the pressure on the return stroke. Try to avoid starting a cut on a sharp edge because it can damage the teeth of the blade. **Figure 2-8** shows the proper use of a hacksaw.

Saw and Knife Safety

- Always wear eye protection when using a hacksaw.
- Always keep blades sharp. The greater the force you have to apply, the less control you have over the cutting action of the knife. The safest knife usually has the sharpest edge.
- Whenever possible, cut away from your body, and keep your hand and fingers behind the cutting edge. Keep handles clean and dry to prevent your hand from slipping onto the blade.
- Never pry with the blade of a knife; the hardened steel can easily break.
- Store knives safely. An exposed blade could cut you severely. Sharp-pointed tools and knives should be kept sheathed while not in use.
- When you are getting to the end of the cut, slow down, reducing the pressure on the forward stroke to avoid injury when the saw cuts through the material. Cut with the saw in one direction only to prevent dulling the blade.
- Never use a damaged blade (cracked, kinked, missing teeth).
- Do not use your thumb to aid in starting a hacksaw. If starting is a problem, use a file to make a starting notch in the work.
- Make as long a stroke as possible to maximize the life of the blade. Short strokes will wear the blade only in the area being used.

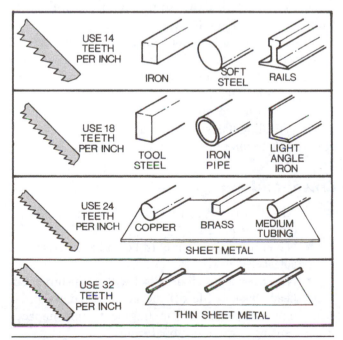

Figure 2-7 Correct hacksaw blade selection should be made for the material to be cut.

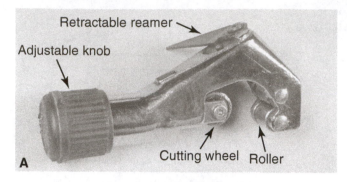

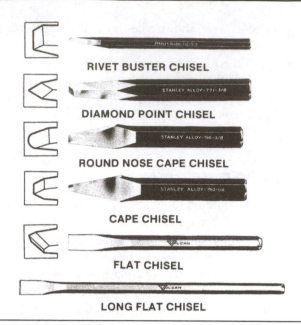

Figure 2-10 Some common chisels with their correct cutting edges.

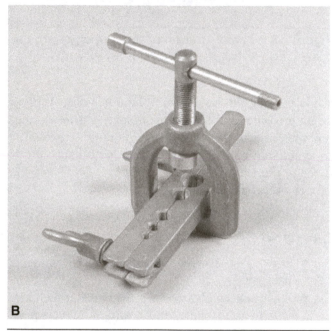

Figure 2-9 Common tubing tools: (A) tubing cutters and (B) a single-loop flaring tool.

TUBE CUTTING, FLARING, AND SWAGING TOOLS

A truck may use tubing manufactured from steel, copper, aluminum, or plastic that at some time may require repairs. Tubing tools are made for such tasks as cutting, **deburring** (removing the sharp edges from a cut), swaging (increasing the size of a piece of tubing, flaring (spreading gradually outward), and bending. **Figure 2-9** shows common tubing tools.

CHISELS AND PUNCHES

Chisels are tools that can be used to cut metal by driving it with a hammer or air-powered hammer. Technicians should have a variety of chisels in their toolboxes for cutting sheet metal, shearing off rivets and bolt heads, splitting rusted nuts, and chipping metal.

When you use a chisel, the blade should be at least as large as the cut being made. Hold the chisel firmly

enough to guide it, but lightly enough to ease the shock of the hammer blows. Hold the chisel just below the area where the hammer makes contact (head) to prevent pinching the hand in the event that the hammer misses striking the chisel. Grip the end of the hammer handle and strike with enough force to cut into the material. Make sure the head of the hammer strikes the chisel squarely, and check your progress every two or three blows. Correct the angle of the chisel until you have cut through what you have intended. **Figure 2-10** shows some common chisels with their correct cutting edges.

Punches are another tool in the same category as chisels. Punches are used to drive out pins, rivets, or shafts. Tapered punches are extremely useful for aligning holes in parts during assembly. Center punches are used to mark the starting point for a drill bit when drilling holes. Punches are designated by their point diameter and punch shape (see **Figure 2-11**).

Chisel and Punch Safety

- Always wear eye protection when working with chisels or punches.
- Never use a punch or chisel on a hardened metal surface.
- Never use a punch or chisel with a mushroomed head. Instead file off the head to redress it. A mushroomed head can fragment upon impact, causing serious personal injury.
- Do not drive a punch too deeply into a bore, or it may become stuck due to its taper.

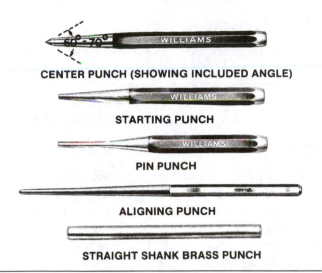

CENTER PUNCH (SHOWING INCLUDED ANGLE)

STARTING PUNCH

PIN PUNCH

ALIGNING PUNCH

STRAIGHT SHANK BRASS PUNCH

Figure 2-11 Punches are designated by point diameter and punch shape.

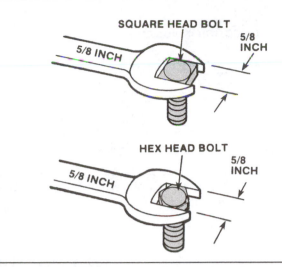

Figure 2-12 The open-end wrench grips only two faces of a fastener.

WRENCHES

Wrenches are one of the most essential tools used in the heavy truck industry. A well-equipped technician will have both standard and metric wrenches in a variety of lengths and sizes. It is important to note that standard and metric wrenches are not interchangeable. For example a 9/16-inch wrench is 0.02 inches larger than a 14-millimeter nut. If a 9/16-inch wrench is used to turn or hold a 14-millimeter nut, there is a good chance the wrench will slip, rounding the **hex** points of the nut and possibly skinning the knuckles of the technician in the process.

The word **wrench** means twist. And this is exactly what a wrench is intended to do: twist and/or hold bolt heads or nuts. The width of the jaw opening determines the size of the wrench. For example, a 5/8-inch wrench has a jaw opening (from face to face) of 5/8 of an inch. **Figure 2-12** The size is actually slightly larger that its nominal size so that the wrench may fit easily around the fastener head of equal nominal size.

Open-End Wrenches

Wrenches are classified by their shape (see **Figure 2-13**). An **open-end wrench** is just as the name implies: it is open at one end so that the wrench may slide onto the fastener from the side.

Box-End Wrenches

The other wrench shape is called the **box-end wrench** and has no opening around its perimeter. The jaws of this style of wrench completely surround the bolt or nut hex, gripping on all six points of the fastener.

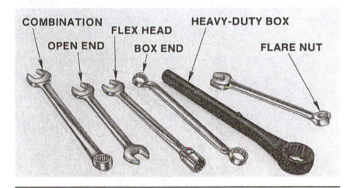

Figure 2-13 An assortment of wrench styles. *(Courtesy of Snap-on Tools Company)*

The box-end wrench is much less likely to slip off the nut while the technician is applying torque to the fastener, making it safer to work with than an open-end wrench (see **Figure 2-13**).

Combination Wrench

A **combination wrench** is basically a combination of an open-end wrench at one end and a box-end wrench on the other end. Both of the ends of this wrench have the same nominal size. Technicians will likely have two sets of combination wrenches, one for holding and the other for turning. A combination wrench is a necessary toolbox item because it complements open-end, box-end, and socket wrench sets (see **Figure 2-13**).

Adjustable Wrenches

An **adjustable wrench** is one that can be adjusted to fit many different nominal sizes. This is accomplished

APPLY FORCE IN DIRECTION INDICATED.

Figure 2-14 Pull the adjustable wrench so that the force bears against the fixed jaw.

by rotating a helical adjusting screw that is mated to the teeth in the sliding jaw. The other jaw is fixed, and all force should be applied only in the direction of the fixed jaw. This type of wrench should really only be used as a last resort as it is more likely to slip than a properly fitting open- or box-end style wrench (see **Figure 2-14**).

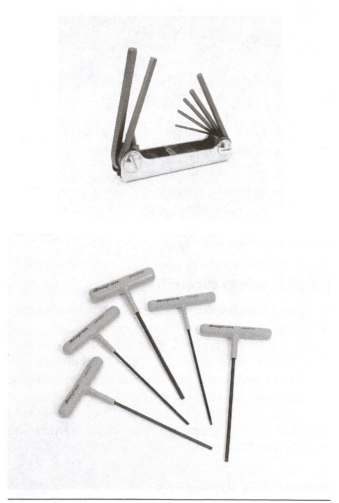

Figure 2-15 Typical Allen wrench sets. *(Courtesy of Snap-on Tools Company)*

Allen Wrenches

Allen wrenches are used on Allen-type set screws and fasteners with a recessed hex-shaped head. This style of fastener is used throughout much of the cab interior and is also used on brake calipers. This is a tool that must be in every technician's toolbox (see **Figure 2-15**).

TORX®

This type of recessed fastener is much like an Allen-type of screw head and is characterized by a six-point star-shaped pattern. Many laypeople refer to this fastener as a "star." The generic name is hexalobular internal driving feature and it is standardized by the **International Organization for Standardization.**

The design of the TORX® prevents it from cam-out better than a Phillips-head or slot-head screw. Phillips heads are designed to cause the tip to cam-out, in order to prevent overtightening of the screw, where the TORX® design prevents cam-out. The reason for this is to develop better torque-limiting automatic screwdrivers for use in the manufacturing industry, unlike the Phillips screw that relies on the driver tip slipping out of the screw head when a torque level is reached (see **Figure 2-16**).

TORX® SIZING

The bit sizing for the TORX® head is described using the capital letter "T" followed by a number. The smaller the number, the smaller the point-to-point dimension of the screw head. Common sizes include T5, T10, T15, and T25, although they reach as high as T100. The proper size bit must be used with a fastener to prevent ruining the driver or the screw. The same series of TORX® drivers is used to drive SAE, metric, and other thread system fasteners, reducing the number of bit sizes required.

Figure 2-16 Standard TORX® bit socket.

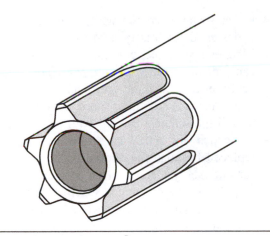

Figure 2-17 Secure TORX® bit socket.

TORX® VARIATIONS

When security is an issue, TORX® produces a tamper-proof fastener. A post is contained in the center of the head, preventing a standard TORX® bit from being inserted into the fastener (see **Figure 2-17**).

An external TORX® version exists, where the screw head had the shape of a screwdriver bit, and a special TORX® socket is used to drive it. These are found primarily on automobile engines.

Socket Wrenches

Socket wrenches are used in many situations and are faster, safer, and easier to use than an open-end wrench. The reason for this is that the technician is not required to remove the socket and realign it every time the limit of rotation is reached. This is accomplished by the use of a **ratchet** (see **Figure 2-18**).

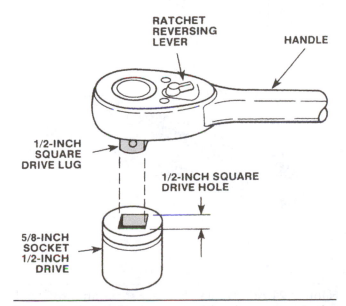

Figure 2-18 A socket square drive lug recess is the same size as the handle drive lug.

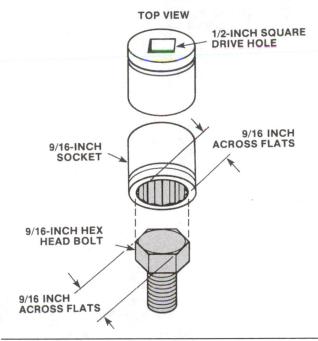

Figure 2-19 The size of a socket is determined by the size of the bolt or nut it will securely hold.

A basic socket wrench set consists of a ratchet handle and several barrel-shaped sockets. The socket fits over and around a given size bolt or hex head. The inside of the socket is machined as either a hex or double-hex box-end wrench. Sockets are available in 6-, 8-, or 12-point configurations. The top side of the socket has a square hole that engages with a square ratchet drive lug (see **Figure 2-19**).

SIZE

Socket sizes are generally referred to by the size of the ratchet that drives the socket. These would be 1/4-, 3/8-, 1/2-, and 3/4-inch drives. Good-quality ratchet drives use a spring-loaded ball that fits into a depression in the socket. This assembly allows the ratchet to retain the socket until the socket is removed by the technician.

Sockets are available in various sizes, lengths, and bore depths. Both standard SAE and metric socket wrenches are required to work on current model trucks. Generally the larger the socket size the deeper the well. Deep-well sockets are made extra long to fit over bolt ends or studs. They are also handy for reaching nuts or bolts in areas (see **Figure 2-20**). of limited access. Deep-well sockets should not be used when a regular-size socket will do the job. This is because the longer socket will develop more torque twist and is more susceptible to slipping off the fastener.

Heavier-walled sockets made of a softer steel are designed to be used with an **impact wrench** and are called **impact sockets.**

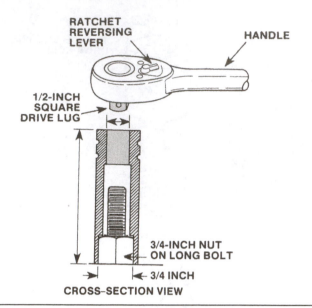

RATCHET
REVERSING
LEVER   HANDLE

1/2-INCH
SQUARE
DRIVE LUG

3/4-INCH NUT
ON LONG BOLT

3/4 INCH

CROSS-SECTION VIEW

Figure 2-20 Deep-well sockets fit over the bolt hex and long studs.

There are many accessories that will enhance the usefulness of a socket wrench set. Technicians should have an assortment of these in their toolboxes. Accessories include screwdriver attachments, crow's foot, flex design sockets, as well as Allen and TORX® head sockets. These accessories can make the difference on hard-to-turn or stubborn fasteners due to the extra torque that can be applied from the leverage of the ratchet handle.

Wrench Safety

- Always use wrenches that fit properly. Loose-fitting wrenches can slip, causing damage to the bolt head or nut, skin the knuckles of the user, or even cause the user to get off balance and fall.
- Always use the wrench that best fits the application, giving you a sure grip and a straight, clean pull. If the tool must be cocked at an angle, it puts intense stress on the contact points, a frequent cause of tool failure. Whenever available, use offset wrenches, an angle head ratchet, or even universal or crow's foot socket to allow you to get a clean pull in tight places.
- Never use an extension on a wrench like a pipe or hooking two wrenches together to obtain more leverage. The handle length is made to apply the maximum safe force that the socket can withstand. The use of excessive force can also cause the wrench to break or break the bolt or cause the socket to round the bolt head. This can also result in the technician falling or skinning his knuckles.

- Never strike a wrench with a hammer unless it is designed for this type of use.
- Whenever possible, position yourself so that you can pull the wrench toward you. This will prevent you from striking your knuckles or injuring your hand on something sharp. If you are unable to pull the wrench, push it with an open palm.
- Wrenches that are worn or cracked should be replaced because they could break at any time, causing serious personal injury. Never try to straighten a bent wrench; it will weaken it.
- Never use an adjustable wrench when a properly fitting combination wrench is at hand. The adjustable wrench is a multipurpose tool and is only a substitute when the proper wrench is unavailable.

PLIERS

Pliers are general-purpose tools used for gripping wires, clips, and pins. Technicians should have an assortment of several different types: standard pliers, needlenose for small, difficult-to-access components, and large adjustable pliers for heavy-duty work with pipes and filters.

Adjustable Pliers

Adjustable pliers, also known by the names pipe pliers, water pump pliers, or even by a manufacturer's name, Channel Locks, have a multiposition slip joint that may be used to grip shafts, pipes, or filters (see **Figure 2-21**).

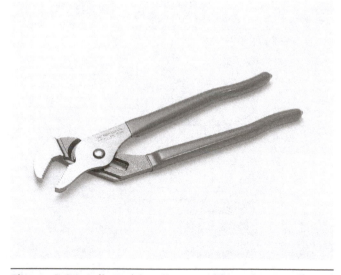

Figure 2-21 Adjustable pliers are sometimes referred to as water pump pliers. *(Courtesy of Snap-on Tools Company)*

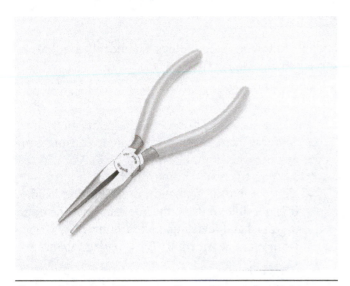

Figure 2-22 Needlenose pliers. *(Courtesy of Snap-on Tools Company)*

Needlenose Pliers

Needlenose pliers are so called because of their long tapered jaws. They are very handy in clutching small components or for reaching into tight spots. Needlenose pliers will commonly incorporate a wire cutting edge at the pivot point of the jaw. These pliers are essential for electrical work (see **Figure 2-22**).

Locking Pliers

Locking pliers, often referred to by a manufacturer's name, **Vice Grips™,** are similar to regular pliers except that they have the ability to lock into a closed position, allowing the operator to stop applying pressure to keep the jaws grasping. Locking pliers can be used to hold or clamp parts together, and are also useful for getting a firm grip on a damaged fastener on which sockets and wrenches are no longer useful. Locking pliers come in several sizes and jaw configurations, including needlenose, and are useful for many mechanical repair jobs (see **Figure 2-23**).

Figure 2-23 Locking pliers, or Vice Grips™. *(Courtesy of Snap-on Tools Company)*

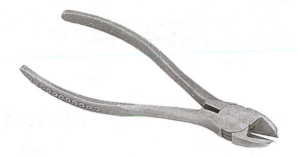

Figure 2-24 Diagonal cutting pliers, often referred as sidecutters.

Diagonal Cutting Pliers

Diagonal cutting pliers, often referred to as **sidecutters,** are used extensively in electrical work for cutting wire or connectors. They are also very useful for removing cotter pins. The jaws of these pliers have hardened cutting edges for increased tool life (see **Figure 2-24**).

Plier and Cutting Safety

- Never use pliers in place of a wrench. They are incapable of holding the work securely and can damage bolt heads and nuts by rounding of the corners.
- Wear eye protection whenever cutting with pliers or cutters. Wire, cotter pins, or connector pieces can fly with much force, causing serious eye injury. Try to cup your hand over the cutters to protect yourself.
- Pliers are designed for pinching, holding, squeezing, or cutting, but not for turning.
- Always observe the following safety precautions:
  1. Select the correct size cutters for the job.
  2. Position the blades at right angles to the stock.
  3. Don't rock the cutters to increase the speed of the cut.
  4. Adjust the cutters to maintain a small clearance between the blades. This prevents the hardened blades from striking each other when the handles are closed.

FILES

The purpose of a **file** is to remove small amounts of metal, for shaping, smoothing, or sharpening parts. Files are classified as single cut, double cut, fine, and coarse. The different types of files are easily identified by the file tooth pattern. Files are available in an assortment of different shapes and sizes.

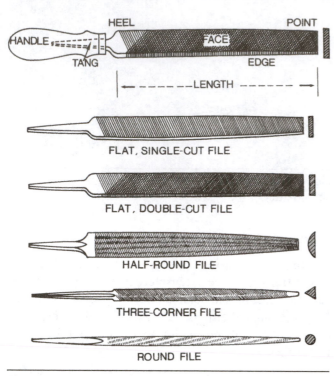

Figure 2-25 A selection of file types.

The teeth of the file are designed to cut in only one direction—the forward stroke. Select a file that is correct for the job being performed. When cutting soft material, use a coarse file to prevent clogging the file teeth with filings. If a slight pressure is applied to the file on the back stroke while cutting soft material, it will clear the file teeth from filings (see **Figure 2-25**).

Filing Technique

With one hand, grasp the file by the handle. With the other hand, hold the point of the file with the thumb and first two fingers of the other hand. Apply a very slight pressure while pushing the file away from your body. The file will cut only on the forward stroke, so don't drag the file back toward you on the return stroke (see **Figure 2-26**).

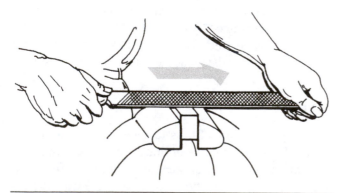

Figure 2-26 A demonstration of the correct method of holding a file.

File Safety

The following safety rules should always be observed when using a file:

- Wear safety glasses or goggles whenever performing any filing work.
- Never strike the file with a hammer or against anything hard. A file can shatter, causing serious personal injury.
- Always file away from your body.
- Never use a file without a proper fitting handle. It is possible to drive the file tang into the palm of your hand, causing severe injury.
- Do not use worn (dull) files; replace them.

SPECIALTY TOOLS

Aside from the tools mentioned so far, a well-equipped technician's toolbox will contain some of the following specialty tools.

Taps and Dies

The **tap** is a hand tool used to cut internal threads (see **Figure 2-27**). An internal thread is the thread that is cut on the inside of a thread, like those on the inside of a nut. A tap can also be used to clean up or restore previously cut threads. A set of hand taps include three types: a tapered tap, a plug tap, and a bottoming tap. The tapered tap is tapered from the tap end to the sixth thread, making the tool easy to screw (get started) into the work piece for the first time. The plug tap is tapered from the tap end to the third or fourth thread so that it is used for a rough finish of the work piece after threading it using the tapered tap. The bottoming tap has only one tapered thread at the end of the tap and is used to finish the threading.

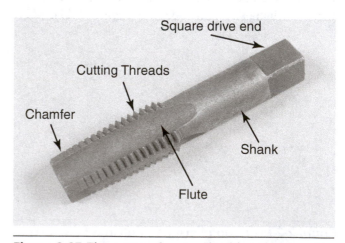

Figure 2-27 The parts of a standard hand tap.

TO USE A HAND TAP

- Drill a pilot hole larger than the minor diameter of the thread. The drill bit diameter should be approximately 75 to 80 percent of the thread diameter. (Consult a tap drill index for exact size tap drill size.)
- Set the tap into the tap wrench handle and position them at a right angle to the work piece.
- Turn the tap clockwise slowly. Be sure to hold the work piece with a vise or equivalent tool.
- After threading the tap in one or two turns, back the tap out a quarter to half turn to break the chip. Repeat the process as you continue tapping.
- Lubricate the tap with cutting oil frequently.
- Be aware that the hole may clog with metal chips, causing the tap to bind. It may be necessary to extract the tap to clean off the filings. Failure to do so could result in breakage of the tap.
- Once threads are cut to the desired depth, turn the handle counterclockwise to remove the tap.
- Clean tap and lightly oil it before placing it in a protective box for storage.

Hand-Threading Dies

Hand-threading dies are used to cut external threads (see **Figure 2-28**). These are the threads that are found on the outside, like on a bolt or a piece of threaded rod. Taps and dies are produced in various sizes depending upon the thread size and pitch required. Dies may be solid (fixed size), split on one side to allow for adjustment, or manufactured in two halves held together in a cullet that provides for individual adjustments. Dies fit into holders called **die stocks.**

TO USE A HAND DIE

- Select the correct size die and place in the die stock.
- Turn the handle clockwise, threading the die onto the work.

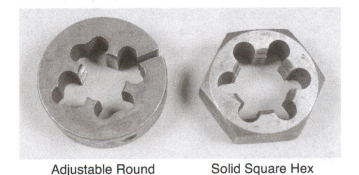

Adjustable Round   Solid Square Hex

Figure 2-28 Common die shapes.

- Use lubricant and back up every one or two turns to keep chips for binding.
- Once threads are cut to the desired dimensions, turn the handle counterclockwise to remove the die.
- Clean die and lightly oil it before placing it in a protective box for storage.

Gear and Bearing Pullers

Often, gears and bearings are installed onto a shaft with what is called an **interference fit** (press or **shrink fit**). What this means is that that either the shaft or housing is of a smaller or larger size. The general rule of press fit allowances = 0.0015 inches. times the diameter of the part in inches. Parts can also be fitted together by making use of the natural tendency of metals to expand or contract when heated and cooled. By heating a part, you can expand it, and it can be then slipped on a mating part. When the heated part cools, it will contract and grip the mating part, often with tremendous force. Parts may also be mated by cooling one or the other so that it contracts, thus making it smaller. As the part warms up in the ambient air, it will expand to meet the mating part (see **Figure 2-29**).

An interference or shrink fit allows no movement between mating components, thus eliminating wear. The removal of gears and bearing must be carefully to avoid damage to the components. Hammering or prying can fracture or bind the mating components. For this reason, a puller with the proper jaws and adapters should be used when applying force to remove gears and bearings. Force may be applied gradually, increasing intensity.

There are many other hand tools that are extremely useful when servicing trucks. including pry bars, wire brushes, scrapers, strap wrenches, propane torch, tire gauges, grease guns, tire irons, and trouble lights. The use of some of these tools will be explained in further detail in appropriate chapters (see **Figure 2-30**).

POWER TOOLS

The use of power tools can make the job of a technician less physically demanding. They operate much faster and develop greater torque than can be achieved using hand tools. The force used to supply power to the tool can come from electricity, air (**pneumatic**), or fluid (**hydraulic** actuated). These tools require the technician to practice safe working techniques, because the tools will keep operating until the tool is turned off.

Figure 2-29 Various gear and bearing pullers.

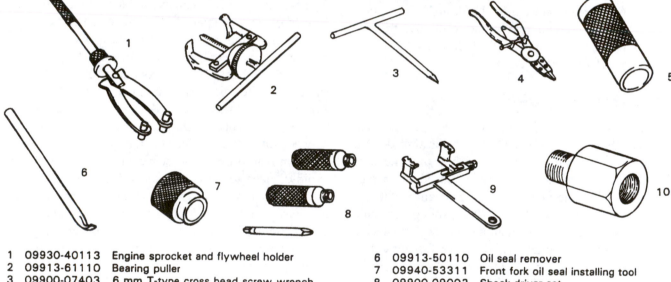

1	09930-40113	Engine sprocket and flywheel holder	6	09913-50110	Oil seal remover
2	09913-61110	Bearing puller	7	09940-53311	Front fork oil seal installing tool
3	09900-07403	6 mm T-type cross head screw wrench	8	09900-09002	Shock driver set
4	09920-70111	Snap ring opener	9	09920-53710	Clutch sleeve hub holder
5	09913-80111	Bearing and oil seal installing tool	10	09930-33710	Rotor remover attachment

Figure 2-30 Manufacturers' special tools make jobs easier and are usually assigned a part number that can be identified in service literature.

Most shops will have electrical power tools such as grinders, drill presses, electric drills, and such, but most of the tools will be pneumatic. Pneumatic tools have four major advantages over electrically powered equipment in a truck shop (see **Figure 2-31**):

- **Flexibility:** Air tools have the advantage of running cooler, and speed and torque are easily controlled; damage from overload or stall is minimized. They also fit in tight spaces.
- **Lightweight:** Air-powered tools are lighter in weight, reducing the effort required by the technician operating them.

- **Safety:** Air tools reduce the risk of arcing that could result in a fire and/or shock hazards.
- **Low-cost operation and maintenance:** Generally, air tools require fewer moving parts and thus require fewer repairs and less preventative maintenance. Also, the purchase cost of air-driven tools is usually less than that of equivalent electrical tools.

Maintenance

Most air tool manufacturers recommend adding a few drops of air tool oil daily. This recommendation

Figure 2-31 A common 3/8-inch air drill on the left and a 3/8-inch electric drill. The air drill weighs 2½ lb; the electric drill weighs 4½ lb.

should be strictly adhered to, as the oil helps to not only lubricate moving parts but also help prevent corrosion within the tool caused by the moisture forced into the tool by the compressor.

The mechanical repair industry was one of the first industries to recognize the advantages of air-powered tools. Today they are necessary tools for the professional truck technician. One of the only real disadvantages of air tools in general is the excessive noise they produce while operating.

Power Tool Safety

Power tools must be used in a safe manner regardless of how the tool is powered. Carelessness or mishandling of power tools can cause serious injury. The following is a list of some general rules that should always be adhered to:

- Wear eye protection.
- Wear hearing protection whenever noise is excessive. This same noise may also be a hazard to others where you are working.
- Wear gloves when operating air chisels or air hammers.
- All electrical equipment should be grounded, unless it is the double-insulated type.
- Never make adjustments, lubricate, or clean a piece of equipment while it is running.
- Compressed air should never be used to clean yourself or others.
- Report any suspect or malfunctioning machinery to the instructor or service manager.
- Know your power tool. Read the operator manual carefully. Learn its applications and limitations as well as the specific potential hazards particular to the tool being used. Do not operate any electrically powered tool in damp or wet locations; keep your work area well lit.

- Never abuse an electrical tool's power cord. Always disconnect by unplugging the cord at the receptacle, and never yank the cord to disconnect it.
- No machine should be started unless guards are in place and in good condition. Defective or missing guards should be reported to the instructor or service manager immediately.
- Check and make all adjustments before applying power.
- Give the machine your undivided attention while you are operating it. Never turn your head to talk to someone.
- Inspect all equipment for safety and for apparent defects before using.
- Whenever safeguards or devices are removed to make repairs or adjustments, equipment should be turned off and the main switch locked and tagged.
- Start and stop your own machine and remain with it until it has come to a complete stop.
- Always allow any machine to reach operating speed before applying a load.
- No attempt should be made to retard rotation of the tool or work.
- Do not attempt to slip broken belts or other debris from a pulley in motion or reach between belts and pulleys.
- Do not wear loose-fitting clothing or use shop rags around operating machinery.
- Use the right tool. Do not force a small tool or attachment to do the job of a heavy-duty tool.
- Do not force the tool. It will do the job better and more safely at the rate for which it was designed. That is, do not attempt to use a machine or tool beyond its stated capacity or for operations requiring more than the rated horsepower of the tool.
- Tools should be maintained with care. Keep tools sharp and clean at all times, for best and safest performance. Follow instructions for lubricating and changing accessories.
- Always remove adjusting wrenches and keys from machinery, making it a habit to check before turning it on.
- Never reach over an operating machine. Keep feet slightly apart to maintain proper footing and to avoid slipping.
- Tools should be disconnected when not in use or when making adjustments or changing attachments. Before plugging in any electric tool or machine, make sure the switch is in the off position to prevent serious personal injury.

- Remove all sharp edges and burrs before completing any job.
- Guards should be kept in place and in good working order.
- Never leave a power tool running unattended. Before leaving, turn off the machine.
- When using power equipment of small parts, never hold the part in your hand. Always mount the part in a bench vise or use locking pliers.

Impact Wrenches

Impact wrenches (see **Figure 2-32**) are used to speed up the wrenching process normally done by hand. An impact wrench is a handheld reversible wrench. Medium-duty models can deliver up to 200 lb of torque. When the trigger is depressed, the output shaft of the wrench, unto which the socket is attached, spins freely at 2,000 to 14,000 rpm, depending on the make and model of the wrench. When the impact wrench senses resistance to rotation, a small spring-loaded hammer, which is situated near the end of the tool, strikes an anvil attached to the drive shaft end where the socket is attached. Each impact moves the socket around a little until torque equilibrium is reached, the fastener breaks, or the trigger is released.

Keep the following safety rules in mind when operating impact wrenches:

Figure 2-32 An impact wrench or gun in use to remove wheel nuts.

Figure 2-33 A typical air ratchet.

- Use only sockets approved for use on impact wrenches. Never use sockets designed for hand tools; they can break, causing damage or injury.
- Ensure the socket is properly secured and snapped in place before using the wrench.
- Hold the wrench firmly with both hands.
- Keep hands clear of moving parts.
- Wear both eye and ear protection.
- Keep face away from work when using an impact wrench.

Air Ratchets

An **air ratchet** is much like a hand ratchet with an ability to get into tight places. Its angle drive reaches in and will tighten or loosen fasteners where other power tools cannot reach. The air ratchet looks like an ordinary ratchet but has a fat handgrip that contains an air vane motor and drive mechanism (see **Figure 2-33**).

Air Drills

Air drills are available in 1/4-, 3/8-, and 1/2-inch chuck sizes and are operated the same as electric drills but are smaller and lighter. This compactness makes them much easier to use for drilling in tight areas of a truck.

Air Chisels and Hammers

Air chisels and hammers (see **Figure 2-34**) are two of the most versatile tools in a mechanic's toolbox. When used with the illustrated accessories in **Figure 2-35**, these tools can perform many functions:

- **Universal joint and tie-rod tool:** This tool helps to shake loose stubborn universal joints and tie-rod ends.
- **Ball joint separator:** The wedge action breaks apart frozen ball joints.
- **Shock absorber chisel:** With this tool, quick work is made of the roughest jobs without the usual bruised knuckles and lost time. It easily cracks frozen shock absorber nuts.
- **Exhaust pipe cutter:** The cutter slices through mufflers and exhaust pipes making disassembly much easier.

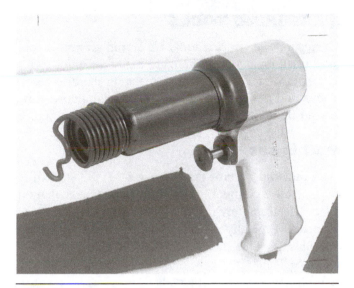

Figure 2-34 A typical air hammer or chisel.

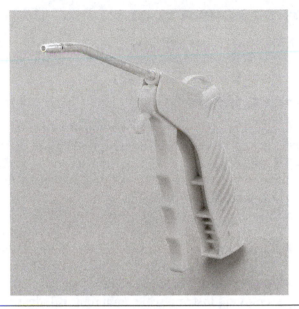

Figure 2-36 A typical blowgun.

- **Tapered punch:** Driving frozen bolts, installing pins, and punching or aligning holes are some of the many uses of this accessory.
- **Rubber bushing splitter:** This accessory is designed to remove all types of bushings to the correct depth. A pilot prevents the tool from sliding.

Blowgun

A blowgun is a tool that can be plugged into the end of an air line. A blowgun directs the flow of air when a button is pressed. Before using a blowgun, be sure it has not be modified to eliminate air bleed holes on the side. Blowguns are used for blowing material off parts during cleaning (see **Figure 2-36**).

CAUTION *Never point a blowgun at yourself or someone else.*

Other safety rules to remember when using air-powered tools:

- Hose and hose connections used to supply compressed air to equipment must be used only for the pressure and service for which they are designed.
- When connecting air tools and lines, check to make sure they are attached securely and properly.
- Do not use compressed air for cleaning, unless it is reduced to a pressure of less than 30 psi (pounds per square inch) and then only with personal protection equipment. (Air tools are usually operated with much greater air pressure.)

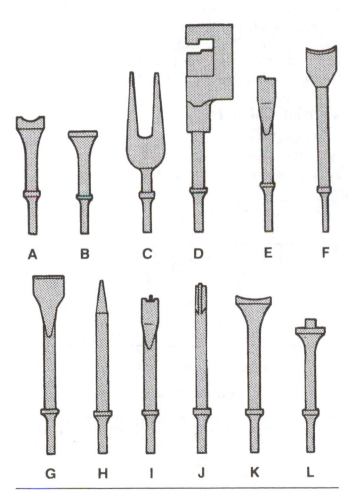

Figure 2-35 Air chisel accessories: (A) a universal joint and tie-rod tool; (B) a smoothing hammer; (C) a ball joint separator; (D) a panel crimper; (E) a shock absorber chisel; (F) a tail pipe cutter; (G) a scraper; (H) a tapered punch; (I) an edging tool; (J) a rubber bushing splitter; (K) a bushing remover; and (L) a bushing installer.

OTHER POWER TOOLS

Most truck shops will have other power tools that the truck technician may use, including the following.

Bench Grinders

A bench grinder will either be bolted to a stand or mounted to a work bench. Bench grinders are classified by wheel size. Wheel sizes from six to ten inches are the most common in truck repair shops. Three types of wheels are available with these types of grinders:

- **Grinding wheel:** For a wide variety of grinding jobs, from sharpening cutting tools to deburring.
- **Wire wheel:** For general cleaning and buffing, removing rust and scale, pain removal, deurring, and so forth.
- **Buffing wheel:** Used for general-purpose buffing, polishing, light cutting.

CAUTION *Never wear gloves when operating a bench grinder. If the glove gets caught, it will pull your hand into the wheel surface and cause severe damage to the hand and fingers. Always stand to the side when starting a bench grinder. If the grinding wheel has been damaged, this is usually the time it will fly apart and could cause serious personal injury or death.*

Presses

Many jobs in a truck shop will require the use of a **press.** A press can exert tremendous force to assemble or disassemble parts that are press fit together. Servicing rear axle bearings, pressing brake drum and rotor studs, transmission assembly work, and frame work are just a couple of examples.

The force used by the press can be hydraulic, electric, air, or hand driven. Capacities range up to 150 tons of pressing force, depending on the size and design of the press. Smaller arbor and C-frame presses can be bench or pedestal mounted, while high-capacity units are free standing or floor mounted.

CAUTION *Whenever a press is in use, the component is subjected to a tremendous force that can easily kick out the work piece. Try to position your body away from the trajectory.*

MEASURING TOOLS

Accurate measuring tools are an indispensable tool for a truck technician. There are many different styles to perform accurate measurement in hard-to-reach places, and it is important that a technician knows how to read the tool, whether it is mechanical or digital.

Machinist's Rule

A **machinist's rule** (see **Figure 2-37**) is a steel ruler not unlike a common ruler, but unlike the common ruler, it is accurately divided into small (1/64-inch) increments. Typically a machinist's rule will be marked on both sides. One side is marked off at 1/16-, 1/8-, 1/4-, 1/2-, and 1-inch intervals, and the other side is marked at 1/32- and 1/64-inch intervals.

Machinist's rules are also available in metric and decimal graduations. Metric rulers are normally divided into 0.5 mm and 1 mm increments. Decimal machinist's rulers are typically divided into 1/10, 1/50, and 1/100 inch (0.10, 0.05, and 0.01 inch). Decimal rulers are convenient when measuring components' dimensions that are specified in decimals.

Vernier Calipers

A **vernier caliper** is a highly precise measuring tool. Proper maintenance of the tool is essential to keep it in good condition. Accurate measurements

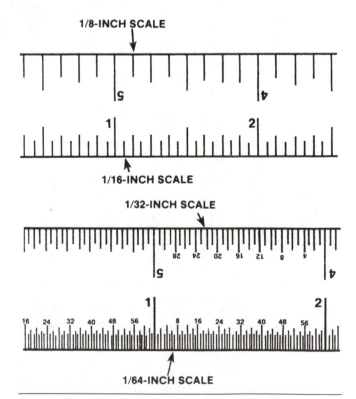

Figure 2-37 Machinist's rule graduations.

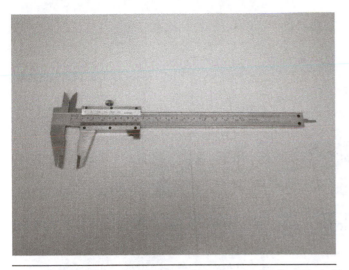

Figure 2-38 A vernier caliper. *(Courtesy of John Dixon)*

cannot be obtained unless the tool is handled properly. The vernier caliper can be used to measure the length, outside diameter, inside diameter, and depth of the material body. Vernier calipers may provide a dial and are called dial calipers. There are three types of vernier calipers, with the minimum measurement being 1/10 mm, 1/20 mm, and 1/50 mm respectively. The vernier calipers with 1/20 mm (0.05 mm) as the minimum measurement are used most often. Some vernier calipers are calibrated in inch (see **Figure 2-38**).

READING INCH VERNIER CALIPERS

Vernier scales are engraved with either 25 or 50 divisions. If you are using a unit with 25 divisions, each inch on the main scale is divided into 10 major divisions numbered from 1 to 9. Each major division is 100 (100 thousandths of an inch). Each major division has four subdivisions with a spacing of 0.025 (25 thousandths of an inch). The vernier scale has 25 divisions with the zero line being an index.

To read the vernier caliper, count all of the graduations to the left of the index line. This would be 1 whole inch plus 2/10 or 0.200, plus 1 subdivision. The value of the partial subdivision is determined by the coincidence of one line on the vernier scale with one line of the true scale, For this example, the coincidence is on line 13 of the vernier scale. This is the value in thousandths of an inch that has to be added to the value read on the beam (see **Figure 2-39**).

	1.000 inch
	0.300 inch
	0.075 inch
	0.024 inch
Total	1.399 Inches

READING A METRIC VERNIER CALIPER

The method of reading a metric vernier caliper is exactly the same as previously described for the inch system vernier caliper.

The discrimination of metric vernier caliper models varies from 0.02 mm, 0.05 mm, to 0.1 mm. The most commonly used type discriminates to 0.02 mm. The main scale on a metric vernier caliper is divided into millimeters with every tenth millimeter mark numbered. The 10 mm line is numbered 1, the 20 mm line is numbered 2, and so on, up to the capacity of the tool. The vernier scale on the sliding jaw is divided into 50 equal spaces with every fifth space numbered. Each number division on the vernier represents one-tenth of a millimeter. The five smaller divisions between the numbered lines represents two-hundredths (0.02 mm) of a millimeter.

To make a reading refer to **Fig 2-40** on the main scale, whole millimeters to the left of the zero on the index line of the sliding jaw. This example indicates 29 mm plus part of an additional millimeter.

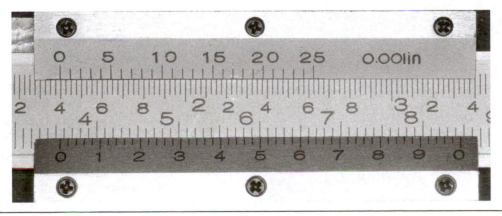

Figure 2-39 The top set of numbers in this picture illustrates an imperial vernier measurement. *(Photo by Al Thompson)*

Figure 2-40 The bottom of this picture illustrates the measurement of a metric vernier measurement. *(Photo by Al Thompson)*

The vernier scale coincides with the main scale at the 43rd vernier division. Since each vernier scale spacing is equal to 0.02 mm the reading on the vernier scale is equal to 43 times 0.02, which equals 0.86 mm. The 0.86 mm is then added to the reading found on the main scale to get a final reading.

Main scale reading	29.00 mm
Vernier scale coincidence line	0.86 mm
Final reading	29.86 mm

Dial Calipers

The **dial caliper** is definitely something that should be in a mechanic's toolbox. It can be used for measuring both inside and outside and depth and step measurements. This measuring tool is capable of measuring from 0 to 6 inches in units up to 0.0005 inch. In metric calibrations it will measure from zero to 150 mm in increments of 0.02 mm (see **Figure 2-41**).

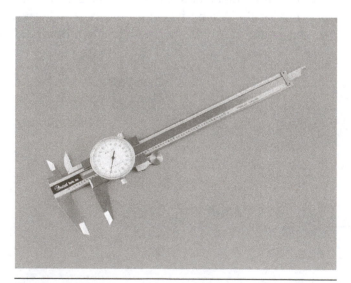

Figure 2-41 A typical dial caliper. *(Courtesy of Central Tools, Inc.)*

The standard dial caliper has a depth scale, bar scale, dial indicator, inside measurement jaws, and outside measuring jaws. The bar scale is divided into one-tenth (0.10) of an inch increments. The dial of the dial indicator is divided into one-thousandth (0.001) of an inch increments. This means that every revolution of the dial indicator needle is equivalent to one-tenth of an inch.

The metric dial caliper is comparable to a standard dial caliper in appearance. The bar scale on this instrument is divided into 2 mm increments. As for the dial of the metric dial caliper, one revolution is equal to 2 mm.

Both standard and metric dial calipers use a thumb-operated knob for fine adjustment. In using a dial caliper, move the measuring jaws together and apart to center the jaws on the work, ensuring that the jaws lie flat on the object being measured. This is important because if the jaws or the work are tilted in any way, the result will be an inaccurate measurement. Even though a dial caliper is a precise measuring instrument, it is only accurate to within ±0.002 inch. The accuracy of this instrument is limited to jaw flatness and feel. A micrometer is a tool better suited for high-precision measurement.

Micrometers

A **micrometer** (often referred to as a mike) is a very precise measuring tool that is commonly used by truck technicians during engine rebuilding procedures. This tool is commonly used for measuring inside or outside diameters of shafts and bores. Both outside and inside micrometers are calibrated and read in the same manner and both are operated so that the measuring points exactly contact the surfaces being measured (see **Figure 2-42**).

The major components and markings of the micrometer include the frame, anvil, spindle, lock nut,

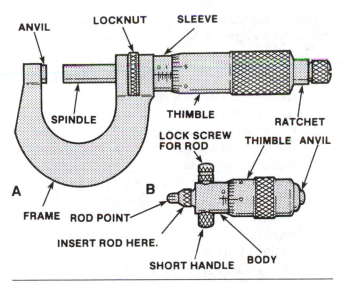

Figure 2-42 Nomenclature and components of (A) an outside and (B) an inside micrometer.

sleeve, sleeve numbers, sleeve long line, thimble marks, thimble, and ratchet. Micrometers are calibrated in standard or metric denominations.

Whether you use an inside or outside micrometer, the thimble is revolved between the thumb and forefinger. Very light pressure is all that is required when bringing the points into contact with the surfaces being measured. Keep in mind that the micrometer is a delicate instrument and that even slight excessive pressure will result in inaccurate measurements.

USING AN OUTSIDE MICROMETER

The first part to making an accurate measurement is to know how to use the tool properly. Before any measurement can be made, the calibration of the mike should be checked. To check calibration, turn the thimble all the way in until the spindle and the anvil are touching each other; if not, follow adjustment procedures to calibrate the tool. The zero on the thimble should be lined up with the zero on the sleeve. The proper method for measuring small objects with an outside mike is to grab the mike with your right or left hand.

READING A STANDARD OUTSIDE MICROMETER

A standard micrometer is made so that each turn of the thimble moves the spindle 0.025 inch (25 thousandths of an inch). This is accomplished by using 40 threads per inch on the thimble. The sleeve long line is marked with sleeve numbers 1, 2, 3, and so on. These numbers on the sleeve signify 0.100 inch, 0.200 inch, 0.300 inch, and so on. The sleeve on the micrometer contains sleeve marks that represent 1 inch in 0.025 inch (25 thousandths of an inch)

increments. Each of the marks on the thimble represents 0.001 inch (one-thousandth of an inch). In one complete turn, the spindle will move 25 marks or 0.025 inch (25 thousandths of an inch). Standard graduated micrometers are available in a range of size of zero to 1 inch, 1 to 2 inch, 2 to 3 inch, 3 to 4 inch, and so on. A micrometer calibrated in one-thousandths of an inch increments are the most common.

To read the micrometer measurement, follow the order provided here:

- First read the last whole sleeve number that is visible on the sleeve long line.
- Second, count the number of full sleeve marks past the number.
- Third, count the number of thimble sleeve marks past the sleeve marks. Add these three measurements together for the measurements. These three readings indicate tenths, hundredths, and thousandths of an inch, respectively. For example, a 2- to 3-inch micrometer that has taken a measurement is used as an example in **Figure 2-43.**

1. Start with 2 inches because you are using a 2- to 3-inch micrometer.
2. The largest sleeve number visible is 4, indicating 0.400 inch (four-tenths of an inch).
3. The thimble is three full sleeve marks past the sleeve number. Each sleeve mark indicates 0.025 inch, so this indicates 0.075 inch (75 hundredths of an inch).
4. Lastly, on the thimble the number 12 mark is lined up with the sleeve long line. This indicates 0.012 inch (12 thousandths of an inch).
5. Now add together the readings from steps 1 through 4. The total of the four readings is the final measurement.

Micrometer size	2.000 inch
Sleeve	0.400 inch
Sleeve marks	0.075 inch
Thimble marks	0.012 inch
Total	2.487 inches

Therefore the final reading would be 2.487 inches.

READING AN OUTSIDE MICROMETER WITH A VERNIER SCALE

In cases where even more accuracy is required, to within 0.0001 inch (one ten-thousandth of an inch), a micrometer with a vernier scale should be used. The micrometer is read in the same way as a standard

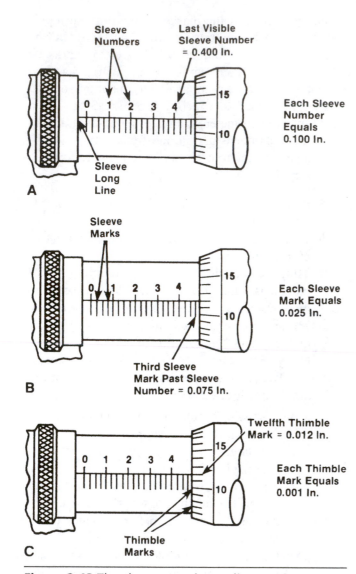

Figure 2-43 The three steps in reading a micrometer: (A) measuring tenths of an inch; (B) measuring hundredths of an inch; and (C) measuring thousandths of an inch.

micrometer. However, in addition to the three scales found on the typical micrometer, this type has a vernier scale on the sleeve. When taking measurements with the micrometer (often referred to as a mike), read it in the same way as you would a standard mike. Then, find the thimble mark that lines up precisely with one of the vernier scale lines. Only one of these lines will match up correctly. All other lines will be mismatched. The vernier scale number that matches up with a thimble mark is the 0.0001 inch (one 10-thousandth of an inch) measurement.

READING A METRIC OUTSIDE MICROMETER

Reading of the metric outside micrometer is performed in the same way as the inch graduated micrometer except that the graduations are in the metric

system of measurement. Readings are performed as follows (see **Figure 2-44**):

- Each number of the sleeve of the micrometer represents 5 millimeters (see **Figure 2-44A**).
- Each of the ten equal spaces between each number, with index lines alternating above and below the horizontal line, represents 0.5 millimeter or five tenths of a millimeter. One revolution of the thimble changes the reading one space on the sleeve scale or 0.5 mm (see **Figure 2-44B**).
- The beveled edge of the thimble is divided into 50 equal divisions with every fifth line numbered 0, 5, 10 . . . 45. Since one completer revolution of the thimble advances the spindle 0.5, each graduation on the thimble is equal to 1/50 of 0.05 or one hundredth of a millimeter (0.01 mm) (see **Figure 2-44C**).

Just like the inch graduated micrometer, the three separate readings are added together to attain the total reading.

1. Start by reading the largest number on the sleeve that has been uncovered by the thimble. In this illustration it is 5, which means the first number in the series is 5 mm.

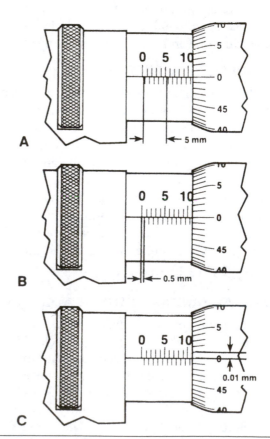

Figure 2-44 Reading a metric micrometer: (A) 5 mm; (B) 0.5 mm; (C) 0.01 mm.

2. Next count the number of lines past the number that the thimble has uncovered. In the example, this is 4, and since each space is equal to 0.5 mm, 4 spaces equal 4 × 0.50 or 2 mm. This added to the figure obtained in step 1 equals 7 mm.

3. Finally read the graduation line on the thimble that coincides with the horizontal line of the sleeve scale and add this to the total obtained in step 2. In this example the thimble scale reads 28 or 0.28 mm. This added to the 7 mm from step 2 gives a total reading of 7.28 mm.

Large number	5.00 mm
Lines past large number 2×0.5	2.00 mm
Thimble scale reading	0.28 mm
	7.28 mm

READING AN INSIDE MICROMETER

An inside mike is a tool used to measure bore sizes. Often they are used with outside mikes to reduce the chance of error.

To make a measurement with an inside mike, place it inside the hole or bore to be measured and extend it till the measuring surfaces of the mike touch the bore surface. If the bore is large, it might be necessary to use an extension rod to increase the measuring range. These extension rods are available in various lengths. The inside micrometer is read in the same manner as an outside micrometer.

To take a precise measurement in either inch or metric graduations, hold the anvil firmly against one side of the bore and rock the inside mike back and forth and from side to side. This will ensure that the mike fits in the center of the work with the correct amount of resistance. Like the outside micrometer, this procedure will require a little practice until you get the feel for the correct amount of resistance and fit of the mike.

Once a measurement has been made with an inside mike, it can be double-checked with an outside mike being used to actually measure the inside mike. This practice reduces the chance of errors and helps ensure an accurate measurement.

MICROMETER CARE

- Micrometers should always be clean before use.
- Do not touch the measuring surfaces.
- The micrometer must be stored properly. The spindle face should not come into contact with the anvil face, or a change in temperature could spring the micrometer.

- Clean the micrometer after every use. Wipe it clean of any oil, dirt, or dust using a lint-free cloth.
- A mike should never be handled roughly. It is a sensitive instrument that must be handled with care. Dropping a mike can ruin it for good.
- Check the calibration of the mike weekly. If the mike is accidentally bumped, jarred, or dropped, its calibration should be checked immediately.

Digital Measuring Tools

Today there are many precision digital measuring tools that are available that can make the life of a technician much easier than the more labor-intensive mathematical additions that must be performed with conventional measuring equipment. But a technician who has mastered the reading of standard micrometers and vernier calipers will have a much better understanding of what a digital reading actually represents. Digital measuring tools can be used to measure anything from brake drum inside diameter to a cylinder head valve guide bore and almost anything in between. Digital measuring tools have the following advantages:

- They are just as accurate as their mechanical counterparts.
- They are faster than their mechanical counterparts.
- They eliminate technician reading and interpretation errors.
- With the push of a button they can convert standard to metric conversions (and vice versa).
- They can be zeroed at any location.
- They are more versatile than mechanical measuring tools: One quality digital caliper can replace a full set of 1-inch through 6-inch micrometers.
- They are often lower in cost than their equivalent mechanical counterparts.

Digital measuring tools compute precise linear and bore measurements with greater accuracy than their mechanical counterparts. Just as the scientific calculator made slide rulers obsolete, digital measuring tools will make the need to interpret the calibrations on a micrometer or vernier scale a thing of the past. The only mystery is why it is taking so long for many experienced technicians to adopt this relatively new technology.

Other Measuring Tools

There are a few more measuring tools that truck technicians may have the opportunity to use.

Figure 2-45 A typical thickness gauge pack.

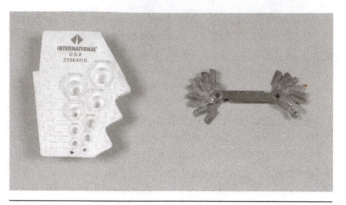

Figure 2-46 A screw pitch gauge.

THICKNESS GAUGES

The **thickness gauge** or **feeler gauge** consists of strips of metal of a known and closely controlled thickness. Several of these metal strips are often combined in a multiple measuring instrument that pivots in a manner similar to a pocket knife. The desired feeler gauge can be pivoted away from the others for convenient use. A standard feeler gauge pack contains leaves of 0.002 to 0.024 inch thickness in step increments of 0.002 inch. Feeler gauges up to 0.050 inch are available (see **Figure 2-45**).

SCREW PITCH GAUGES

In order to identify the thread pitch of fastener, a **screw pitch gauge** can be used. This handy little tool provides a quick and accurate method of checking the threads per inch. The leaves of this tool are marked with the various pitches. The technician just holds the teeth of the leaf up to the threads of the fastener until a match is found; the pitch can be read directly from that leaf.

Screw pitch gauges are available for the various types of fastener thread used in the truck industry: American National coarse and fine threads, metric threads, and International Standard threads (see **Figure 2-46**).

TORQUE WRENCHES

Torque wrenches are used to apply a specific amount of turning force to a fastener (bolt or nut). Conventional torque wrenches have scales there are usually read in foot pounds (ft. lb) or inch pounds (in. lb) and metric scales in Newton-meters (N-m).

Every truck manufacturer lists torque specifications for fasteners on its vehicles. This is evidence of this tool's importance to the truck technician when tightening nuts or bolts.

There are three general types of torque wrenches, which are the flex bar, dial indicator, and sound-indicating (click) types. The flex bar type is fairly accurate and inexpensive. The dial indicator type is very accurate but, like the flex bar type, can be hard to read in tight quarters. The sound-indicating type torque wrench is fast and easy to use. It makes a pop or click sound when a preselected torque value is reached. With this type of torque wrench, it is not necessary to watch an indicator needle while torquing.

A torque wrench makes it possible to apply the correct clamping pressure without overstressing either the tool or the fastener. A torque wrench should never be used to remove or disassemble fasteners but only for tightening purposes (see **Figure 2-47**).

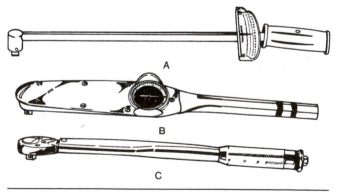

Figure 2-47 Three types of torque wrenches: (A) a flex bar; (B) a dial indicator; and (C) a click or sound-indicating wrench. *(Courtesy of Snap-on Tools Company)*

Summary

- Tools give the technician the ability to put his thoughts into action.
- Screwdriver size is determined by the length of the shank and the size of the tip.
- A hammer is generally classified by the weight and material of the head.
- The hacksaw can be used to cut bolts, angle iron tubing, and so on.
- Tubing tools are made for such tasks as cutting, swaging (increasing the size of a piece of tubing), flaring, and bending.
- Chisels are a tool that can be used to cut metal by driving it with a hammer or air-powered hammer.
- Wrenches are one of the most essential tools in the industry. A well-equipped technician will have both standard and metric wrenches in a variety of lengths and sizes.
- Socket wrenches are used in many situations and are faster, safer, and easier to use than an open-end wrench.
- Pliers are a general-purpose tool used for gripping wires, clips, and pins.
- The purpose of a file is to remove small amounts of metal, for shaping, smoothing, or sharpening parts.
- A tap is a hand tool used to cut internal threads.
- Dies are used to cut external threads.
- Power tools can make the job of a technician less physically demanding.
- Blowguns are used for blowing material off parts during cleaning. Never point a blowgun at yourself or someone else.
- Accurate measuring tools are an indispensable tool to a truck technician.
- A machinist's rule is a steel ruler not unlike a common ruler.
- A dial caliper can be used for measuring both inside and outside and depth and step measurements.
- A micrometer is commonly used for measuring inside or outside diameters of shafts and bores.
- Digital measuring tools are available that can make the life of a technician much easier than the more labor-intensive tools requiring mathematical addition.
- Thickness gauges (or feeler gauges) are strips of metal of a known and closely controlled thickness.
- In order to identify the thread pitch of fastener, a screw pitch gauge can be used.
- Torque wrenches are used to apply a specific amount of turning force to a fastener (bolt or nut).

Review Questions

1. From the following, which is **not** a safety requirement for using a hammer safely?

   A. Keep the hammer face parallel with the surface being struck.

   B. Choose the correct size hammer for the job.

   C. Always wear eye protection.

   D. Grasp the hammer as closely as possible to the head.

2. When using a hacksaw, why should you slow down when approaching the end of the cut?

   A. To prevent dulling the blade

   B. To prevent the blade from breaking

   C. To prevent the material from slipping

   D. To prevent injury to the user's hands

3. From the following list items, which would be considered a specialty fastener?

   A. TORK®

   B. Posi-Drive™

   C. Phillips

   D. Robertson

   E. Both A and B are correct.

4. You should never strike a punch or chisel upon what material? _____

5. Which of the following is recommended for the maintenance and care of a screwdriver?

   A. Shape the tip with a grinder.

   B. Reshape the tip of the screwdriver with a hammer.

   C. Periodically harden the screwdriver tip with a torch.

   D. Reshape the tip of the screwdriver with a hand file.

6. From the following wrenches, which one allows the technician to work faster?

   A. Open-end wrench

   B. Box-end wrench

   C. Combination wrench

   D. Socket wrench and ratchet

7. From the following list of pliers, which is best suited for gripping small components?

   A. Needlenose pliers

   B. Locking pliers

   C. Diagonal cutting pliers

   D. Combination pliers

8. If you use a coarse cut chisel on soft material, it will prevent _____.

   A. dragging

   B. injury

   C. damaging the file

   D. clogging

9. Hand taps are different from dies in the fact that:

   A. they remove, rather than cut, threads.

   B. they increase thread pitch.

   C. they cut external threads.

   D. they cut internal threads.

10. What tool should be used to remove precision bearings or gears?

    A. Swing press (hammer)

    B. Prybar

    C. Puller

    D. Gear and bearing wrenches

11. Which of the following accessories, used with an air chisel or hammer, can be used to install pins and drive seized bolts?

    A. Universal joint and tie-rod separator

    B. Ball joint separator

    C. Exhaust pipe cutter

    D. Tapered punch

12. From the following list of measuring tools, which one has the greatest precision?

    A. Metric machinist's rule

    B. Decimal machinist's rule

    C. Thickness gauge

    D. Micrometer

13. What part of a micrometer is used between the thumb and forefinger to bring the measuring points into contact with surfaces being measured?

    A. Thimble

    B. Spindle

    C. Sleeve

    D. Frame

14. When should a micrometer's calibration be checked?

   _____

   _____ .

15. Which of the following would a screw pitch gauge be used for?

   A. To determine the number of threads per inch

   B. To measure thickness

   C. To measure turning force

   D. To determine clamping force

CHAPTER

3 Fasteners

Objectives

Upon completion and review of this chapter, the student should be able to:

- Explain how fasteners are graded.
- Describe what a Huck fastener is and where it is used in truck assembly.
- Demonstrate the proper use of a torque wrench.
- Explain why proper torque is essential.
- List three groups that twist drills are divided into.
- Demonstrate how to check a drill bit for proper cutting angle and lip length.
- Describe how threads may be repaired.
- Describe ways in which broken studs or fasteners may be extracted.
- Explain procedures for using thread locking compounds.
- List procedures for performing buck riveting.

Key Terms

aerobic	Heli-Coil™	stud removers
anaerobic	Huck fastener	thread chaser
brazier head	rivet	torque-to-yield
bucking bar	rivet set	yield point
elastic	screw extractor	

INTRODUCTION

Technicians working in truck shops require a general knowledge of the fasteners that they use daily to perform service tasks on today's trucks. Use of the correct fastener and torque can make the difference between a successful repair and a catastrophic failure, causing bodily injury or even death.

Because of damage or corrosion, fasteners' threads can become damaged and may need to be repaired in order to save a costly component. For this reason, technicians must know how to recondition threads while not sacrificing any of the intended strength or clamping force. Technicians must also be able to tighten fasteners to their intended load without going over the prescribed torque and damaging the fastener.

SYSTEM OVERVIEW

This chapter explains how a technician can identify the many different fasteners that most truck components and shop equipment use. Depending on the manufacturer of the vehicle, the fasteners can be standard SAE or metric-size fasteners. These fasteners are also available in different grades. The technician must be able to distinguish the type, size, and grade of fasteners so that exact replacement fasteners can be selected when replacement becomes necessary.

The chapter also explains the reasons why a technician must have the skills necessary for tightening fasteners to their recommended torque without overtorquing.

Thread repair and stud removal are also discussed and are challenging skills that all technicians will be faced with sooner or later.

Finally, the technician is introduced to another fastening technique called buck riveting. This riveting technique is used extensively in truck and coach cabs as well as trailers.

FASTENERS

All trucks and trailers on our roads use a combination of metric and standard fasteners. Metric fasteners are common but not universal in late model vehicles. The engine and major components use metric fasteners almost exclusively (diameter and pitch are measured in millimeters).

The majority of fasteners on a truck are hex head design. Hardened flat washers are used under the bolt head, between the clamped components and the hex nut, to distribute the load; this prevents localized stress. The washers are cadmium or zinc plated and have a hardness rating of 38 to 45 Rockwell "C" hardness (HRC).

Some fasteners, often those smaller than half-inch diameter, have integral flanges that fit against the clamped surfaces. These flanges eliminate the requirement for washers.

Fastener Grades and Classes

Fasteners are grouped into grades based on the tensile strength of the fastener. These grades are established by the SAE or the International Fastener Institute (IFI). The higher the number (or letter), the stronger the fastener. The grade on a bolt (capscrew) can be identified by the number of pattern or radial lines forged on the bolt head (see **Figure 3-1**).

The grade of hex nuts and lock nuts can also be identified by the number of pattern or axial lines and dots on various surfaces of the nut (see **Figure 3-2**).

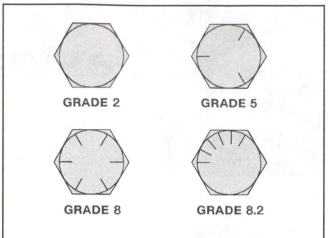

NOTE: Grade 2 bolts have no grade markings; grade 2 bolts are rarely used by Freightliner.

These grade markings are used on plain hex–type and flanged bolts (capscrews). In addition to the grade markings, the bolt head must also carry the manufacturer's trademark or identification.

Figure 3-1 Bolt (capscrew) identification. *(Courtesy of Freightliner, LLC)*

Most bolts used on heavy duty vehicles are grades 5, 8, and 8.2. Matching grades of hex nuts are used with grade 5 bolts; grade 8, grade C, or grade G (flanged) hex nuts are used with grade 8 or 8.2 bolts.

Metric fasteners are divided into classes adopted by the American National Standards Institute (ANSI). The higher the number of the class, the stronger the fastener. The class of the bolt can be identified by numbers forged on the bolt head (see **Figure 3-3**). The grade of hex nuts and lock nuts can be identified by the lines or numbers on various surface of the nut (see **Figure 3-4**) Class 8 hex nuts are always used with class 8.8 bolts, and class 10 hex nuts with class 10.9 bolts. The threads of a bolt can be measured with a thread pitch gauge.

Frame Fasteners

Most components that are attached to the frame are done so with threaded fasteners; grade 8 and 8.2 phosphate and oil-coated hex head bots and grade C cadmium-plated and wax-coated prevailing torque lock nuts are used. Prevailing torque lock nuts have distorted sections of threads to provide torque retention. For attachments where clearance is minimal, low-profile hex head bolts and grade C prevailing torque lock nuts are used (see **Figure 3-5**).

Many truck frames today are assembled using a fastener with constant clamping force such as a

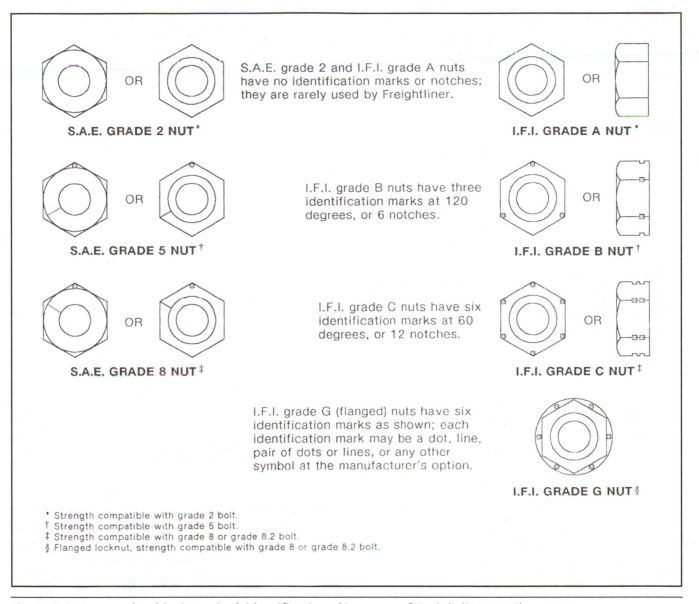

Figure 3-2 Hex nut (and lock washer) identification. *(Courtesy of Freightliner, LLC)*

Huck fastener. A Huck fastener is installed with the use of hydraulic clamping forces. The bolt is under great hydraulic tension while a collar is hydraulically squeezed to lock the bolt threads. These fasteners are a one-time application and must be destroyed to be disassembled.

Tightening Fasteners

When a bolt or capscrew is torqued to the fastener's specified torque, or a nut is tightened to its torque value on a bolt, the shank of the bolt or capscrew stretches slightly. This stretching (tensioning) results in a preload. OEM-specific torque values are calculated to provide enough clamping force on bolted components and the correct tension on the fastener to maintain the clamping force.

Overtensioning of a fastener causes permanent stretching of the fastener, which can result in breakage of the component or fastener. Overtensioning can be avoided by using a torque wrench to tighten fasteners.

When torquing a fastener, typically 80 to 90 percent of the turning force is used to overcome thread, cap, and nut-face friction; only 10 to 20 percent results in capscrew or bolt-clamping force. About 40 to 50 percent of the turning force is needed to overcome the friction between the underside of the capscrew head or nut and the washer. An additional 30 to 40 percent is needed to overcome the friction between the threads of the capscrew and their threaded hole, or the friction between the threads of the nut and bolt.

To some extent, all metal has some elastic properties. This means that the material can be both

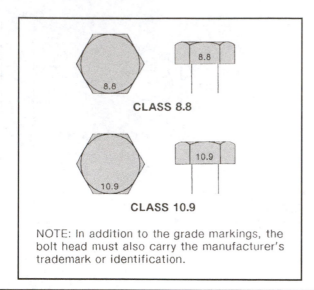

Figure 3-3 Bolt classes can be identified by the numbers forged on the head of the bolt. *(Courtesy of Freightliner, LLC)*

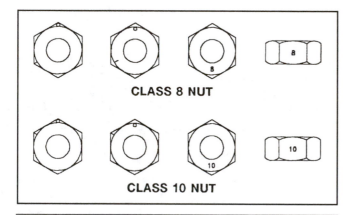

Figure 3-4 Identification markings on Class 8 and Class 10 nuts. *(Courtesy of Freightliner, LLC)*

stretched and compressed to a certain point. It is this **elastic** or springlike property that provides the clamping force when a bolt is threaded into a tapped hole or when a nut is tightened. As the bolt is stretched, clamping force is created due to the tension of the bolt.

Just like a spring, the more a bolt is stretched the tighter it will become. That being said, a bolt can also be stretched too far, in which case it will lose its elasticity and therefore lose its clamping load. A bolt that has been overstretched will no longer safely clamp the load it was designed to support.

As previously mentioned, the elasticity of a bolt allows it to be stretched to a certain point, and when the tension placed on the bolt is reduced, the bolt will return to its original, normal size. If a bolt is stretched too much, the tension will cause to bolt to get to its

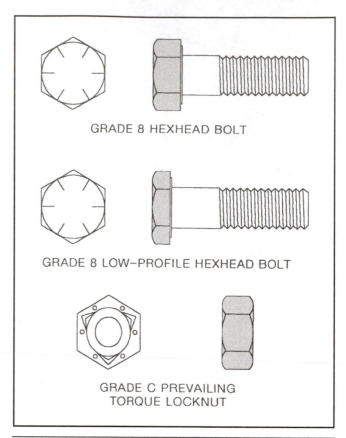

Figure 3-5 A Grade C prevailing torque lock nut. *(Courtesy of Freightliner, LLC)*

yield point. When a bolt is stretched to its yield point, it will no longer return to its original, normal size after the load on the bolt has been removed. A bolt will continue to stretch a little more each time it is used; just like bending an aluminum can back and forth, it will stretch until it finally breaks (see **Figure 3-6**).

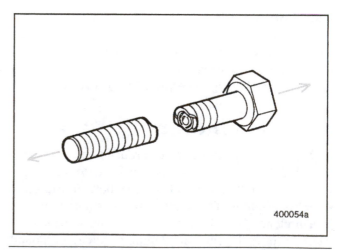

Figure 3-6 This bolt has been torqued beyond its yield point, resulting in shear.

The proper use of a torque wrench will eliminate the overstretching of fasteners to a position beyond their yield point. When a torque value is calculated, it is done so at 25 percent below the fastener's yield point for a safety factor. There are some fasteners that are intentionally torqued just barely into a yield condition, although not quite enough to create the classic coke bottle shape of a necked-out bolt. This type of fastener, known as a **torque-to-yield** (TTY) bolt, produces close to 100 percent of its tensile strength, compared to around 75 percent on a regular fastener when both are torqued to specification. A TTY fastener should never be reused unless otherwise specified. Engine heads and main bearings often use a TTY fastener.

The torque required to tighten a fastener is reduced when friction is reduced. If a fastener is dry (unlubricated) and plain (unplated), thread friction is high. If a fastener is wax-coated or oiled, or has a zinc phosphate coating or cadmium plating, friction forces are reduced. Each of these coatings and combinations of coatings has a different effect. Using zinc-plated hardened flat washers under the bolt (capscrew) head and nut reduces friction. Dirt or other foreign material on the threads or component surface of the fastener or component can increase friction to the point where the torque specification is met before a clamping force is produced.

Although there are many conditions that will affect the amount of friction, a different torque value cannot be given for each. Most OEMs recommend that all fasteners be lubricated with oil (unless specifically instructed to install them dry), and then torqued to the values for lubricated and plated fasteners. If a locking compound or an antiseize compound is recommended for a fastener, the compound acts as a lubricant, and oil is not required.

Impact wrenches can easily strip bolt thread, so care must always be exercised. These tools can also put a bolt beyond its yield point within a split second. Some friction is required to prevent a nut from spinning. When a nut is lubricated, there is no friction left to stop the impact wrench from hammering the nut beyond the bolt yield point and/or stripping the threads.

While using an impact wrench, never run a nut at full speed onto the threads of a bolt. The nut should be run up slowly until it comes into contact with the work, and then make note of the socket position, observing how far it turns. Small air-powered tools like air ratchets do not produce the severe turning forces that an impact wrench produces and are much safer. This procedure should be followed when using a torque-modulated air wrench as well.

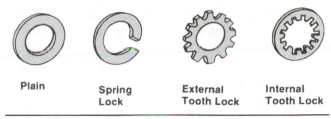

Figure 3-7 Washers are sometimes used to lock fasteners to keep them from coming loose.

When it comes to lock washers, a rule of thumb is, if the fastener assembly did not come with one, do not add one. Lock washers are extremely hard and tend to break under severe pressure. Use lock nuts with hard, flat washers. Properly torqued, this type of fastener should never loosen even when lubricated (see **Figure 3-7**).

Fastener Replacement

The following points should be kept in mind when installing replacement fasteners:

- When replacing fasteners, use only identical bolts, washers, and nuts that are the same size, strength, and finish as originally specified.
- When replacing graded (or metric) bolts and capscrews, use only fasteners that have the manufacturer trademark or identification on the bolt head. Don't settle for inferior fasteners.
- When using nuts with bolts, use a grade (or class) of nut that matches the bolt.
- If installing nonflanged fasteners, use hardened steel flat washers under the bolt (capscrew) head and under the hex nut or lock nut.
- For bolts 4 inches (100 mm) or less in length, make sure that at least one and a half threads and no more than 5/8 inch (16 mm) bolt length extends through the nut after it has been tightened. For bolts longer than 4 inches (100 mm), allow a minimum of one and a half threads and a maximum of 3/4 inch (19 mm) bolt length protrusion.
- Never hammer or screw a bolt into place. Align the holes of the mating components so that the fastener surfaces are flush with the washers and the washers are flush with the clamped surfaces.
- When installing a fastener into threaded aluminum or plastic components, start the fasteners by hand, to prevent cross-threading into the soft material.
- Do not use a lock washer (split or toothed) next to aluminum surfaces.
- When installing studs that do not have an interference fit, install them with thread locking compound.

- When installing components mounted on studs, use free-spinning (nonlocking) nuts and helical spring (split) lock washers or internal-tooth lock washers. Avoid using lock nuts because they tend to loosen the studs during removal. Do not use flat washers.
- Do not use lock washers and flat washers in combination (against each other); each defeats the other's purpose.
- Use stainless steel fasteners against chrome plating, unpainted aluminum, or stainless steel.

Tightening Fasteners

When tightening fasteners, keep in mind the following procedures:

- Clean all fasteners, threads, and all surfaces before installing them.
- To ensure fasteners are torqued accurately, fasteners should be lubricated with oil (unless you are specifically instructed to install them dry), and then torqued to the values for lubricated- and plated-thread fasteners. When locking compound or antiseize compound is used, oil is not needed.
- Hand torque fasteners until they contact before using a torque wrench to tighten them to their final torque value.
- Whenever possible, tighten the nut, not the bolt head. This gives a truer torque reading by eliminating bolt body friction.
- Always use a torque wrench to tighten fasteners, and use a slow, smooth, even pull on the wrench. Do not use a short, jerky motion, or inaccurate readings can result.
- If using a bar-type torque wrench, try to look straight down on the scale. Viewing the scale from an angle can give a false reading.
- Only pull on the handle of a torque wrench. Do not allow the beam of the wrench to touch anything.
- Bolts should be tightened up in stages. First torque to one-half the specified torque, then to three-quarters, full torque value, and full torque value a second time.
- Never overtorque fasteners. Overtightening causes permanent stretching of fasteners, which can result in breakage of parts or fasteners.
- If specific torque values are not given for countersunk bolts, use the torque value for the corresponding size and grade of regular bolt.
- Follow the torque sequence when provided to ensure that clamping forces are evenly distributed and mating parts and fasteners are not distorted.

TWIST DRILLS

A twist drill is a tool used extensively by technicians and comes in many different styles to accommodate the drilling of holes in almost any surface.

Twist drills are available in three different sizing types:

- Fractional in sizes 1/6 inch to 1/2 inch (1.5 mm to 13 mm)
- Numbers (wire gauge) #1 (0.228 in.) to #60 (0.040 in.). The larger the number the smaller the drill bit.
- Letters in sizes A to Z. The further down the alphabet the larger the drill bit, A being the smallest and Z being the largest.

For general shop use, a fractional and number drill index should be all that is required. Twist drills with a straight shank are used in portable drills and light-duty drill presses. For heavy-duty power drills, the tapered shank drill is used.

Drill bits are commonly manufactured from carbon steel or high-speed steel. A carbon drill bit will require more frequent sharpening and it will not last nearly as long as a high-speed drill bit.

Starting the Drill Bit

Before drilling, it is important that the intended hole be marked out exactly. Once the hole is marked, a center punch should be used to locate the drill bit. This will prevent the drill from skating away from the intended location once the bit is spinning. Apply reasonable pressure until the drill bit cuts a dimple in the material and again check the position. If it is good, proceed; if not, center punch again in the correct spot and start again. If it is possible, use a cutting lubricant while drilling. This will keep the drill bit cool and allow it to remain sharper longer.

Securing the Work Piece

Many preventable accidents occur each year as a result of using a drill press with the work piece unsecured. What generally happens is that the technician thinks he can hold the work piece firmly on the table of the drill press, but if the bit grabs the work it can spin the work, tearing it out of the grasp of the worker. The work piece will now freely spin at the rpm of the drill bit, slamming into everything, including the technician. This will usually occur just as the bit goes through the other side of the work piece.

Sharpening Drill Bits

A skill that every technician should possess is the ability to sharpen a drill bit. If a drill bit were thrown away each time it became dull, it would be expensive and a waste of a drill bit. Although a drill bit can be sharpened freehand on a grinder, the results are generally not all that impressive. Drill grinding jigs can be used with better results. When properly sharpened, each lip of a drill is the same length and has the same angle in relation to the axis of the drill. Use a drill bit gauge to check the correct angle of 59 degrees as well as the lip length.

A cone-pointed drill of two lips relies on four factors for it to work efficiently:

1. The cutting lips have the same inclination to the axis of the drill.
2. Cutting lips are all exactly the same length.
3. Surface back of the cutting edges having the proper lip clearance
4. Correct angle of lip clearance

After sharpening a drill bit freehand, install the bit in a drill press and feed the bit into a piece of scrap steel and observe:

- The chips made by the cutting
- The size of the hole

If the cutting lips have been sharpened to the proper clearance, the chips will curl as they start from the cutting edge. If the cutting lips lack the proper clearance, the chips will look more like they have been ground off rather that cut off. If the cutting lips are not equal in length, the hole will be enlarged over the diameter of the drill.

THREAD REPAIR

A technician must be able to make repairs to threads quickly and correctly (see **Photo Sequence 1**). Threaded holes in parts can become damaged, requiring repairs to salvage the parts.

Minor Thread Repairs

Minor thread damage consists of things like nicks, and partial flattening of the threads. Minor thread damage can usually be repaired with a **thread chaser** or threading tool. A thread chaser is available for "cleaning up" slightly damaged internal or external threads. The chase is run through or over the threads, much like a tap or die (see **Chapter 2**) to restore them.

Major Thread Repair

Major thread damage generally includes badly smashed threads or stripped threads that cannot be repaired easily. Sometimes major thread damage can be repaired with a tap or die. Bolts and studs with major damage should be replaced if possible. Stripped threads are common problems with threaded fasteners. This can be caused by high torque or by cross-threading. Threads can sometimes be replaced with the use of thread inserts. Several types of threaded inserts are available, but the helical coil insert (**Heli-Coil**^{TM}) is the most common (see **Figure 3-8**).

To install this or other reconditioning inserts, proceed as follows:

1. Determine the size, pitch, and length of the threads required to make the repair. Refer to the insert manufacturer for correct size drill for the threaded tap to be used for the repair.
2. Using the specified drill bit, drill out the damaged threads. Clean out the drill swath and chips from the hole.
3. Tap the new threads in the hole using the specified tap. Lubricate the tap while threading the hole; back out the tap every quarter turn or whenever the tap feels like it's binding. When the hole is threaded to the required depth, remove the tap and all metal cuttings from the hole.
4. Select the proper size insert and install it by screwing it in with the special installing mandrel or tool. Make sure the tool engages with the tang of the insert. Screw the insert in the hole by

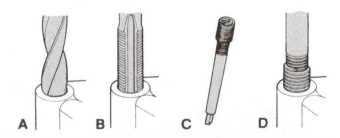

Figure 3-8 Steps in the installation of a helical screw repair coil: (A) Drill the damaged threads using the correct size drill bit. Clean all metal chips out of the hole. (B) Tap new threads in the hole using the specified tap. The thread depth should exceed the length of the bolt. (C) Install the proper size coil insert on the mandrel provided in the installation kit. Bottom it against the tang. (D) Lubricate the insert with oil and thread it into the hole until it is flush with the surface. Use a punch to sidecutters to break off the tang.

PHOTO SEQUENCE 1 Repairing Damaged Threads

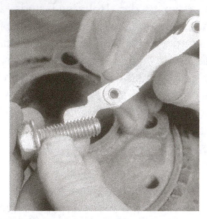

P1-1 Using a threaded pitch gauge, determine the thread size of the fastener that should fit into the damaged internal threads.

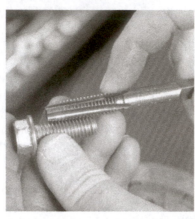

P1-2 Select the correct size and type of tap for the threads and bore to be repaired.

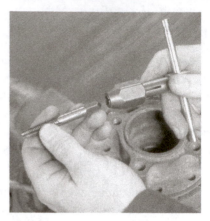

P1-3 Install the tap into a tap wrench.

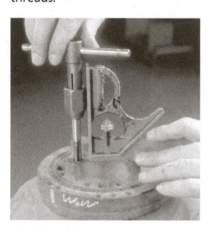

P1-4 Start the tap squarely in the threaded hole using a machinist square as a guide.

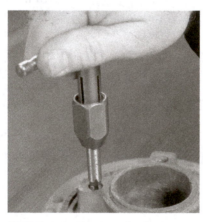

P1-5 Rotate the tap clockwise into the bore until the tap has been run through the entire length of the threads.

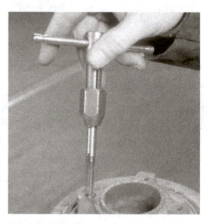

P1-6 Drive the tap back out of the hole by turning it counterclockwise.

P1-7 Clean the metal chips left by the tap out of the hole.

P1-8 Inspect the threads cut by the tap to be sure they are acceptable.

P1-9 Test the threads by threading the correct fastener into the threaded hole.

For more information on Heli-Coils, visit http://www.emhart.com/products/helicoil.asp

Figure 3-9 Two types of stud removers.

turning the installing tool clockwise. Lubricate the thread insert with engine oil if it is installed in cast iron. (Do not lubricate if installing into aluminum.) Turn the thread insert into the tapped hole until it is flush with the surface or one turn below. Remove the installer.

Screw/Stud Removers and Extractors

Two types of **stud removers** are shown in **Figure 3-9.** A stud remover can be used for either installing or removing studs. Stud removers have a large, hardened, grooved eccentric roller or jaws that grip the stud tightly when turned. Stud removers/installers may be turned with a 3/8- to 1/2-inch ratchet.

If a bolt or stud has broken below the surface, it will require the use of an extractor. Twist drills, fluted extractor, and hex nuts are included in a **screw extractor** set (see **Figure 3-10**). This type of extractor reduces the tendency to expand a screw or stud that has been drilled out by providing gripping contact through the full length of the stud.

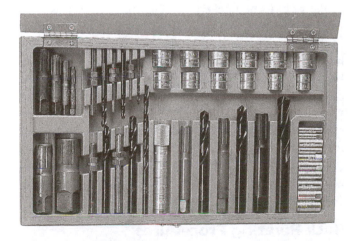

Figure 3-10 A screw extractor set. *(Courtesy of Snap-on Tools Company)*

Shop Talk: For some fasteners, penetrating fluid will help to remove a stubborn fastener. In some applications, heat may be used to aid in the removal of a broken fastener. If it is possible to heat the area around a broken bolt (not the bolt itself), the part may expand enough to allow the broken stud or bolt to turn. Great care must be taken to perform this task. In other cases where an extractor fails to perform the task, skilled technicians weld a nut to the broken stud and unwind the stud using the nut. Another method if all else fails is to use a small chisel. If you can get the chisel on the outer edge of the fastener, drive the fastener in a counterclockwise direction. The impact of the hammer can sometimes free up the threads.

Thread Locking Compound Application

When it is necessary to use a thread locking compound, follow the safety precautions given on the locking compound container. Then continue as follows:

1. Clean both the male and female threads of the fastener to remove dirt, oil, and other contaminants. Use solvent to ensure threads are completely clean, and allow drying for 10 minutes. Ensure solvent is completely evaporated before applying thread adhesive.
2. Apply a small amount of locking compound from the container to the circumference of three or four fastener threads.
3. Install right away and torque the fastener. Retorquing of the fastener is not possible after installation without destroying the adhesive locking bond.

Shop Talk: Thread locking compounds are available in different strengths, so choose the compound that suits your situation.

CAUTION *Thread locking compounds are powerful adhesives. They are color coded, so use only the color code recommended by the OEM.*

In order to remove fasteners that have been held together with a thread locking compound, it may be necessary to heat the bond line to 400°F (204°C) before the nut may be removed. Every time the fasteners are disassembled, replace them. If mating components are damaged by overheating, replace them.

RIVETING

A very common fastener used on tractor cabs, on buses, and on many unibody-style trailers is the **rivet.** This fastener is commonly manufactured from aluminum alloy, mild steel, or stainless steel. A rivet is a clamping style of fastener and must have two heads to clap the materials together. One head is preformed by the manufacturer and is referred to as the manufactured head. The other head is one that is formed when the rivet is being driven through the material to be clamped. The formed head is referred to as a bucked head and is produced when the rivet gun drives the rivet into the face of the **bucking bar.**

A **rivet set** is installed into an air-driven rivet gun or air hammer. The rivet set is driven against the manufactured head and must be the proper style and size so as not to damage the manufactured head. An incorrect size rivet set will damage either the rivet or the skin panels. Too large a set will flatten the rivet head and mar the panels. The rivet gun delivers high-frequency hammer blows that form the bucked head when backed up against a bucking bar. This is a two-person operation: one to operate the rivet gun and the other to hold the bucking bar on the opposite side of the work pieces.

Rivet Types

There are two different styles of rivets that may be used in the manufacturing process for truck cabs and trailer bodies: the **brazier head** and the flush head (see **Figure 3-11**). The rivet set that is used to drive a brazier head rivet is cup-shaped to fit the convex head of

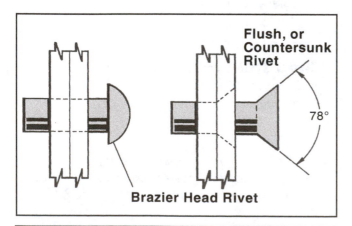

Figure 3-11 Two common types of rivets. *(Courtesy of Freightliner, LLC)*

the rivet. The flush type gets its name from the angle between the shoulders of the rivet. The rivet requires the use of a "mushroom" flush-type rivet set. The rivet set is the component that transmits the hammering force from the rivet gut to the rivet head. The rivet set end must precisely fit the manufactured head of the rivet. If they do not fit, damage to the rivet or the material being fastened can occur, and the result will appear unprofessional. A rivet set has the nominal size of the rivet it is designed to buck stamped on the head.

Rivet Gun

An air hammer or pneumatic rivet gun produces the hammering force required to drive the rivet. Rivet guns are produced in varying sizes. Each rivet gun size has a defined range of rivet sizes that it can drive. Gun sizes are as follows: 2X, 3X, 4X, 5X, 9X, and so on. The larger the number of the gun, the larger the rivet it will be able to buck. In general, rivets 3/16-inch diameter or less can be driven with any one of the following gun sizes: 2X, 3X, or 4X.

Usually a quick-connect air line is attached to the gun, with some models using an air regulator in between. An air regulator allows the technician to adjust the volume of air delivered to the gun to adjust the hitting power of the tool. A retainer spring on the rivet gun barrel is a safety feature that prevents the rivet set from flying out of the gun during operation.

Bucking Bar

A bucking bar is a tool made of hardened steel and is available in a variety of sizes and shapes. Many technicians make their own bucking bars for components found in their shops. (A discarded brake S-cam is used by many trailer technicians.) The required weight of a bucking bar depends on the size, strength, and position of the rivet to be bucked. Bucking bars vary in weight from approximately 1 to 15 lb (0.4 to 6.8 kg) with most in the 5 lb range (2.2 kg). One end of the bucking bar has a polished end known as the face. The face of the bucking bar is held firmly against the protruding shank of the rivet opposite the rivet head and parallel to the work being fastened. When the technician operates the rivet gun, the rivet shank will flatten against the face of the bucking bar producing a uniformly bucked head (with practice). When using a rivet gun, be sure to follow the gun manufacturer's instructions to the letter. Always wear eye and hearing protection.

Buck Riveting Procedures

It is important that all rivets be both driven and bucked properly. Improper use of the tools can cause

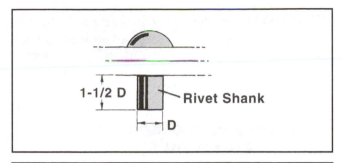

Figure 3-12 The correct rivet shank length before bucking. *(Courtesy of Freightliner, LLC)*

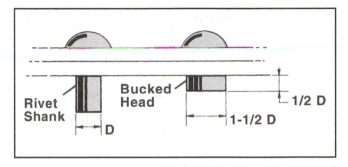

Figure 3-14 A properly bucked rivet. *(Courtesy of Freightliner, LLC)*

damage to the truck's cab, the trailer panels, or the rivet itself. If one rivet in a row of rivets fails, additional load will be placed on the remaining rivets. This additional stress can cause the entire seam to fail.

Practice the following rules while riveting:

1. Start by connecting the air line to the riveting gun. Adjust the air regulator until the desired hitting power is obtained by pressing the set into a block of wood and opening the throttle. This will take some practice on the first few occasions. Experienced technicians can tell by the sound of the rivet gun which way to adjust the air regulator.

2. Drill a hole to the exact size for the rivets being used. If the hole is even slightly too big, clamping forces will be sacrificed. Place the rivet through the drilled holes of the skin panels to be fastened. As a general rule, the projection of the rivet shank, before driving, should be about one and a half times larger than the diameter of the rivet shank (see **Figure 3-12**).

3. Align the rivet gun and bucking bar with the rivet as shown in **Figure 3-13**. Apply moderate pressure to the rivet gun perpendicular to the rivet head to keep rivet set from slipping off the rivet

and damaging the panel. Ensure that the face of the bucking bar is pressed against the rivet shank and that it is kept exactly parallel to the panel while the rivet is being bucked, and this will form the bucked head. If the face of the bucking bar is canted to one side or the other, damage to the panel skin and a deformed bucked rivet head will result.

4. Drive the rivet by opening the throttle of the rivet gun. The application of the gun should only take about 1 second of hitting force if the gun is properly set. The set must be held firmly against the manufactured head and driven carefully to avoid dimpling the panel skin or "eye-browing" the rivet head. These imperfections will be highly visible even after vehicle painting.

5. Inspect the bucked head of the rivet. It should be one and a half times the diameter of the original rivet shank and half the thickness (see **Figure 3-14**).

Rivet Removal

At some point in time it may be necessary to remove a rivet. If it is required to remove a rivet, you may choose one of the following methods:

1. Using a cold chisel, cut off the manufactured head of the rivet (see **Figure 3-15A**). Be careful not to damage the surrounding panel material.

2. Using a drill bit of the same diameter as the rivet shank, drill the exact center of the manufactured head to the depth shown in **Figure 3-15B**. Using a drill punch, pry off the rivet head (see **Figure 3-15C**). Then drive out the shank with the drill punch (see **Figure 3-15D**).

3. Many trailer technicians will enter the trailer and cut off the rivet's bucked head with an air chisel. This prevents possible damage to the finished panel. An air punch is then inserted into the air hammer and the shank of the rivet is driven out of the panel.

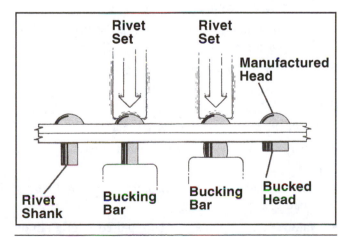

Figure 3-13 The riveting procedure. *(Courtesy of Freightliner, LLC)*

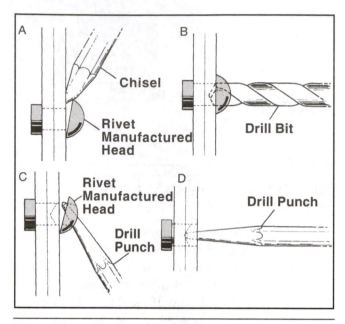

Figure 3-15 The steps to removing a buck rivet. *(Courtesy of Freightliner, LLC)*

ADHESIVES AND CHEMICAL SEALING MATERIALS

Chemical adhesives and sealants may provide added holding force and sealing ability when two components are joined. Sealants are applied to threads where fluid contact is frequent. Chemical thread retainers are either **aerobic** (cures in the presence of air) or **anaerobic** (cures in the absence of air).

When using a chemical adhesive or sealant, follow the manufacturer's instructions. Note that some adhesives bond molecularly to the surface of metals, destroying the materials on removal.

Summary

- The correct fastener and torque can make the difference between a successful repair or a catastrophic failure.
- All trucks and trailers on our roads use a combination of metric and standard fasteners.
- The majority of fasteners on a truck are hex head design.
- Fasteners are grouped into grades based on the tensile strength of the fastener.
- The grade on a bolt can be identified by the number of pattern or radial lines forged on the bolt head.
- The grade of hex nuts and lock nuts can also be identified by the number of pattern or axial lines and dots on various surfaces of the nut.
- Many truck frames today are assembled using a fastener with constant clamping force such as a Huck fastener.
- When a bolt or capscrew is tightened to its specified torque, it stretches slightly.
- Proper use of a torque wrench will eliminate the overstretching of fasteners to a position beyond their yield point.

- When a bolt is stretched to its yield point, it will no longer return to its original size after the load is removed.
- There are some fasteners that are intentionally torqued just barely into a yield condition and are called torque-to-yield fasters.
- The torque required to tighten a fastener is reduced when friction is reduced with the use of a lubricant.
- When replacing fasteners, use only identical bolts, washers, and nuts that are the same size, strength, and finish as originally specified.
- Clean all fasteners, threads, and all surfaces before installing them.
- A twist drill is a tool used to cut holes in almost any surface.
- Before drilling, mark the exact location and use a center punch to give the bit purchase.
- Always secure the work piece when operating a drill press.
- Minor thread damage can usually be repaired with a thread chaser.
- Major thread damage can be repaired with a tap or die.

- Threads can be replaced with the use of thread inserts.
- A stud remover can be used for either installing or removing studs.
- A rivet is a clamping style of fastener and must have two heads to clap the materials together.

- An air hammer or pneumatic rivet gun produces the hammering force required to drive the rivet.
- A bucking bar is used to produce a bucked head on the opposite side of the manufactured rivet head.

Review Questions

1. Which type of fastener requires the use of hardened flat washers?

   A. Any fastener with metric threads

   B. Flanged fasteners

   C. Hex-type fasteners

   D. Rivets

2. If a bolt is tightened beyond its torque rated value, what can result?

   A. The amount of friction will be reduced.

   B. The bolt's elasticity will be reduced.

   C. The bolt may be permanently stretched or even break.

   D. The bolt's elasticity will be increased.

3. Many truck manufacturers use a fastener called a Huck fastener because of the fastener's reusability as well as ease of assembly and disassembly.

   A. True

   B. False

4. What tool can prevent fasteners from being loaded to their yield point?

   A. Dial indicator

   B. Impact gun

   C. Torque wrench

   D. Air ratchet

5. When torquing a fastener that requires the use of an antiseize compound, lubricant is still required.

   A. True

   B. False

6. With regard to twist drills, a technician should be able to perform most shop procedures if he possesses a:

   A. fractional and letter drill set.

   B. a numerical drill set.

   C. a fractional and numerical drill set.

   D. and letter drill set.

7. When using a drill press, it is extremely important that the technician:

   A. use a shape drill bit.

   B. select the correct drill bit speed.

   C. secure the work piece to the drilling platform.

   D. use lubricant to cool and prolong drill bit life.

8. What is required to cure an anaerobic sealant?

   A. Heat

   B. Freezing

   C. Presence of air

   D. Absence of air

9. Threads can be repaired using:

    A. thread chasers.   D. all of the above.

    B. taps or dies.   E. none of the above.

    C. inserts.

10. Which size rivet gun can drive the largest size rivets?

    A. 2X   C. 5X

    B. 4X   D. 9X

4 Preventive Maintenance Inspection

Objectives

Upon completion and review of this chapter, the student should be able to:

- Explain in detail the positive aspects of a well-implemented preventive maintenance program.
- Explain the differences between an A, B, C, D, and L inspection.
- Follow through the steps to perform a pre-trip inspection.
- Describe maintenance issues that would require deadlining a vehicle and an out-of-service (OOS) sticker applied.
- Apply a policy of preventive maintenance scheduling that conforms to federal inspection regulations.
- Explain the responsibilities of the inspecting person and the record-keeping requirements.

Key Terms

Commercial Vehicle Safety Alliance (CVSA)

deadline

Department of Transportation (DOT)

driver's inspection report

gladhand

National Highway Traffic Safety Administration (NHTSA)

original equipment manufacturer (OEM)

out of service (OOS)

pencil inspection

PM form

pre-trip inspection

repair order

Technical and Maintenance Council (TMC)

INTRODUCTION

A preventive maintenance program should allow a vehicle owner to perform the least amount of repair maintenance while performing the routine maintenance that is necessary. Often if problems are caught and repaired early, it can save equipment owners considerable funds. If problems are ignored, the vehicle can become unsafe to operate, jeopardizing the lives of others sharing the road with these unfit vehicles. As well, making repairs on the road can be extremely costly to the owner.

SYSTEM OVERVIEW

A well-planned preventive maintenance program offers some very attractive benefits to the commercial vehicle owner. The following is a list of some of the advantages that a good program has to offer:

- The lowest possible maintenance costs
- Maximum vehicle availability
- Reduced roadside failures resulting in better customer dependability
- A reduction in possible accidents that can be caused by an unsafe vehicle
- Increased customer confidence, resulting in better public relations
- Fewer drivability complaints from drivers

There are many factors that must all work together for a PM program to be completely successful. Careful planning is key to the success of any program, and this is the job of the maintenance manager. You, the technician, are the hinging factor to success. Unless you perform the inspection properly and as scheduled and accurate records are kept, the manager cannot adjust the program for maximum efficiency.

Even though careful planning is important, that alone is not enough. Unless the maintenance manager and technicians implement the preplanned PM program through closely controlled scheduling of the vehicles for inspection and maintenance, the best plans become worthless.

Communication between drivers and technicians is extremely important as well. Because the purpose of both the driver's daily inspection and the technician's PM inspection is to detect failures before they become breakdowns, their importance to a preventive maintenance program is obvious. A driver responsibility is to perform daily pre- and post-trip inspections. These inspections are largely determined by laws with a focus on safety. The inspections performed by maintenance personal are managed in house and are performed at specified intervals determined by the maintenance manager. Today, many maintenance programs use special computer software to schedule and log maintenance inspections.

The responsibility of the technician goes beyond the inspection of the vehicle because they are also required to service and perform repairs to the vehicle. Servicing includes cleaning and lubricating as well as adjusting and tightening of components. Repairs are performed in response to any problems noted by the vehicle operator or found during the technician's inspection.

The instructions on a PM form can be very general, and the technician must know what is involved at each stage of the procedure. For example, the PM specifies to check the cooling system. The technician should know that the system should be pressure tested to 15 psi and the pressure reading checked after 10 minutes, during which time the technician can be checking the condition of the radiator hoses, cab heater water pump, tanks,

radiator core, and head gasket for leaks; then "OK" the system if the pressure indicates a loss of 1 psi or less. If a technician does not know what is required for each aspect of the checklist, he should verify what is expected of him by the maintenance manager. Once the technician knows the specifics of each step of the checklist, he should perform each check conscientiously. It is important to bear in mind that the goal of the PM program is to eliminate any potential problem through careful inspection and proper maintenance.

Both drivers and technicians must understand what is involved in each inspection. They must know how to execute the details of each step, and then must perform them properly and honestly. A major cause of PM program failure is a **"pencil inspection."** This happens when either the driver or the technician responsible for the inspections checks off the item on the PM checklist without actually verifying the condition. This type of inspection is worse than useless: the vehicle can be unsafe for the road and also waste time and resources.

Neat and accurate records are another important aspect of a successful program. Drivers must fill out driver inspection reports. Technicians work with **PM forms, repair orders,** vehicle files, and major component histories. Computer data records are making these tasks easier for both the driver and technician. The following is a list of reasons why accurate records must be kept:

- In many operations they are required by the **Department of Transportation (DOT).**
- Permanent records are invaluable from a performance standpoint. A study of failures of individual parts or components can alert the manager of the need to take corrective action. It also highlights components that have performed well.
- A maintenance manager cannot develop a systematic, preplanned preventive maintenance program without the use of these records.
- In the event of a serious accident, a vehicle's file, which contains written proof that maintenance inspections were performed and repairs were made as required, is usually beneficial in the defense of a lawsuit.

IMPLEMENTING A PM PROGRAM

A successful PM program must be tailored to fit the operation of the fleet it is to service. No single maintenance program can apply to all operations. Some of the variables that must be factored into the program are the age and type of equipment, percentage of units under warranty, distances traveled per year, and how often the unit returns home for service (some long-haul units may be away from home base for a month or more). A basic PM program is not difficult to set up. It is a good idea to follow the servicing recommendations outlined in the preventive maintenance checklist of each major **original equipment manufacturer (OEM).** Another source for maintenance is government information. The Department of Transportation has for years influenced the maintenance practices and mandatory record-keeping practices of those involved in interstate trucking. Rules and guidelines are covered in a manual called Federal Motor Carrier Safety Regulations. Not all truck operations are governed by the regulations; however, even if compliance is not mandatory, following them is a good basic guideline for any maintenance program.

The PM form is extremely important to the success of a maintenance program. This form provides the technician with an orderly list of items to be checked or inspected, and instructions on those items that should be cleaned, lubricated, tightened, adjusted, and replaced. It is the responsibility of the technician to know what must specifically be done at each step of the form. The following is a list of the basic types of inspections in alphabetical order:

- Schedule "A" is a light inspection.
- Schedule "B" is a more detailed check.
- Schedule "C" is a detailed inspection, service, and adjustment.
- Schedule "D" is a comprehensive inspection and adjustment.
- Schedule "L" is a chassis lubrication.

A road test or driver's report should be performed before any inspection is performed.

DRIVER'S INSPECTION

Generally it is the driver who first identifies the need for repairs to be performed while he does his pre-trip and post-trip inspections. A form that is vital to a well-managed maintenance program is the **Driver's Inspection Report.** This report serves two purposes: It is useful to both the repair technician and the maintenance manager in keeping the vehicle in a safe and operable condition, and it is also required by law.

As mentioned earlier, the driver is also responsible for performing a post-trip inspection, as spelled out by the Department of Transportation's Federal Motor Carrier Safety Regulations, which require preparation of a vehicle inspection report after each day's work and each vehicle driven. The form is almost always printed on the back of the driver hours of service log

and completed before drivers turn in their daily logs. Listed are the items that must be addressed:

- Service brakes
- Frames
- Parking brakes
- Sliding subframes
- Brake drums
- Tire and wheel clearance
- Brake hoses and tubing
- Tire low air pressure warning
- Wheels and rims
- Tractor protection valve
- Windshield glazing
- Air compressor
- Wipers
- Hydraulic brake systems
- Vacuum brake systems
- Fifth wheel\exhaust system
- Fuel system
- Lighting
- Cargo securement
- Steering components
- Suspensions

After the checks, the driver is also required to note any defects or problems that could affect the safe operation of the vehicle, and then to sign the report. The driver's report makes up the operator part of the maintenance program and places direct responsibility on the drivers to report malfunctions that might arise on a daily basis. If this system is followed to the letter, there should be no reason for a defective vehicle to be in service.

A driver's pre-trip inspection is less specific. In addition to ensuring that the vehicle is safe, drivers are required to review the last vehicle inspection report. If the driver identifies defects listed by the other driver that have not yet been repaired, both the technician and the new driver must sign off on the work before the vehicle can be dispatched. If the problem noted did not require repair, that too should be noted. In addition, a copy of the latest inspection report must remain with the vehicle. Carriers are required to keep the original copy of each report and the certification of repairs for at least 3 months.

While the regulations do place more emphasis on the post-trip inspections, most drivers think that the pre-trip inspection is more important. Because the driver doing the inspection will be operating the vehicle, the incentive to ensure that the vehicle is safe is probably greater than for a driver who has returned from a long trip and just wants to get home. Repairs are less expensive and faster if the driver finds them on the pre-trip before inspectors charge fines or impose penalties.

INSPECTION PROCEDURE

Both the driver and the technician should perform a **pre-trip inspection** in the same way every time to learn all the steps. If the inspection procedures are performed in the same order every time, they will become routine and there will be much less chance of overlooking or forgetting to check something altogether. The following procedure is an adaptation of the **Commercial Vehicle Safety Alliance (CVSA)** standard inspection and should be a useful guide. Sometimes this procedure is referred to as a circle check (see **Figure 4-1**):

- **Step 1 (Vehicle Overview).** Observe the general condition of the vehicle. If the truck is leaning to one side, there may be a suspension-related problem, improper load distribution, or flat or mismatched tires. Look on the ground under the vehicle for signs of fluid leaks such as engine coolant, engine oil, or transmission fluid. Review the last vehicle inspection report, noting any defects reported by the prior driver. Inspect the vehicle and confirm that any necessary repairs were performed.

- **Step 2 (Engine Compartment).** Before checking the engine compartment, ensure that the parking brakes are applied and the wheels are chocked. Ensure that there is adequate space in front of the vehicle to tilt the hood or cab and that there is nothing unsecured inside the cab that could fall and break. Tilt the cab or hood to expose the engine and check the following items:

Note: Not all vehicles will include all the options listed here.

1. Coolant level in the radiator or overflow tank: condition of the coolant hoses and fittings
2. Power steering fluid level and condition of the power steering lines
3. Windshield washer fluid level
4. Battery fluid level or indicator on maintenance-free batteries and the battery connections and hold-downs (batteries may be in other areas of the vehicle).

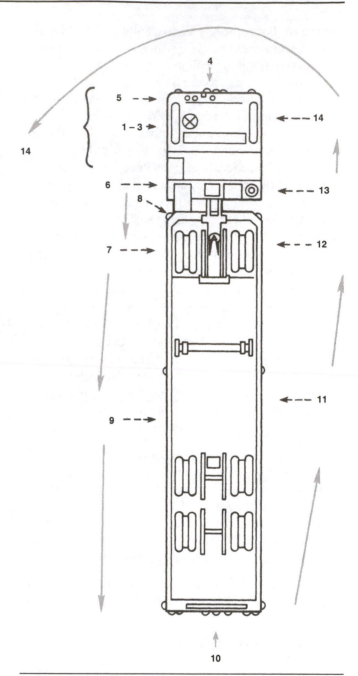

Figure 4-1 Make inspection a habit by following a circular, walk-around sequence (numbers correspond with the text). Most jurisdictions require the use of this or a variation of this, pre-trip circle inspection. *(Courtesy of Heavy Duty Trucking)*

5. Automatic transmission fluid level (engine will probably have to be running to perform this task)
6. Inspect all drive belts for proper tension and wear.
7. Check for any fluid leaks in the engine compartment, this includes fuel, coolant, oil, power steering fluid, hydraulic fluid, and battery fluid.

8. Visually inspect wiring and wiring harnesses for abrasions, rubbing, or cracked insulation.

Lower and secure the engine hood or cab.

- **Step 3 (Cab Interior).** Ensure that the parking brakes are applied and that the gearshift is in neutral (if transmission is manual) or in park (if transmission is automatic). Start the engine, listening for any unusual sounds. Check the dash instruments and gauges for the following:

  1. Oil pressure should rise up to operating pressure within seconds of the engine starting.
  2. Ammeter and/or voltmeter should indicate that the alternator is charging the batteries once the vehicle starts.
  3. Engine coolant temperature should start to rise gradually up to normal operating temperature (obtaining full engine temperature may require running the engine with some load applied to it).
  4. Engine oil temperature should start to gradually rise into the normal operating range.
  5. Warning indicators (lights/buzzers) for oil, coolant, and charging circuit should stop immediately once the engine has started.

Inspect the cab interior seats and seat belts. Check windshield and mirrors for cracks, cleanliness, stickers, or anything else that could obstruct the driver's view. Clean and adjust as necessary. Check for excessive play in the steering wheel. Manual steering play should not exceed about 2 inches, and power steering play should not exceed about 2-1/2 inches. Check the following components for looseness, sticking, damage, or improper setting:

1. Steering wheel
2. Clutch
3. Throttle pedal
4. Brake controls
5. Foot brake
6. Trailer brake
7. Parking brake
8. Retarder controls (if equipped)
9. Transmission controls
10. Interaxle differential lock (if equipped)
11. Horn(s)
12. Windshield wiper/washer
13. Heater fan and controls

14. Lights, including headlights dimmer switch, turn signal, four-way flashers, clearance, identification, marker light switch(s)

Shop Talk: The National Highway Traffic Safety Administration (NHTSA) issued a rule that standardizes the buckle-release mechanism that requires the belts to have either an emergency locking retractor or an automatic locking retractor. The rule also requires that retractors be attached to the seat so that they will move along with the air suspension system.

Besides the components of the truck's interior that have been checked, the truck must also have certain safety equipment that must be checked. There should be three red reflective triangles, and a properly charged and rated fire extinguisher within arm's reach of the driver's seat. As well, there should be spare electrical fuses, flares, lanterns, and flags.

Other optional equipment that may be required are tire chains if the vehicle is operated in winter conditions (in some areas this is mandatory equipment), tire-changing equipment, a list of emergency phone numbers, and an accident reporting kit.

- **Step 4 (Front of Cab).** Complete a visual inspection on the steering system looking closely for any defects. Keep an eye out for missing nuts, bolts, cotter pins, or other parts. Look for bent, loose, or broken parts such as the steering column, steering gearbox, or tie-rods.

Check the operation of the headlights for operation of both high and low beams, turn signals, and emergency flashers (four ways) for proper color and operation. Inspect the front suspension on both sides, looking at:

1. Spring hangers that allow the axle to move out of alignment
2. Cracked or broken spring hangers and condition of swinging shackle
3. Missing or broken leaves in a leaf spring. If one-fourth or more are missing, it will put the vehicle out of service, but any defect could be dangerous.
4. Make sure that, if a leaf is broken, it can't shift, possibly hitting the tire or other parts of the steering system.
5. Shock absorbers for signs of leakage or broken mounts or bolts and bushings

6. Torque rod or arm, U-bolts, or other axle positioning parts that are cracked, damaged, loose, or missing

7. Air suspension systems that are damaged and/or leaking

8. Any loose, cracked, broken, or missing frame members

9. Inspect headlamps for high and low beams. Check turn signals, emergency flashers, marker lights, and reflectors. Ensure that all lamps are clean and the correct color.

Perform an inspection of the front brakes, paying attention to the lining and making sure that all components are attached and operational. While looking at the lining, check for cracks, missing pieces of lining, and wetness that would indicate a leaking or weeping wheel seal. Check brake lines for leaks, damage, or abrasions as well as brake chambers for cracks or insecure mounting. Make sure that the pushrods, slack adjusters, and clevis and pins are secure and operational. Test for any audible leaks in the brake chamber by having a helper apply the brakes while you watch and listen, and then have them release the brakes. Check for too much pushrod travel. Inspect brake drums for cracks that open upon brake application.

- **Step 5 (Left Side of Cab).** Examine the front left wheel for:

  1. Defective welds, cracks, or breaks (especially between handholds or stud holes)
  2. Unseated locking rings on multi-piece rims
  3. Broken, missing, or loose lugs, studs, or clamps
  4. Bent or cracked rims
  5. Check for "bleeding" rust stains, defective nuts, or elongated stud holes.
  6. Spoke wheels should be checked for cracks across the spokes, and look for any scrubbed or polished areas on either side of the lug to indicate the rim has slipped on studs. Also check the rims for any cracks or bends and that the valve stem is centered between the spokes.
  7. Check the condition of the front left tire for imperfections like bulges, separation, or cracks in the sidewall. Measure tire tread depth and inflation pressure.
  8. Check for tire contact with any part of the vehicle.
  9. Inspect the frame for cracked, sagging rails. Look for any broken or loose bolts or brackets.

- **Step 6 (Left Fuel Tank Area).**

  1. Check the level of the fuel in the tank and compare it to the fuel gauge reading during the cab inspection.
  2. Pay attention for signs of fuel leakage, as well as any damage to the fuel tank.
  3. Check that the tank is mounted securely to the vehicle frame and that there are no components of the hold-down missing.
  4. Check that the fuel cap is present and seals the tank properly.
  5. Look at the electrical connection for the fuel gauge and verify that the fuel crossover line is secure.
  6. Check the air lines for the trailer and the condition of the seven-way connection. Watch out for tangles, crimps, chafing, or dragging.
  7. Check the connections to the trailer and listen for any air leaks; if so, inspect the **gladhand** rubbers and replace as necessary.
  8. Check the truck frame for any sign of damage, cracks, or sags. Ensure that all bolts and brackets are tight.

- **Step 7 (Left Rear Tractor Area).**

  1. Check the drive tires, wheels, and rim as previously described at step 5. Be sure to examine the inside wheel assembly, because these are the most neglected wheels and are often overlooked.
  2. Make sure the tires are of the same type and size. Check for any foreign material that may have become wedged between the tires and remove as necessary.
  3. Make sure there is an air space between the tires (at the bottom where they budge).
  4. Check the suspension and braking system on the rear axle as described in step 4.
  5. Inspect the fifth wheel top plate for cracks (may need to be cleaned to identify cracks) and for cracks in the mounting surfaces as well. Check the operation of the locking mechanism and adjust as necessary. Check for loose or missing mounting nuts and bolts.
  6. If the fifth wheel is mounted to a slider assembly, make sure that the slider is locked.
  7. If tractor is coupled to a trailer, make sure the jaws are properly secured around the trailer's kingpin.
  8. Check the tractor's stop, turn, emergency flashers, license plate, and running lights for proper color and operation.

9. Check the frame for any sign of damage, cracks, sags. Ensure that all bolts and brackets are tight.

■ **Step 8 (Front Side of Trailer).**

1. Check the body (van trailer) condition for breaks that could allow water to damage the load.
2. If trailer is a flat bed, check mounting of head board and that any load securing equipment is secured itself. If a tarp is used, inspect for rips/tears and that it is tied down properly.
3. Check for proper placarding.
4. Inspect trailer seven-way electrical receptacle box, ensuring all pins are clean and not damaged.
5. Check operation of top left clearance light.

■ **Step 9 (Left Side of Trailer).**

1. Check the side of the body (van trailer) for breaks that could cause water damage.
2. Check condition of trailer upper coupler for damage.
3. Check for any missing rivet or fastener on trailer upper and lower rails as well as posts.
4. Make sure the trailer landing gear is fully raised and the crank handle is secured.
5. Check operation of midway marker/turn signal.
6. Check for proper placarding.
7. Inspect conspicuity tape to ensure that there are no sections missing.
8. Inspect the load (flatbed trailer) and check to see that the load is secure and has not shifted.
9. If a tarp is used, inspect for rips/tears and that it is tied down properly.

■ **Step 10 (Rear of Trailer).**

1. Check operation of the rear tail lights, turn signals, brake lights, emergency flashers, clearance, license plate, and marker lights. Inspect reflectors on conspicuity tape. Inspect for cleanliness and proper color.
2. Check the braking system and suspension as described in step 4.
3. Check wheels, tires, and rims as described in step 5.
4. Inspect the rear underside guard for damage, broken or cracked welds, and any deformities that would put the guard out of compliance.

5. Check rear doors for missing or broken hinges and that they are securely closed or locked.
6. For flat beds, inspect as described in step 8.
7. Check for proper placarding.
8. Make sure the license place is legible.
9. Inspect the frame for cracked, sagging rails and check for broken or loose bolts or brackets.

■ **Step 11 (Right Side of Trailer).** Check the same items that you checked on the left side of the trailer in step 9. In addition:

1. Check the spare tire carrier for proper mounting and the condition and air pressure.
2. Make sure the trailer landing gear is fully raised and the crank handle is secured.
3. Check for missing, bent, or damaged parts.
4. Check condition of trailer upper coupler for damage.

■ **Step 12 (Right Rear Tractor Area).** Check the same items that were checked on the lift side at step 7.

■ **Step 13 (Right Fuel Tank Area).** Check the same items that were listed in step 6. In addition, check the exhaust system:

1. Check that the exhaust is securely mounted to the cab and to the underside of the cab.
2. Make sure that there is nothing in contact with the exhaust, like fuel or air lines or electrical wires.
3. Look for carbon deposits around seams and clamps signifying exhaust leaks.

■ **Step 14 (Right Side of Cab).** Check the same items as listed in step 5.

■ **Step 15 (Road Test).**

1. Watching the air pressure gauges, pump the brake pedal. The reading should drop as you pump the brakes.
2. Keep pumping the brakes to exhaust the air system. The warning light/buzzer should activate at about 55 psi (379 kPa) or above.
3. Ensure that your seat belt is fastened and release the brakes. Move the vehicle ahead slowly and trigger the parking brakes to make sure they are operational.
4. Start moving again and at about 5 mph (8 kph), apply the service brakes. Note any unusual pulling, delay, or play in the brake pedal.

☐ N   **DRIVER'S VEHICLE INSPECTION REPORT**

Date: _____

Carrier's
Name: _____

Address: _____
CHECK ANY DEFECTIVE ITEM WITH AN X AND GIVE DETAILS

TRACTOR/TRUCK NO. _____ Odometer _____

☐ Air Compressor   ☐ Horn   ☐ Suspension
☐ Air Lines   ☐ Lights   System
☐ Battery   Head - Stop   ☐ Starter
☐ Body   Tail - Dash   ☐ Steering
☐ Brake Accessories   Turn Indicators ☐ Tachograph
☐ Brakes, Parking   ☐ Mirrors   ☐ Tires
☐ Brakes, Service   ☐ Muffler   ☐ Tire Chains
☐ Clutch   ☐ Oil Pressure   ☐ Transmission
☐ Coupling Devices   ☐ Radiator   ☐ Wheels and
☐ Defroster/Heater   ☐ Rear End   Rims
☐ Drive Line   ☐ Reflectors   ☐ Windows
☐ Engine   ☐ Safety Equipment ☐ Windshield
☐ Exhaust   Fire Extinguisher   Wipers
☐ Fifth Wheel   Reflective Triangles ☐ Other
☐ Frame and Assembly   Flags - Flares - Fusees
☐ Front Axle   Spare Bulbs & Fuses
☐ Fuel Tanks   Spare Seal Beam

TRAILER(S) NO.(S) _____

☐ Brake Connections   ☐ Hitch   ☐ Tarpaulin
☐ Brakes   ☐ Landing Gear   ☐ Tires
☐ Coupling Devices   ☐ Lights - All   ☐ Wheels and
☐ Coupling (King) Pin   ☐ Roof   Rims
☐ Doors   ☐ Suspension System ☐ Other

REMARKS: _____
_____
_____
_____
_____

☐ Condition of the above vehicles is satisfactory

Driver's Signature: _____
☐ Above defects corrected
☐ Above defects need not be corrected for safe operation of vehicle

Mechanic's Signature: _____ Date: _____

Driver Reviewing Repairs: Signature: _____ Date: _____

131-FS-E3 Rev. (5/03)

ORIGINAL

Figure 4-2 A typical driver's report form as issued by the American Trucking Association. *(Courtesy of © and used by permission of JJ. Keller & Associates Inc.: www.jjkeller.com)*

If any problems are noted during the pre-trip inspection, it should be noted on the pre-trip form. This allows the technician to make a judgment call as to whether something needs to be repaired immediately or if the vehicle may be placed back in service. The technician is ultimately responsible for verifying that the vehicle is safe (see **Figure 4-2**).

OUT-OF-SERVICE OR DEADLINING A VEHICLE

It can be a hard decision as to whether or not to take a vehicle **out of service (OOS)** (see **Figure 4-3**). You must first determine the possibility that a mechanical problem will cause an accident or breakdown against the need

Figure 4-3 Vehicle out of service. *(Courtesy of John Dixon)*

to get the load moved. The question is, how does one determine when it is necessary to **deadline** a vehicle?

The Federal Motor Carrier Safety Regulations, Part 393, state that a mechanical system that can either cause or prevent an accident is classified as a safety item. Potential safety problems should be rectified, no matter how minor they appear. Regulations prohibit dispatching a vehicle that is likely to break down. Regardless of whether or not a fine might be imposed for a safety violation, dispatching a truck likely to break down is not a good business practice. A broken-down vehicle on the highway is always a traffic hazard regardless of whether the driver actually makes it off the shoulder of the road. The vehicle remains a hazard. Adding up the costs of driver down-time, unhappy shippers/receivers, and a towing and/or road repair bill, the person who dispatched the unfit vehicle will be annoyed by their lack of foresight.

Management does have the ability to minimize road-side breakdowns by listening to the opinions of both drivers and technicians.

It is essential that both drivers and technicians be familiar with the CVSA out-of-service criteria. Refer to **Table 4-1.** This table represents the guidelines being used by law enforcement officials in practically every state in the United States and Mexico and in every Canadian province in increasingly frequent truck inspections.

A vehicle considered likely to cause an accident or breakdown because of mechanical of mechanical conditions or improper loading will be taken out of service, tagged with an out-of-service (OOS) sticker. The vehicle cannot be moved under its own power until all required repairs are completed. A "restricted service" designation (indicated by an asterisk (*) in **Table 4-1**) makes the provision for a vehicle to travel

up to 25 miles (40 km) to a repair facility if the inspector judges that the truck continuing is less hazardous than requiring it to remain at the inspection site.

If it has been decided to deadline a vehicle, the next step would be for the technician to repair the problem. Many problems that could put a vehicle out of service (brakes out of adjustment, air line replacements, headlamp replacement) can be repairs taking less than an hour. If the problem is of a more serious nature, a manager may have to bring in a substitute vehicle and switch the load over to make the delivery. After the repair has been performed to the downed vehicle, the technician is required to certify in writing that the required repair has been performed.

VEHICLE PM SCHEDULING

Scheduling preventive maintenance for vehicles at regular intervals is important for any maintenance program. This can be accomplished by timetabled servicing of equipment. This regular servicing of the vehicle will allow you to spot potential (and repair) problems long before they become roadside breakdowns. Implementing a program requires good scheduling and controls to ensure that all vehicles are serviced on their set cycle whether it be miles traveled, hours of operation, gallons of fuel consumed, or monthly (see **Figure 4-4**).

PM services are categorized into three different groups. (see **Figure 4-5**). The first, an "A" PM, is the easiest to perform and is done at the shortest interval. The "B" PM is more involved and performed less often. The "C" PM adds more steps to the "B" PM. It is performed the least often of the three inspections. A recent **Technical and Maintenance Council (TMC)** report indicated that:

- "A" PM service averaged 2.4 hours but might vary from 45 minutes all the way up to 5-1/2 hours.
- "B" PM service averaged 5 hours but might vary from 2 hours to 8 hours.
- "C" PM (major inspections) averaged about 1 day, but might be longer depending on the servicing required.
- There are also some fleets the use a more detailed "D" PM.
- The miles between unscheduled repairs averaged about 20,000. Some supervisors have created their own "Mini-A" at 6,000 miles (9,659 km) specifically to look for items that can cause an unplanned shop visit. A mini check takes about half an hour and can eliminate unscheduled maintenance.

TABLE 4-1: CVSA VEHICLE OUT-OF-SERVICE CRITERIA

Inspected Item	Out of service if

BRAKE SYSTEM

Defective Brakes

20 percent or more of the brakes on a vehicle are defective. A brake is defective if:
1. Brakes are out of adjustment. (Measured with engine off and reservoir pressure at 80–90 psi with brakes fully applied.)

CLAMP-TYPE BRAKE CHAMBER			ROTOCHAMBER		
Type	Outside Dia.	Max. Stroke	Type	Outside Dia.	Max. Stroke
6	$4^1/_2$"	$1^1/_4$"	9	$4^9/_{32}$"	$1^1/_2$"
9	$5^1/_4$"	$1^3/_8$"	12	$4^{13}/_{16}$"	$1^1/_2$"
12	$5^{11}/_{16}$"	$1^3/_8$"	16	$5^{13}/_{32}$"	2"
16	$6^3/_8$"	$1^3/_4$"	20	$5^{15}/_{16}$"	2"
20	$6^{25}/_{32}$"	$1^3/_4$"	24	$6^{13}/_{32}$"	2"
24	$7^7/_{32}$"	$1^3/_4$" (2" for long strock)	30	$7^1/_{16}$"	$2^1/_4$"
30	$8^3/_{32}$"	2"	36	$7^5/_8$"	$2^3/_4$"
36	9"	$2^1/_4$"	50	$8^7/_8$"	3"

	BOLT-TYPE BRAKE CHAMBER			
Type	Effective Area	Outside Dia.	Max. Stroke	WEDGE BRAKE
A	12 sq. in.	$6^{15}/_{16}$"	$1^3/_8$"	Movement of the scribe mark on the lining shall not exceed $1/_{16}$".
B	24 sq. in.	$9^3/_{16}$"	$1^3/_4$"	
C	16 sq. in.	$8^1/_{16}$"	$1^3/_4$"	
D	6 sq. in.	$5^1/_4$"	$1^1/_4$"	
E	9 sq. in.	$6^3/_{16}$"	$1^3/_8$"	
F	36 sq. in.	11"	$2^1/_4$"	
G	30 sq. in.	$9^7/_8$"	2"	

2. On application of service brakes, there is no braking action (such as brake shoe(s) failing to move on a wedge, S-cam, cam, or disc brake).
3. Mechanical components are missing, broken, or loose (such as shoes, linings, pads, springs, anchor pins, spiders, cam rollers, pushrods, and air chamber mounting bolts).
4. Audible air leak at brake chamber is present.
5. Brake linings or pads are not firmly attached or are saturated with oil, grease, or brake fluid.
6. Linings show excessive wear.
7. Required brake(s) are missing.

Steering Axle Brakes

1. On vehicles required to have steering axle brakes, there is no braking action on application (includes the dolly and front axle of a full trailer).
2. Air chamber sizes or slack adjuster lengths are mismatched on tractor steering axles.
3. Brake linings or pads on tractor steering axles are not firmly attached to the shoe; are saturated with oil, grease, or brake fluid; or lining thickness is insufficient.

Parking Brakes

Upon actuation, no brakes are applied, including driveline hand-controlled parking brake.*

TABLE 4-1: (*continued*)

Inspected Item	Out of service if
Brake Drums or Rotors	1. External crack(s) on drums open on brake application. 2. A portion of the drum or rotor is missing or in danger of falling away.
Brake Hose	1. Hose damage extends through the outer reinforcement ply. 2. Hose bulges/swells when air pressure is applied or has audible leak at other than proper connection. 3. Two hoses are improperly joined and can be moved or separated by hand at splice. 4. Hoses are improperly joined but cannot be moved or separated by hand.* 5. Air hose is cracked, broken, or crimped and airflow is restricted.
Brake Tubing	Tubing has audible leak at other than proper connection or is cracked, heat-damaged, broken, or crimped.
Low-Pressure Warning Device	Device is missing, inoperative, or does not operate at 55 psi and below, or half the governor cutout pressure, whichever is less.*
Air Loss Rate	If air leak is discovered and the reservoir pressure is not maintained when governor is cut in, pressure is between 80 and 90 psi, engine is at idle, and service brakes are fully applied.
Tractor Protection	Protection valve on tractor is missing or inoperable.
Air Reservoir	Mounting bolts or brackets are broken, missing, or loose.*
Air Compressor	1. Drive belt condition indicates impending or probable failure.* 2. Mounting bolts are loose. Pulley is cracked, broken, or loose. Mounting brackets, braces, and adapters are cracked or broken.
Electric Brakes	1. Absence of braking on 20 percent or more of the braked wheels. 2. Breakaway braking device is missing or inoperable.
Hydraulic Brakes	1. No pedal reserve with engine running, except by pumping the pedal. 2. Master cylinder is less than one-quarter full. 3. Power-assist unit fails. 4. Brake hoses seep or swell under application. 5. Hydraulic fluid is visibly leaking from brake system and master cylinder is less than one-quarter full.* 6. Check valve is missing or inoperative. 7. Hydraulic fluid is visibly leaking from brake system. 8. Hydraulic hoses are worn through outer cover-to-fabric layer. 9. Fluid lines or connections are restricted, crimped, cracked, or broken. 10. Brake failure light/low fluid warning light is on and/or inoperative.*
Vacuum System	1. Vacuum reserve is insufficient to permit one full brake application after engine is shut off. 2. Hoses or lines are restricted, excessively worn, crimped, cracked, broken, or collapsed under pressure. 3. Low-vacuum warning light is missing or inoperative.*
COUPLING DEVICES Fifth Wheels	1. Mounting to frame—more than 20 percent of frame mounting fasteners are missing or ineffective. 2. Movement between mounting components is observed. Mounting angle iron is cracked or broken. 3. Mounting plates and pivot brackets, more than 20 percent of fasteners on either side are missing or ineffective. Movement between pivot bracket pin and bracket exceeds $3/8$". Pivot bracket pin is missing or not secured.

(*continued*)

TABLE 4-1: CVSA VEHICLE OUT-OF-SERVICE CRITERIA (*continued*)

Inspected Item	Out of service if
	4. Cracks in any weld(s) or parent metal are observed on mounting plates or pivot brackets.*
	5. Sliders—more than 25 percent of latching fasteners, per side, are ineffective. Any fore or aft stop is missing or not secured. Movement between slider bracket and base exceeds $3/8$".
	6. Cracks are observed in any slider component parent metal or weld.*
	7. Lower coupler—horizontal movement between the upper and lower fifth wheel halves exceeds $1/2$". Operating handle is not closed or locked in position. Kingpin is not properly engaged. Cracks are observed in fifth wheel plate. Locking mechanism parts are missing, broken, or deformed to the extent the kingpin is not securely held.
	8. Space between upper and lower coupler allows light to show through from side to side.*
Pintle Hooks	1. Mounting to frame—fasteners are missing or ineffective. Frame is cracked at mounting bolt holes. Loose mounting. Frame cross member providing pintle hook attachment is cracked.
	2. Integrity—cracks are anywhere in the pintle hook assembly. Section reduction is visible when coupled. Latch is insecure.
	3. Any welded repairs to the pintle hook are visible.*
Drawbar Eye	1. Mounting—any cracks are visible in attachment welds. Fasteners are missing or ineffective.
	2. Integrity—any cracks are visible. Section reduction is visible when coupled. Note: No part of the eye should have any section reduced by more than 20 percent.
Drawbar/Tongue	1. Slider (power/manual)—latching mechanism is ineffective. Stop is missing or ineffective. Movement of more than $1/4$" between slider and housing. Any leaking air or hydraulic cylinders, hoses, or chambers.
	2. Integrity—any cracks. Movement of $1/4$" between subframe and drawbar at point of attachment.
Safety Devices	1. Missing or unattached, or incapable of secure attachment.
	2. Chains and hooks worn to the extent of a measurable reduction in link cross section.*
	3. Improper repairs to chains and hooks including welding, wire, small bolts, rope, and tape.*
	4. Kinks or broken cable strands. Improper clamps or clamping on cables.*
Saddlemounts (method of attachment)	1. Any missing or ineffective fasteners, loose mountings, or cracks in a stress or load-bearing member.
	2. Horizontal movement between upper and saddlemount halves exceeds $1/4$".
EXHAUST SYSTEM	1. Any exhaust system leaking in front of or below the driver/sleeper compartment and the floor pan permits entry of exhaust fumes.*
	2. Location of exhaust system is likely to result in burning, charring, or damaging wiring, fuel supply, or combustible parts.
FUEL SYSTEM	1. System with visible leak at any point.
	2. A fuel tank filter cap missing.*
	3. Fuel tank not securely attached due to loose, broken, or missing mounting bolts or brackets.*
LIGHTING DEVICES When Lights Are Required	1. Single vehicle without at least one headlight operative on low beam and without a stoplight on the rearmost vehicle.

TABLE 4-1: (*continued*)

Inspected Item	Out of service if
	2. Vehicle that does not have at least one steady burning red light on the rear of the rearmost vehicle (visible at 500 feet).
	3. Projecting loads without at least one operative red or amber light on the rear of loads projecting more than 4 feet beyond vehicle body and visible from 500 feet.
At All Times	1. No stop light on the rearmost vehicle is operative.
	2. Rearmost turn signal(s) do not work.*
SAFE LOADING	1. Spare tire or part of the load is in danger of falling onto the roadway.
	2. Protection against shifting cargo—any vehicle without front-end or equivalent structure as required.*
STEERING MECHANISM Steering Wheel Free Play	When any of the following values are met or exceeded, the vehicle will be taken out of service.

Steering Wheel Diameter	Manual System Movement	Power System Movement
	30 Degrees or	45 Degrees or
16"	$4^1/_2$" (or more)	$6^3/_4$" (or more)
18"	$4^3/_4$" (or more)	$7^1/_8$" (or more)
20"	$5^1/_4$" (or more)	$7^7/_8$" (or more)
21"	$5^1/_2$" (or more)	$8^1/_4$" (or more)
22"	$5^3/_4$" (or more)	$8^5/_8$" (or more)

Inspected Item	Out of service if
Steering Columns	1. Any absence or looseness of U-bolts or positioning parts.
	2. Worn, faulty, or obviously repair-welded universal joints.
	3. Steering wheel not properly secured.
Front Axle Beam	1. Includes all steering components other than steering column, including hub.
	2. Any cracks.
	3. Any obvious welded repair(s).
Steering Gear Box	1. Any mounting bolt(s) loose or missing.
	2. Any cracks in gear box or mounting brackets.
Pitman Arm	Any looseness of the arm or steering gear output shaft.
Power Steering	Auxiliary power-assist cylinder loose.
Ball and Socket Joints	1. Movement under steering load of a stud nut.
	2. Any motion (other than rotational) between linkage member and its attachment point over $^1/_4$".
Tie-Rods and Drag Links	1. Loose clamp(s) or clamp bolt(s) on tie-rods or drag links.
	2. Any looseness in any threaded joint.
Nuts	Loose or missing nuts in tie-rods, pitman arm, drag link, steering arm, or tie-rod arm.
Steering System	Any modification or other condition that interferes with free movement of any steering component.
SUSPENSION Axle Parts/Members	Any U-bolt(s), spring hanger(s), or other axle positioning part(s) cracked, broken, loose, or missing resulting in shifting of an axle from its normal position.

(continued)

TABLE 4-1: CVSA VEHICLE OUT-OF-SERVICE CRITERIA (continued)

Inspected Item	Out of service if
Spring Assembly	1. One-fourth of the leaves in any spring assembly broken or missing. 2. Any broken main leaf in a leaf spring. 3. Coil spring broken. 4. Rubber spring missing. 5. Leaf displacement that could result in contact with a tire, rim, brake drum, or frame. 6. Broken torsion bar spring in torsion bar suspension. 7. Deflated air suspension.
Torque, Radius, or Tracking Components	Any part of a torque, radius, or tracking component assembly that is cracked, loose, broken, or missing.

FRAME

Frame Members	1. Any cracked, loose, or sagging frame member permitting shifting of the body onto moving parts or other condition indicating an imminent collapse of the frame. 2. Any cracked, loose, or broken frame member adversely affecting support of functional components such as steering gear, fifth wheel, engine, transmission, or suspension. 3. Any crack $1\frac{1}{2}$" or longer in the frame web that is directed toward the bottom flange. 4. Any crack extending from the frame web around the radius and into the bottom flange. 5. Any crack 1" or longer in bottom flange.
Tire and Wheel Clearance	Any condition, including loading, causes the body or frame to be in contact with a tire or any part of the wheel assemblies at the time of inspection.
Adjustable Axle (sliding subframe)	1. Adjustable axle assembly with more than one-quarter of the locking pins missing or not engaged. 2. Locking bar not closed or not in the locked position.

TIRES

Any Tire on Any Steering Axle of a Power Unit	1. With less than $\frac{2}{32}$" tread when measured in any two adjacent major tread grooves at any location on the tire. 2. Any part of the breaker strip or casing ply is showing in the tread. 3. The sidewall is cut, worn, or damaged so that ply cord is exposed. 4. Labeled "Not for Highway Use" or carrying markings that would exclude use on steering axle. 5. A tube-type radial tire without the stem markings. These include a red band around the tube stem, the word "radial" embossed in metal stems, or the word "radial" molded in rubber stems.* 6. Mixing bias and radial tires on the same axle.* 7. Tire flap protrudes through valve slot rim and touches stem.* 8. Regrooved tire except on motor vehicles used solely in urban or suburban service.* 9. Visually observable bump, bulge, or knot related to tread or sidewall separation. 10. Boot, blowout patch, or other ply repair.* 11. Weight carried exceeds tire load limit. This includes overloaded tire resulting from low air pressure.* 12. Tire is flat or has a noticeable leak. 13. So mounted or inflated that it comes in contact with any part of the vehicle.

	TABLE 4-1: *(continued)*
Inspected Item	**Out of service if**
All Tires Other Than Those Found on the Steering Axle of a Powered Vehicle	1. Weight carried exceeds tire load limit. This includes overloaded tires resulting from low air pressure.* 2. Tire is flat or has noticeable leak (that is, one that can be heard or felt). 3. Bias ply tire—when more than one ply is exposed in the tread area or sidewall, or when the exposed area of the top ply exceeds 2 sq. in. Note: On duals, both tires must meet this condition. 4. Bias ply tire—when more than one ply is exposed in the tread area or sidewall, or when the exposed area of the top ply exceeds 2 sq. in.* 5. Radial ply tire—when two or more plies are exposed in the tread area or damaged cords are evident in the sidewall, or when the exposed area exceeds 2 sq. in., tread or sidewall. Note: On dual wheels, both tires must meet this condition. 6. Radial ply tire—when two or more plies are exposed in the tread area or damaged cords are evident in the sidewall, or when the exposed area exceeds 2 sq. in., tread or sidewall.* 7. Any tire with visually observable bump or knot apparently related to tread or sidewall separation. 8. So mounted or inflated that it comes in contact with any part of the vehicle. (This includes any tire contacting its mate in a dual set.) 9. Is marked "Not for Highway Use" or otherwise marked and having similar meaning.* 10. So worn that less than $1/32$" tread remains when measured in any two adjacent major tread grooves at any location on the tire. Exception: On duals, both tires must be so worn.*
WHEELS AND RIMS	1. Lock or side ring is bent, broken, cracked, improperly sealed, sprung, or mismatched. 2. Rim cracks—any circumferential crack except at valve hole. 3. Disc wheel cracks—extending between any two holes. 4. Stud holes (disc wheels)—50 percent or more elongated stud holes (fasteners tight). 5. Spoke wheel cracks—two or more cracks more than 1" long across a spoke or hub section. Two or more web areas with cracks. 6. Tubeless demountable adapter cracks—cracks at three or more spokes. 7. Fasteners—loose, missing, broken, cracked, or stripped (both spoke and disc wheels) ineffective as follows: For 10 fastener positions: 3 anywhere, 2 adjacent. For 8 fastener positions or less (including spoke wheel and hub bolts): 2 anywhere. 8. Welds—any cracks in welds attaching disc wheel disc to rim. Any crack in welds attaching tubeless demountable rim to adapter. Any welded repair on aluminum wheel(s) on a steering axle. Any welded repair other than disc to rim attachment on steel disc wheel(s) mounted on the steering axle.
WINDSHIELD GLAZING	Any crack over $1/4$" wide, intersecting cracks, discoloration not applied in manufacture, or other vision-distorting matter in the sweep of the wiper on the driver's side.*
WINDSHIELD WIPERS	Any power unit that has inoperable parts or missing wipers that render the system ineffective on the driver's side.

The PM scheduling in most fleets is determined by the amount of vehicle usage. The servicing frequency of the different inspections is determined by the service manager according the specific needs of the trucking operation.

FEDERAL INSPECTION REGULATIONS

The **National Highway Traffic Safety Administration (NHTSA)** has set up a periodic minimum inspection standards program, under which the following vehicles must be inspected:

| VEHICLE MAINTENANCE RECORD & SCHEDULE |||||||
| Truck # _____   Make _____   Model _____   Serial # _____ |||||||
Month or mileage	Due date	Service due	Mileage serviced	Date serviced	Repairs performed	RPO#
January		(1) O&F				
February		(1)				
March		(1) O&F				
April		(1) WF				
May		(1) O&F				
June		(1)				
July		(1) O&F				
August		(1) WF				
September		(1&2) O&F				
October		(1)				
November		(1) O&F				
December		(1) WF				

O = OIL
F = OIL FILTER
WF = WATER FILTER

Figure 4-4 Typical monthly PM schedule.

- Any vehicles involved in interstate commerce with a gross vehicle weight over 10,000 lb (454 kg) with or without power, including trucks, buses, tractor/trailer, full trailer, semi-trailer, converter dollies, container chassis, booster axles, and jeep axles
- Any vehicle, regardless of weight, that is designed to carry more than 15 passengers, including the driver
- Any vehicle, regardless of weight, carrying hazardous materials in a quantity requiring placarding

Each vehicle must carry proof that an inspection was completed. Proof can be either a copy of the inspection form kept in the vehicle or a decal. If a decal is used, a copy of the inspection form must be kept on file and the label must indicate where an inspector can call or write to get a copy of the form.

Size and shape of the decal is not specified. It may be purchased from a supplier or even be made by the PM shop. The only requirement is that it remain legible. Each "vehicle" must be inspected separately. This means that a tractor trailer combination is actually two vehicles and each must be inspected separately and requires its own decal; a converter dolly is also a separate vehicle.

The decal must show the following information:

- Date (month and year) the vehicle passed the inspection
- Name and address to contact about the inspection record. The record can be stored anywhere as long as the decal states where to contact the fleet.
- Certification that the vehicle has passed the inspection
- Vehicle identification information (fleet unit number or serial number sufficient to identify the vehicle)

The NHTSA website can be found at nhtsa.dot.gov

RECORD-KEEPING REQUIREMENTS

There are many fleets that have modified their "B" preventive maintenance form to include all the federal safety checks. If this is done every time a "B" service is performed, the vehicle also passes the safety inspection, and a new decal can be affixed. This procedure simplifies the task of tracking each vehicle to perform the annual inspection. Many fleets stagger their annual inspections throughout the year to prevent shop scheduling congestion.

Experience has shown that it is a good idea to keep work orders in the vehicle file folder, and current fleet practice indicates that maintaining

PM INSPECTION LINEHAUL TRACTOR

Unit # _____   □ "A" PM   □ "B" PM   □ "C" PM   Mechanic _____

Date _____   Check-High Oil/Fuel Checklist □

Supervisor _____

Mileage _____

On D/L, Check and Inspect

	Ck	Remarks
Verify PM on History Card		
Start Engine		
Gauges, Warning Lights		
Low Air Buzzer & Light on at 60#		
Air Buildup & Cutoff		
Air Dryer Cutoff		
Air Loss Test, 3# Per Min.		
Heater/Defroster Operation		
A/C Operation		
Horns, Air & Electric		

Moving to Shop, Check and Inspect

	Ck	Remarks
Park Brake Application		
Clutch Operation, Free Travel		
Transmission Shift, Hi-Lo		
Steering Free Travel, Bind, Pull		
Speedometer Operation		
Any Unusual Noise or Vibration		
Windshield Wiper/Washer Operation		
Foot Brake Application		
Headlights & Driving Lights		

In Shop, Check and Inspect

	Ck	Remarks
Turn Signals, Marker Lights		
Brake Lights–Reflectors		
Dome Light (Rear Light)		
Floor Mats, Boots, Coat Hook, DH Seal		
Sun Visors, Dash Screws		
All Glass and Mirrors		
Door Locks, Regulators		
Safety Equipment		
Seat Belts, Retractors		
(Sleeper Compartment Items)		
Trailer Cord Test		
Air Hoses and Hangers		
Lube: Door Latches, Hinges		
Seat Rails, Pivot Points		
Brake/Clutch Peddle Pivot		
Accelerator Peddle Pivots		
Heater/Defroster Cables		
Clean & Lube Floor Mtd. Foot Valve		

Engine Compartment, Check and Inspect

	Ck	Remarks
Hood Condition, Cracks, Damage		
Hinges, Bug Screen, Brackets, Wiring		
(Cab Over, Ck Jack Lift Cyl and Lines)		
Air Cleaner Ducts, Hoses, Brackets		
Air Restriction, Repl. Element @ + 20 in.		
All Belts, Tension-Condition		
A/C Compressor, Condenser,		
Receiver-Dryer, Wiring, Pressure Lines		
Pressure Test Cooling System		
Radiator, Leaks, Mounts, & Brackets		
Cooling Hoses, Any Leaks		
Water Pump Leaks		
Antifreeze, –20°		
DCA Check, Add Nalcool if Required		
Change Water Filter if Equipped		
Alternator Mounting, Wiring		
Starter Mounting, Wiring		
Spray Protectant on All Wiring Term.		
Fuel System, Leaks, Lines Rubbing		
Air Comp. Lines, Mtgs., Air Gov. Screws		
Lube Linkage (Must Have 2 Return Springs)		
Drain Fuel Tank Sumps of Water		
Drain Fuel Heater of Water		
Exhaust System, Clamps, Leaks		
Inspect Fan Hub		
Engine and Transmission Mounts		
Cab Mounts		
Drain Air Tanks of Water		

Under Vehicle

	Ck	Remarks
Change Oil, Oil Filters and Fuel Filters		
Check Trans Lube ("C" PM Change)		
Check Diff(s) Lube (C' PM Change)		
Inspect Axle Breather		
Inspect Pinion Seals, Wheel Seals		
Inspect Dr. Shaft, U-Joints Yokes, Lube		
Inspect Clutch Linkage, Lube as Req.		
Insp. Exh. Pipe, Muffler, Hangers, Leaks		
Insp. Fuel Tanks, Lines, Hangers, Leaks		
Insp. Air Tanks, Hangers, Lines Rubbing		
Record Oil Pressure Hi-Lo		
Check Dipstick Full Mark		
Lube All Required Points		

Used by a major carrier, this PM form provides space for remarks and special instructions, is clearly sequenced, notes specific requirements.

Figure 4-5A An example of a PM form.

electronic vehicle records on the chassis data bus is effective.

The same regulations require that a driver's vehicle condition report be retained for at least 3 months from the date of the report.

Computerized record keeping plays an important role in today's PM programs. Increasingly sophisticated maintenance software with automated data entry facilitates tracking and scheduling. Wireless data exchangers at fuel islands that read the chassis data bus and

PM INSPECTION LINEHAUL TRACTOR

Unit # _____   ☐ "A" PM   ☐ "B" PM   ☐ "C" PM   Mechanic _____
Date _____   Check-High Oil/Fuel Checklist ☐   Supervisor _____
                                                                         Mileage _____

Axles and Chassis

	Ck	Remarks
Vehicle Jacked Up		
Insp. Frt. Spgs., Pins, Hanger, U-Bolts		
Ins. Frt. Brake Chamber, Slacks,		
Lining, Air Lines, Brake Adjustment		
Insp. Frt. End. Tie-Rod Ends, Drag Link		
Pitman Arm, Kingpin, Steering		
Box, Steering Shaft, Hub Lube		
Level, Lube All Points.		
Lower Vehicle		
Insp. Rear Spgs., Pins, Hangers, U-Bolts		
Torque Rods, Rayco Equalizer Wear		
Insp. Rear Brake Chambers, Slacks,		
Lining, Air Lines, Brake Adjustment		
Insp. Frame and Cross Members		
Insp. 5th Wheel, Legs, Brackets, Ground		
Strap, Lube Slider Assembly		
Insp. Battery Box Mounts		
Clean Battery and Post		
Check Battery		
Insp. Battery Cables, Spray Prot.		
Insp. All Chassis Wiring, Spray Prot.		
on All Terminals		
Check Lisc., Regis, Permits, State Insp.		
Clean Glass & Cab Interior		

Add for 'C' PM

	Ck	Remarks
See Foreman for Compuchek		
Replace Gladhand Rubbers		
Check Air Shield Angle		
Check Toe-in		
Check Tandem Alignment		
Change Trans and Diff Lube		
Change Rear Center Pump Filter/Screen		
Check U-Bolt Torque		
Change Steering Box Oil, As Required		
(Delvac SHC Synthetic)		
Clean Heater Core & A/C Evap.		
Clean Trans. Air Valve Filter		
Air System Check		
Clean & Check 7 Way Cord		
Use UYK Comp. Reverse Ends		

"A" PM ONLY
On D/L & Moving to Shop, Check & Insp.

	Ck	Remarks
Verify PM on History Card		
Start Eng.: Ck Gauges, Warning Lights		
Low Air Buzzer and Light On @ 60#		
Air Build Up & Cut Off		
Air Dryer Cut Off		
Heater, Defroster and A/C Operation		
Horns: Air & Electric		
Clutch Operation, Free Travel		
Steering Free Travel, Bind, Pull		
Foot & Park Brake Application		

In Shop, Check and Inspection

	Ck	Remarks
All Lights		
Floor Mats, Boots, Coat Hook		
Safety Equipment		
All Glass and Mirrors		
Trailer Cord Test		
Check Engine Oil, Coolant Level		
Change Fuel Filter		
Check Front End, Lube All Points		
Check Trans and Diff Lube Levels		
Lube Drive Shaft U-Joints		
Lube 5th Wheel and Sliders		
Drain Air Tanks		
Check Batteries		
Clean All Glass, Inside Cab		

Wheels and Tires, Check and Inspect

Air Pressure	LF	RF	RFI	RFO	RRI	RRO	LRI	LRO	LFI	LFO
Tread Depth										
Loose Lugs										
Cracked Rims										
Valve Caps										

Used by a major carrier, this PM form provides space for remarks and special instructions, is clearly sequenced, notes specific requirements.

Figure 4-5B (*continued*)

diagnostic recorders help the management of driver, vehicle, and trip information. The maintenance managers of today must be very computer savvy and understand complicated programs in order to track and schedule regular maintenance. Understanding these computerized systems allows the manager access to much more information about the condition of a fleet than ever before. This knowledge is important when determining warranty claims, specifying equipment, and cost per mile reduction.

PM SOFTWARE

Many medium to large shops are now using specialized computer software to enhance their PM programs. This software prints, controls, and tracks work orders. This software can be adapted to the specific needs of a fleet. There is a good selection of programs available for the purpose of PM Service in today's market. "Dossier" by Arsenault Associates (http://www.truckfleet.com/) is an example of a program that will do everything from scheduling and reordering parts for inventory to tracking costs for every vehicle in a fleet. (see **Figure 4-6**).

Advantages

Some of the advantages of using this type of software are reduced maintenance costs. The following is a list of items that will reduce maintenance costs:

- Scheduled repair work becomes approximately 80 percent of all work performed.
- Vehicles are able to operate from PM to PM service without a breakdown.
- Provide clear instructions to mechanics for each unit serviced.
- Identify and reduce rework by mechanics and vendors by approximately 85 percent.
- Increase warranty recapture dollars by 50 percent.

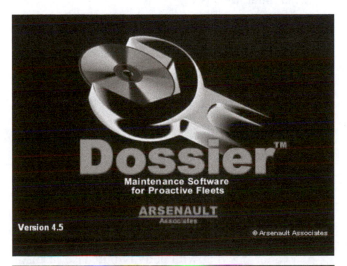

Figure 4-6 Commonly used fleet maintenance software program. (*Courtesy of Arsenault Associates*)

- Identify high cost units and why these units are more expensive.
- Make maintenance decisions based upon facts instead of gut feeling.
- Provide an easy method to schedule and track problem drivers.

REDUCE PARTS COSTS AND INVENTORY

- Reduce the quantity of on-hand parts by as much as 30 percent.
- Identifying nonused parts and obsolete parts can save 10 percent.
- Reduce vehicle down-time caused by out-of-stock parts.
- Reduce parts cost through comparison price shopping.
- Physically counting inventory becomes easier.
- Stop duplicate orders by knowing where to find existing parts.
- Stop excessive parts inventory by using parts cross-referencing.
- Track and control fuel, oil, and other fluids as inventory items.

INCREASE LABOR PRODUCTIVITY

- Identify total actual hours worked versus total hours paid.
- Identify percentage of productivity by each mechanic.
- Work-pending files to track deferred work or PM follow-up work.
- Measure each mechanic by comparison of shop averages.
- Reduce routine paperwork by as much as 8 hours per week.
- 100 percent increase in reporting accuracy and accountability.

INCREASE FUEL, TIRES, ADMINISTRATION EFFICIENCY

- Establish your own fleet benchmarks for costs and operations.
- Identity poor fuel mpg units as compared to fleet averages.
- Track tire cost by unit and measure cost per mile against fleet average.
- Make new vehicle buying decisions based on performance comparisons.
- Centralize control with consistent fleet reporting using local data access.

- Save up to 5 percent on insurance costs with good reports and PM process.
- Never have to pay fines for out-of-date licenses and permits.
- Identify cost trends and make operating budget projections.
- Export your fleet data to various departments by file.

PROGRAM FEATURES

Fleet maintenance programs may be available with more or fewer features depending upon the complexity of the fleet. Dossier fleet management systems are available in a standard format, which is also featured in their Platinum Edition.

The following is a list of features in Dossier's Standard Fleet Management Edition:

- Fleet inventory with user definable specifications
- Complete maintenance and repair histories (see **Figure 4-7**).
- Prints, controls, and tracks work orders (see **Figure 4-8**).
- Automatic PM Inspection scheduling
- Customizable PM checklists for each unit
- Automatically captures unit and component warranty
- Unit and component warranty notices and reporting
- Prints warranty tracking and labels at time of repair
- Parts, labor, tire, and component costing.
- Cost per mile, hour, or your customized measurement
- Tracks work pending, complaints, and campaigns
- Text notes pages for all areas of the program
- Records unlimited serial numbers for fast lookups
- Unlimited license and permit tracking and scheduling
- Automated fleet budget control and comparison
- Down-time tracking and analysis
- Fuel and fluids inventory tracking, automatic reordering, costing, and inventory control
- One-click repair order template to reduce data entry
- Attach unlimited photos, documents, or scanned images
- Automated license and permit renewal notices
- Fuel and oil consumption and mpg with histories
- Utilization measured by up to three different methods
- Track fixed-cost items like depreciation and more
- Automatic posting of reoccurring costs

- Daily "Work in Progress" reminder feature (see **Figure 4-9**).
- Easy unit look up by "Search 'n' Select" feature
- VMRS coding compliant. Today, equipment users worldwide use VMRS to capture and report their vehicle maintenance activities. Equipment manufacturers and maintenance software suppliers use VMRS coding for parts, thus providing additional impetus for fleets to adopt this universal coding scheme.
- Tracks vehicles at multiple sites from one location
- Tracks one or multiple on the same program
- Each vehicle can have its own costing method.
- Fast and easy equipment search and sorting by specifications (see **Figure 4-10**).
- Automated benchmarking and costing averages (see **Figure 4-11**).

PARTS INVENTORY AND VENDOR FEATURES IN PLATINUM EDITION

The following are extra features in addition to standard format:

- Unlimited number of parts can be entered and tracked.
- Detailed description of each part in inventory
- Automated parts histories of each use and transaction
- Automated parts warranty tracking and reporting
- Warranty notices at time of repair order entry
- Produces warranty labels at time of repair order entry
- Automatic reorder based on minimum and maximum levels
- Parts cross-referencing and multiple sites
- Prints and tracks purchase orders
- History of use, received, returns, and adjustments
- Tracks and records part transfers between fleet shops
- Attach unlimited photo, schematic, and document images
- Parts back order tracking
- Obsolete parts reporting with amount and values
- Parts inventory turns reporting
- Freeform note pages for each part
- Automated parts inventory valuation tracking
- Multiple part pricing methods and markups
- Search and report by any part specification
- Vendor tracking with activity/purchase history
- Track parts and service dollars spent by vendor
- Vendor detail files with services provided look-ups
- Track parts and service expenses by vendor

Figure 4-7 Computer screen shot of "Work in Progress" reminder feature. *(Courtesy of Arsenault Associates)*

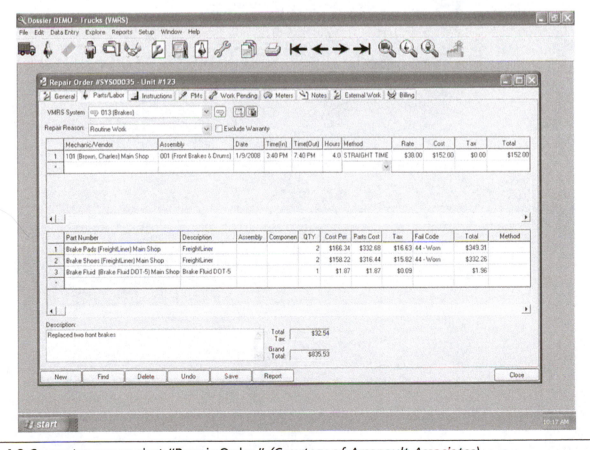

Figure 4-8 Computer screen shot "Repair Order." *(Courtesy of Arsenault Associates)*

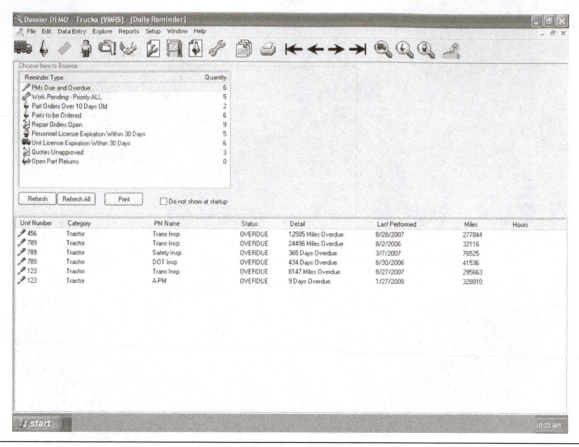

Figure 4-9 Computer screen shot "Daily Reminder." *(Courtesy of Arsenault Associates)*

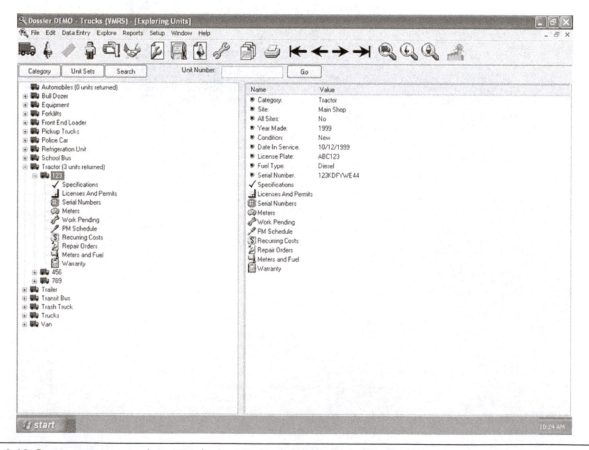

Figure 4-10 Computer screen shot "Exploring Units." *(Courtesy of Arsenault Associates)*

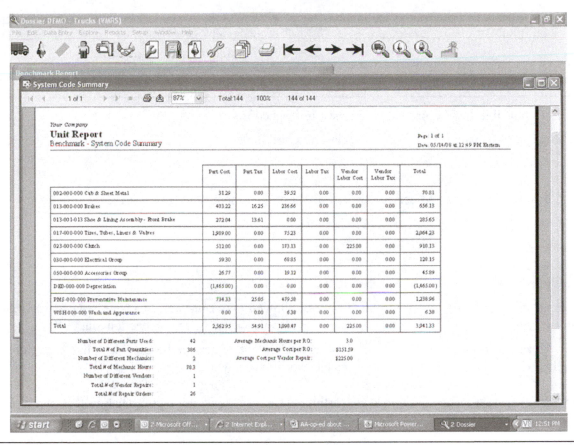

Figure 4-11 Computer screen shot "Automatic Benchmarking." *(Courtesy of Arsenault Associates)*

PERSONNEL TRACKING FEATURES IN PLATINUM EDITION

- Tracks any person attached to your fleet operations
- Complete mechanic and driver files and records
- Fast access to emergency contacts
- Automatically documents which employee made which repairs
- Will track up to six labor rates per mechanic
- Automatically notifies personnel of physical exam renewals
- Automatic labor productivity tracking
- Tracks date/time in and out of each repair order
- Tracks assigned vehicles by drivers with histories
- Commercial drivers' licenses and physical exam exploration and tracking
- Unlimited certification records and renewal notices
- Attach photo images to each person's file

Summary

- A preventive maintenance program allows a vehicle owner to perform the least amount of repair maintenance while performing routine maintenance.

- A well-planned preventive maintenance program offers attractive benefits to the commercial vehicle owner.

- Careful planning is key to the success of any program.

- Communication between drivers and technicians is extremely important.

- The technician's responsibility goes beyond the inspection of the vehicle; they are also required to service and perform repairs.

- The driver identifies many repairs when performing the pre-trip and post-trip inspections.

- A vehicle considered likely to cause an accident or breakdown because of mechanical of mechanical conditions or improper loading will be taken out of service and, tagged with an out-of-service (OOS) sticker.

- It is essential that both drivers and technicians be familiar with the CVSA out-of-service criteria.

- Maintenance managers and truck technicians are expected to be computer literate because most service tracking is computerized.

Review Questions

1. Who is responsible for the pre-trip and post-trip inspection?

   A. The vehicle inspector

   B. The maintenance manager

   C. The truck technician

   D. The driver

2. Of the different types of PM service, which one takes the most time to perform?

   A. Category A

   B. Category B

   C. Category C

   D. Category L

3. A deadlined vehicle would be described as a vehicle that:

   A. has not been properly inspected.

   B. has broken down on the highway.

   C. has been allowed by an inspector to have repairs performed within a 25-mile radius of the inspection site.

   D. has been taken out of service because it is likely to cause an accident or break down.

4. How many separate inspections must be performed on a tractor double combination that also uses a converter dolly?

   A. Two

   B. Three

   C. Four

   D. Five

5. While performing a pre-trip inspection, which of the following is part of the inspection procedure?

   A. Checking to make sure all necessary emergency equipment is on board the vehicle

   B. Checking under the vehicle for sign of fluid leakage

   C. Starting the truck engine and listening of any unusual noises

   D. All of the above

6. Whenever a driver or technician performs a pencil inspection, the vehicle is in jeopardy of being unsafe or even breaking down.

   A. True

   B. False

7. The driver's inspection report serves three purposes. Which of the following is **not** one of them?

   A. Helps the technician keep a vehicle in a safe and operable condition

   B. Helps drivers to maintain their estimated time of arrival (ETA)

   C. Helps the maintenance manager keep a vehicle in a safe and operable condition

   D. Required by law

8. NHTSA has set up a periodic minimum inspection standards program. Which of the following vehicles are exempt?

   A. Vehicles carrying 15 people or more including the driver

   B. Converter dollies

   C. Passenger automobiles

   D. Container chassis

9. Regulations require that a driver's vehicle condition report be retained for at least _____ month(s) from the date of the report.

   A. 1

   B. 2

   C. 3

   D. 4

10. Which of the following is **not** a typical feature offered by PM software programs?

    A. Track and cost labor hours on specific jobs or trucks

    B. Schedule and prioritize unplanned breakdown repairs

    C. Perform a driver's pre-trip inspection

    D. Schedule PM operations by date, mileage, or vehicle data bus faults

5

Cooling System Service and Inspection

Objectives

Upon completion and review of this chapter, the student should be able to:

- List the three types of antifreeze used in today's diesel engines and the advantages and disadvantages of each.

- Describe the need for a supplemental cooling additive package.

- Determine the freezing and boiling points of a coolant mixture based on antifreeze and water ratios.

- Properly mix coolant using the correct proportions of water, antifreeze, and supplemental coolant additives (SCAs) according to the OEM's recommendations and ambient temperature conditions.

- Measure the coolant strength (freeze level) using the appropriate instrument.

- Test the SCA level and maintain it at the desired level.

- Test coolant for contamination.

- Remove and replace a coolant filter and check it for leaks.

- Pressure test a radiator cap and determine its serviceability.

- Test a cooling system thermostat and determine its serviceability.

- Check the condition of the water pump and drive belt tension and condition.

- Test the various forms of cooling fans.

- Diagnose basic cooling system malfunctions.

- Inspect the radiator condition and mounting as well as pressure test the cooling system for leaks and restrictions to proper air flow.

- Inspect coolant lines, hoses, and clamps.

- Inspect the coolant recovery system and determine its serviceability.

Key Terms

aqueous

coolant

ethylene glycol (EG)

extended life coolant (ELC)

hydrometer

propylene glycol (PG)

refractometer

supplemental coolant additive (SCA)

INTRODUCTION

Cooling systems in the trucking industry are one of the most neglected areas of maintenance. Yet if maintenance personal understood that this one system is responsible for nearly half the cost of repairs to a heavy-duty engine, chances are that the system would soon become a priority.

According to a study by Fleetguard, 40 percent of engine repair cost can be traced back to cooling system problems. Now with new exhaust-gas-recirculation (EGR) engines producing more heat than ever, the cooling system maintenance is even more critical.

New EGR engines produce roughly 30 percent more heat that must be handled by the cooling system. This means there must be some changes by manufacturers to the cooling system to increase its efficiency as well. In some applications, the size of the cooling system module has been increased. In cases where space restrictions prevented upsizing, some manufacturers have increased the air flow through the system by increasing fan speed and using a more efficient fan and fan shroud. Other options used by some manufacturers are coolants with a higher boiling point and raising the pressure limit on the radiator cap. The Mack Vision has mounted the radiator in front of the charge air cooler. International is using a side-by-side configuration of the radiator and the charge air cooler on its 8000 and 9000i series.

SYSTEM OVERVIEW

This chapter will deal with all aspects for maintaining an efficient and chemically stable cooling system. It is important that the technician understand the chemistry of the system and just how important it is to the life of the system.

Cooling systems can fail for a number of reasons, the four main causes being:

- Corrosion
- Scale buildup
- Cavitation erosion
- Inhibitor drop-out

Note: The chemical condition of the **coolant** should never be overlooked by the technician. Cooling system preventive maintenance is generally overlooked until someone complains to the maintenance staff that the truck is overheating or there is a puddle of green or red coolant under the vehicle. By this time, much damage may have taken place, and the expenses will start to mount up. This condition might have been avoided if regular cooling system preventive maintenance had been performed.

COOLANT

Engine coolant is a chemical mixture of water, antifreeze, and a **supplemental coolant additive (SCA)** package. If the only requirement of the coolant were to act a heat-transferring medium, water would work more efficiently than any presently used antifreeze mixture. The problem with water only in a cooling system is that it has a low boiling point, a high freezing point, and poor lubrication properties, and it promotes oxidation and scaling activity.

Most current truck engines use a mixture of **ethylene glycol** (EG), **propylene glycol** (PG), or carboxylate-type extended-life antifreeze, plus water to form what is known as coolant. Alcohol-based solutions are not used any more as they evaporate at low temperatures. Engine coolant should always be maintained at the correct proportions of water, antifreeze, and coolant additives. When EG or PG is used as the antifreeze component of the coolant, the additive package must be monitored and routinely replenished.

Advantages

A coolant mixture of water, antifreeze, and additive package should provide the following benefits to the cooling system:

- Corrosion inhibitors in both the antifreeze and additive package provide corrosion protection for the metal, plastic, and rubber compounds found within the cooling system.
- The antifreeze component protects the coolant from freezing and is directly related to the proportion of antifreeze in the mixture.
- The antifreeze component also raises the boiling point of the mixture and is again directly related to the proportion of antifreeze in the mixture.
- The additive package should also contain an antiscaling additive to prevent hard water mineral deposits from adhering to the walls of the cooling system. This scale can contribute to poor heat transfer as well as reduced coolant flow.
- Inhibit the formation of acid in the coolant by the use of a pH buffer. Acid, if present within the cooling system, will cause internal corrosion.
- Foaming suppression of the coolant prevents aeration of the coolant caused by the action of

the water pump and the flow through the cooling system.

- An anti-dispersant prevents insoluble particles from coagulating and plugging the small passages of the cooling system.

Glycol-Based Antifreeze

The freeze protection and antiboiling characteristic of glycol-based antifreeze do not break down over time, unlike the SCA package, which depletes with engine operation (see **Figure 5-1**). This additive package must be tested and restored at regular intervals in order to retain the protection level required for an efficient cooling system. Ethylene glycol and propylene glycol are both petrochemical products. Ethylene glycol has been the standard antifreeze for many years, but the Federal Clean Air Act and OSHA regard it as a toxic hazard. When propylene glycol is in its virgin state, it is said to be less toxic than ethylene glycol. Consequently, propylene glycol is becoming more popular as a base antifreeze ingredient. This being said, leaks and spills of both ethylene and propylene glycol

are to be regarded as dangerous to mammals (including humans) and plant life. The longer the coolant remains in the engine the more toxic the mixture will become.

When ethylene or propylene glycols are mixed with water, they may be described as **aqueous,** meaning they mix easily and don't separate.

ACCIDENTAL MIXING

Mixing of ethylene and propylene glycol-based coolants should not be done. Mixing of the two will not cause harm to the engine or cooling system but measuring the strength of the coolant mixture will be impossible using either a **refractometer** or a **hydrometer.** If it is known that ethylene and propylene glycols have been mixed in a cooling system and it is impossible to flush the system immediately, measure using a refractometer with a ethylene glycol scale and a propylene scale. Then take the average of the two readings. In order to avoid problems down the road, the system should be drained completely and replaced with a mixture of ethylene glycol and water or propylene glycol and water.

Mixing Heavy-Duty Coolant

It is preferable not to mix coolant within the vehicle's cooling system. The best method is to premix the coolant in an external container and then pour or pump the coolant into the vehicle's cooling system. Clean water should be added to the container first. This water should not be excessively hard or have any iron content. Next, the correct amount of antifreeze is added, and finally the SCA package. Combine the contents of the container thoroughly before adding to the vehicle's cooling system.

Note: Tap water should not be used in a cooling system because the magnesium and calcium found in most tap water can cause scaling on cooling system components. Corrosion of engine and cooling system parts can also result from the presence of sulfates in tap water. For these reasons, distilled water should always be used when mixing coolant solutions.

CAUTION Whenever contact with the skin is made with either antifreeze or a coolant solution, the affected area should be washed thoroughly without delay.

Figure 5-1 Typical glycol-based antifreeze container. (Courtesy of John Dixon)

When topping up a cooling system, some OEMs recommend the use of commercially premixed coolant. This shifts the responsibility of mixing coolant from the technician because the water, antifreeze, and SCA are already premixed in the correct proportions. It also eliminates the problems involved with poor-quality water.

Extended Life Coolants

Extended life coolent (ELC) is low maintenance, as the life of the coolant is 6 years, with only the SCA package being replenished once (see **Figure 5-2**).

Antifreeze is an extremely important component of the coolant, as water must not be allowed to freeze within the cooling system. Water expands about 9 percent in volume when it freezes and can distort or break the container holding it as it freezes. This can even occur if the container is a cast iron engine block. Water takes up the least amount of space when it is in the liquid state, close to its freezing point of 32°F (0°C). When water is heated from its near frozen state to close to its boiling point 212°F (100°C), it will expand approximately 3 percent. If the water is contained in a 50/50 mixture of antifreeze, it will expand even more than straight water to approximately 4 percent through the same temperature range. For this reason, the cooling system must be able to accommodate the expansion and contraction rate of the liquid coolant within the system. Antifreeze also increases the boiling point of the mixture.

Extended life coolants are claimed to have a lower toxicity than ethylene- or propylene-based coolants.

Extended life coolant is sold only in a premixed container and is dyed red in color. Only an extended life premix should be added to the cooling system, or an extended life concentrate may be added in extremely cold conditions. Extended life coolant is not compatible with ethylene or propylene glycol antifreeze.

Some of the advantages of extended life coolants:

- Significantly extended service life, 6 years or 600,000 miles (960,000 km)
- No SCA testing required
- Longer water pump life because of reduced TDS content (TDS can be abrasive)
- Less hard water scaling
- Better corrosion protection
- Better cavitation protection
- Increased ability to transfer heat
- No gelling problems because no silicates are used (EG and PG get sludge buildup due to silicate drop-out)

Figure 5-2 Extended life coolant. *(Courtesy of John Dixon)*

- Better protection for aluminum components
- Improved high temperature performance over EGs and PGs

High Silicate Antifreeze

When aluminum is one of the internal components that must be protected, a coolant with high silicate concentrations should be used. However, many OEMs require the use of low silicate coolant to be used in their engines. High silicate and low silicate antifreeze should not be mixed. High silicate antifreeze should not be used unless specified by the OEM. ELCs (extended life coolants) do not use silicates and other chemicals to reduce scaling but instead use a carboxylate base that, according to the manufacturer, drastically outperforms the complicated cocktail required of an ethylene or propylene glycol coolant.

Note: In severe winter conditions, a running truck circulates coolant through the engine and radiator, which is exposed to wind chill factors well below ambient temperatures. The coolant can ice up if the freeze protection is not sufficient. Make sure the coolant freeze protection is adequate for running wind chill factors.

If an engine is using an EG or PG coolant, it is permissible to change it over to an ELC. All that needs to be done is to drain the old EG or PG and dispose of it correctly. Then the cooling system is flushed with fresh water. The ELC can then be added to the system. If the cooling system used a coolant filter, it must be replaced with a SCA-free filter. The freeze protection of ELC may be strengthened for users in northern states and Canada. Check with the manufacturer of the ELC for a concentrate additive.

Testing Coolant Strength

It is very important that technicians regularly test the strength of the engine coolant. Frequently, the cooling system will be topped up with straight water while the vehicle is on the road, and this will have a major effect on the freeze protection of the coolant mixture. Standard antifreeze hydrometers are calibrated to measure the strength of ethylene glycol mixtures and tend to be fairly inaccurate, requiring calculations to correct for temperature. A refractometer is a more accurate measuring instrument to test coolant strength (see **Figure 5-3**). These devices are recommended by most OEMs and the TMC (Truck Maintenance Council). Refractometers should be accurate to within 7° of the actual freeze point of the coolant throughout the temperature range that the vehicle will be expected to operate in. Refractometers designed for use with truck diesel engine coolant will have a calibration scale for both ethylene and propylene glycol-based coolants. Some refractometers can also be used for checking the strength of electrolyte solution in batteries.

The technician is responsible for ensuring that the correct scale for the antifreeze base is being used, or

measurement will be inaccurate. The refractometer measures how far light is bent as it passes through a liquid. The term used to indicate this is refraction. The refraction index of coolant increases as the concentration of antifreeze increases.

Refractometer Operation

The refractometer is used by taking a small sample of coolant and placing a drop of it on the prism surface (see **Figure 5-4**). The lid is then lowered, and the technician looks through the eyepiece while holding the lens up to the light. Ambient light is usually sufficient to get a good view. Focusing the eyepiece provides a sharp, easy-to-read scale, which is designed with a semitransparent background. The view inside the eyepiece will appear like a shade being pulled down on a window. The intersecting point of the light and the shade will correspond with a temperature scale indicating the freeze protection of the coolant mixture. Although most refractometers can be used in this manner, always follow the operating procedures outlined by the manufacturer of the instrument (see **Photo Sequence 2**).

Scaling

Scaling is caused by the minerals found in hard water that adhere to the surface of the cooling systems where temperatures are highest. Scaling buildup insulates engine components designed to transfer heat. If scaling conditions are left unimpeded, engine overheating and consequent engine failure will occur.

De-scaling agents are commercially available to remove minor scale buildup, followed by flushing the cooling system. If the engine has already developed an overheating problem caused by scaling, the descaling agent will have little effect on the cooling system. In this case, the engine will have to be disassembled and the cylinder block and heads boiled in a soak tank.

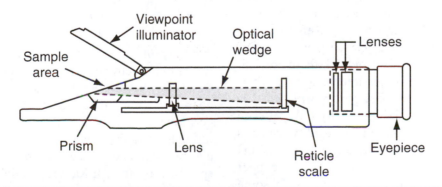

Figure 5-3 The parts of a typical handheld refractometer.

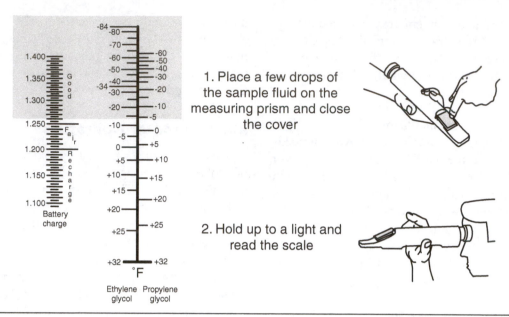

1. Place a few drops of the sample fluid on the measuring prism and close the cover

2. Hold up to a light and read the scale

Figure 5-4 Using a refractometer.

Testing Supplemental Coolant Additives

The SCA should be tested on regular PM services or when there is a substantial loss of coolant and the system has to be replenished. SCA testing is recommended by OEMs to ensure the coolant is protecting the internal components of the cooling system. The SCA is determined by the engine manufacturer and is specific to the materials used within the cooling system. Therefore, test kits of one manufacturer's engine cannot generally be used for other OEM products. The test kits usually contain test strips that must be stored in airtight containers. These test strips are also manufactured with an expiration date that, if not adhered to, may lead to inaccurate test results. The coolant test kits allow the technician to test the concentration level of the SCA, the pH level, and **total dissolved solids (TDS).** The pH level determined indicates the relative acidity or alkalinity of the coolant. When coolant comes in contact with degrading metals (ferrous and copper based) or combustion gasses, acids may form in the coolant. The test is known as a **litmus test** and is performed by immersing a test strip in a coolant sample. Once the test strip is removed, the color will change. The strip is then compared to a color chart provided with the kit. The best possible pH level is determined by each OEM, but generally falls between 7.5 and 11.0 on the pH scale. Higher acidity readings of the test coolant (readings below 7.5 on the pH scale) indicate the corrosion of ferrous and copper metals and exposure to combustion gasses. In some cases, high acidity can indicate the breakdown of the

coolant itself. Higher than normal alkalinity levels indicate aluminum corrosion and the possibility that a low silicate antifreeze has been used where a high silicate antifreeze is required. From the tables, the amount of additive to add to the system can be determined. The additive can be installed in the cooling system in a number of ways. It may be installed in the engine coolant filter, although this method generally results in higher than normal additive levels. The SCA may also be mixed with the coolant outside the cooling system and then reinstalled. Most OEMs recommend testing additive levels in the coolant, then adding to adjust to the required additive levels. Unmeasured quantities of coolant additive should not be dumped into a cooling system at each PM service, as engine damage could occur due to higher than normal acidity.

The SCA may also be adjusted to suit a specific operating environment or set of conditions. Unusually hard water, for instance, requires a greater amount of antiscale protection.

Testing for total dissolved solids (TDS) requires the use of a TDS probe. This probe measures the conductivity of the coolant by conducting a current between two electrodes. Distilled water does not conduct electricity, but as the total dissolved solids build up, so does the ability to conduct electricity. This test is performed by inserting the probe into the coolant through the radiator cap. The TDS is measured in parts per million, and if the reading is higher than that specified by the OEM, the condition of the coolant may be conducive to scale buildup.

2 Testing Coolant Strength

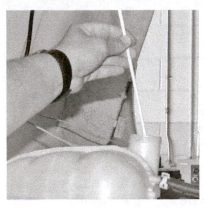

P2-1 Take a sample of coolant from either the recovery bottle or from the top of the radiator.

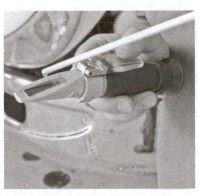

P2-2 Apply coolant to the prism surface.

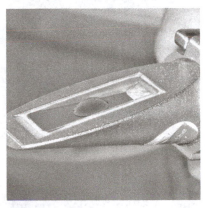

P2-3 One drop is all that is required.

P2-4 Close the lid on the refractometer.

P2-5 Hold the refractometer up to the light and make a reading. It will appear as if a shutter has been pulled down.

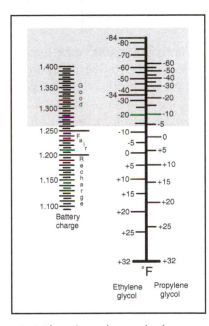

P2-6 The view through the refractometer.

Coolant Recycler

This is a compact unit that can be used for all your coolant service work (see **Figure 5-5**). It recycles on or off the vehicle, drains and refills a system, has a reverse flow option that provides back flushing, and includes a gauge to check system pressure for diagnostics:

- Drain and fill for simple coolant exchange.
- Pressure test for leaks.
- Vacuum fill an empty system.
- Reverse coolant flow to clean system (optional adapter required).

COOLANT FILTERS

Engine coolant filters are usually of the spin-on cartridge type and are connected in parallel to coolant flow. Some filters are charged with the SCA package,

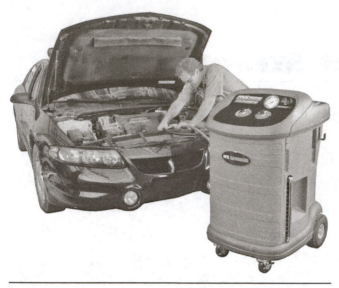

Figure 5-5 An engine coolant recycler. *(Courtesy Robinair, a Business Unit of Service Solutions)*

Figure 5-7 A typical engine coolant filter with hand shutoff valves. *(Courtesy of John Dixon)*

so it is important that the correct filter be used along with the correct coolant base as specified by the OEM. This is also a good reason to avoid overservicing of the filter.

Fluid flow through the filter is consistent with that of most other engine filters in that the coolant enters the canister through the outer ports and exits the center single port. Some coolant filters are equipped with a zinc oxide electrode to cancel out the electrolytic effect of the coolant. The electrode is sacrificed to the coolant, protecting internal metal components within the cooling system. These electrodes are more commonly found in marine applications (see **Figure 5-6**).

Coolant Filter Replacement

When it is time to change a coolant filter, the technician should familiarize himself with the type of shutoff valve used on the vehicle. Some are automatic, while others are of the manual type. These valves permit changing of the filter with minimal coolant loss. Once the filter is changed and valves are reopened, the filter will prime itself. Check for any leaks at the filter before allowing the vehicle to go back into service (see **Figure 5-7**).

COOLING SYSTEM COMPONENTS

The cooling system is made up of components that store, pump, condition, and manage engine temperature and coolant flow rate. These components may be very similar from one diesel engine OEM to the next. Nevertheless, when servicing and repairing cooling system components, always consult the appropriate service literature.

RADIATORS

A radiator is a form of heat exchanger. The engine horsepower rating will usually determine the area the radiator will take up in the front of the truck. On average, this is in the range of 3 to 4 square inches per BHP unit. The cooling medium for the radiator is ram air, that is, ambient air forced through the radiator core as the truck is driven down the road. The radiator holds a large volume of coolant in close contact with a large volume of air, so heat will transfer from the coolant to the air. The efficiency of the radiator is determined by the ambient temperature and the speed of the vehicle.

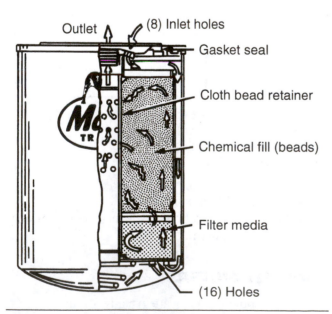

Figure 5-6 The flow of coolant through a typical engine coolant filter. *(Courtesy of Mack Trucks, Inc.)*

Note: The rate of heat transfer depends on the temperature difference between coolant temperature and ambient temperature.

A fan shroud will also increase the heat transferring efficiency of the radiator as well as the efficiency of the fan.

Radiator Components

A typical radiator incorporates the following components (see **Figure 5-8**):

- **Core:** The core is the center section of the radiator. It is made up of tubes and cooling fins.
- **Tanks:** The metal or plastic ends fit over the core tube ends to provide storage for coolant and fittings for the hoses.
- **Filler neck:** The opening for adding coolant, it is also designed to accept a pressure cap and overflow tube.
- **Petcock:** Fitting located at the bottom of the tank is used for draining coolant from the system.
- **Radiator hoses:** Radiator hoses are used to direct coolant between the engine water jackets

and the radiator. These hoses must be flexible to withstand the engine movement resulting from engine torque. The upper radiator hose normally connects the radiator inlet to the thermostat housing on the intake manifold or cylinder head. The lower radiator hose connects the radiator to the water pump inlet.

Radiator Servicing

Radiators are generally neglected by most technicians until the cooling system experiences some sort of failure because of leakage or overheating. The condition of the radiator should be checked on every PM service to ensure that it is clean and that there is no buildup of road dirt or insects that can restrict air flow and in turn hinder the radiator's heat transferability. Radiators should frequently be washed with either a low-pressure steamer or a garden hose and detergent and a soft nylon-bristle brush.

Note: High-pressure washers should never be used to clean a radiator as the high-pressure water can damage the delicate fins and can impede air flow through the radiator.

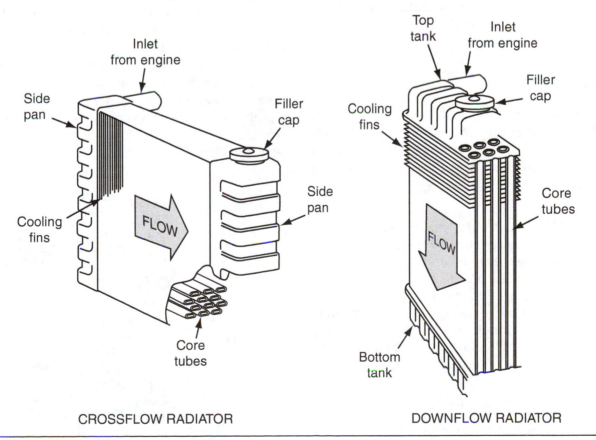

CROSSFLOW RADIATOR   DOWNFLOW RADIATOR

Figure 5-8 Typical radiator components.

While servicing the radiator, the technician should visually inspect all mounting hardware and hose clamp connections and ensure that other components are not permitted to rub against any surface of the radiator.

Radiator Testing

Frequently, coolant leaks are the result of damage and not corrosion. Some leaks may be located by white or reddish stains at the leakage point. To locate problem leaks, the radiator may also be pressure tested to around 10 percent above its normal operating pressure, but OEM recommendations must be followed, especially when plastic tanks are used in their construction. (See cooling system pressure testing later in this chapter.) It is important that repairs to radiators be performed promptly to protect the engine from possible damage. If a leak is found in the radiator caused by external damage and the radiator is in good condition, it may be repaired by shorting out the affected tubes. This procedure usually involves removing the tanks and plugging the damaged tube at the header. If the leaking tube is accessible, it may be repaired by soldering with a 50/50 lead-tin solder, but the preferred method would be to braze the repair with silver solder, which requires considerably more heat that may affect other soldering joints within the radiator core. Frequently, radiators are removed and sent out to shops with equipment and technicians that specialize in the repair of heating and cooling system components.

Radiator Cap

A very important component of cooling system is the radiator pressure cap. The radiator cap locks onto the filler neck with a quarter turn. Usually the radiator cap must be pushed down before it can be turned on or off (see **Figure 5-9**).

> **CAUTION** *The radiator cap should never be removed while the system is under pressure. Hot pressurized coolant may be forced out the filler neck, causing serious burns to anyone close by. Most filler necks are fitted with double-cap lock stops preventing the removal of the radiator cap in a single counterclockwise motion. Never attempt to remove a radiator cap until the system has cooled and the pressure equalized.*

The radiator cap performs four functions:

- Seals the top of the radiator's filler neck, preventing coolant leaks

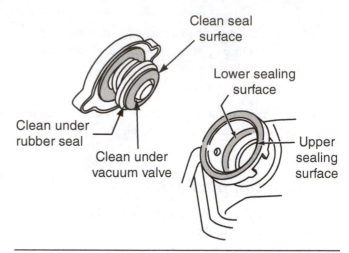

Figure 5-9 A typical radiator pressure cap and radiator filler neck.

- Pressurizes the cooling system, thereby raising the boiling point
- Relieves excess system pressure to protect the system from damage
- Allows coolant to flow to and from the overflow tank to maintain a constant volume of coolant

Radiator caps allow for the thermal expansion of the coolant to pressurize the cooling system. For each 1 psi (7 kPa) above atmospheric pressure, the boiling point raises by 3°F (1.67°C) at sea level. System pressures will hardly ever be designed to operate above 25 psi (172 kPa) and are more commonly designed to operate between 7 psi (50 kPa) and 15 psi (100 kPa).

A radiator cap is identified by the pressure the system is designed to maintain before the spring pressure of the cap is overcome, unseating the seal. When this occurs, the coolant is routed to an overflow tank (see **Figure 5-10**).

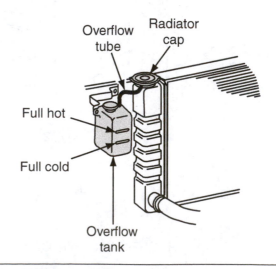

Figure 5-10 An overflow tank.

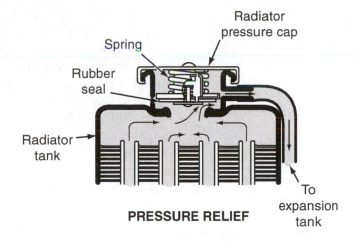

PRESSURE RELIEF

Spring

Radiator
pressure cap

Rubber
seal

Radiator
tank

To
expansion
tank

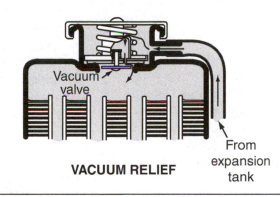

Vacuum
valve

VACUUM RELIEF

From
expansion
tank

Figure 5-11 Flow through a radiator cap.

The overflow tank will be at its highest level when the engine is operating at its warmest temperature. When the engine is shut off, it will begin to cool. This will cause the coolant to contract (thermal contraction), which will decrease the pressure to the point where the system is in a partial vacuum. When this occurs, the vacuum valve within the cap will unseat, allowing coolant from the overflow tank to be pushed by atmospheric pressure back into the radiator (see **Figure 5-11**).

Some cooling systems do not use a coolant overflow tank but use a coolant expansion tank (see **Figure 5-12**). This tank has a sight glass that the technician can watch to make sure that coolant is at the correct level. This allows for space at the top of the tank for the coolant to expand. If pressure exceeds safe limits, the radiator cap will bleed pressure from the tank to maintain the intended pressure setting. As the system cools and the coolant contracts, the vacuum portion of the pressure cap will allow atmospheric pressure to enter the cooling system, preventing the collapsing of radiator and heater hoses.

Note: For the vacuum valve in the pressure cap to open, the atmospheric pressure applied to the surface of the coolant in the reservoir tank must be higher than that of the cooling system.

Expansion tank

Sight glass

Figure 5-12 A coolant expansion tank. (*Courtesy of John Dixon*)

Radiator Cap Testing

The radiator cap may be tested to ensure that it will contain the required pressure and also relieve any pressure above its rated capacity. This can be done with a standard cooling system pressure testing kit. The radiator cap is first installed on the correct size adapter and then connected to the hand pump. The hand pump is used to pressurize the cap to 1 psi (7 kPa) above its rated capacity. The cap should open, allowing pressure to bleed off. Next relieve the pressure and again pump until the exact pressure rating of the cap is achieved. Stop pumping and observe the pressure gauge. A properly sealing cap should not drop more than 2 psi (15 kPa) in 1 minute (see **Photo Sequence 3**).

WATER PUMP

The water pump, as the name implies, circulates coolant throughout the cooling system. Most water pumps are non-positive centrifugal pumps directly driven by gear or by belt. Basically, pumps that discharge liquid in a continuous flow are referred to as non-positive displacement pumps and do not provide a positive seal against slippage like a positive displacement pump (see **Figure 5-13**).

When the water pump is turned by the engine, an impeller spins inside the pump housing. This creates low pressure at its inlet, usually located at or close to the center of the impeller. The vanes of the impeller throw the coolant outward, and centrifugal force accelerates it into the spiraled pump housing and out toward the pump outlet. The fluid leaving the pump flows first

through the engine block and cylinder head, then into the radiator, and finally back to the pump.

Due to the fact that the pressure of the coolant is at its lowest as it enters the water pump, boiling will always take place at this point first. This condition can speed up an overheating condition, as the impeller will not efficiently pump vapor. Water pumps are the main reason that coolants should contain some lubricating properties. The **impeller** is susceptible to the abrasive wear of the coolant when TDS levels are high. Some of the reasons why water pumps fail are listed below:

- Overloading of the pumps bearings and seals caused by too much belt tension or belt misalignment
- Erosion of the impeller caused by high TDS levels in the coolant
- Mineral scale buildup on the pump housing
- Overheating of the water pump caused by coolant boiling at the pumps inlet. If this happens during a hot shutdown, a vapor lock may occur.

Water Pump Replacement and Inspection

The water pump is responsible for circulating the coolant through the engine and may be referred to as the heart of the cooling system. When a water pump from a reputable manufacturer is properly installed in a well-maintained cooling system, it will last for an extremely long time, and they are often the victim of misdiagnoses and premature replacement. The presence of a small amount of coolant on the water pump housing does not necessarily mean that the water pump is ready to fail. In fact it is often part of normal operation, depending on the application and temperature conditions, and should not raise much concern. Water pump bearings are also routinely misdiagnosed as having too much play. Bearings require clearance to turn, so some play is normal. Judging how much is too much by hand is generally not easy.

A water pump will need replacement if the seal fails. This condition is easily diagnosed, since it is usually accompanied by a large volume of coolant loss that pools under the engine.

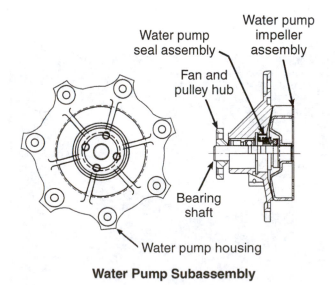

Water Pump Subassembly

Figure 5-13 Cutaway of a typical water pump assembly. *(Courtesy of Navistar International Corp.)*

CAUTION *Before removal and replacement of a water pump, ensure that the coolant has been drained according to environmental requirements in your area.*

Pressure Testing for a Radiator Cap

PHOTO SEQUENCE 3

P3-1 Check radiator cap and make note of the pressure setting of the cap. Ensure it is the correct cap for the system you are working on.

P3-2 Using the hand pump of the cooling system pressure tester, raise the pressure within the cooling system to system pressure, as indicated on the pressure cap. Note: it may be permissible to raise the pressure 10 percent above the rating of the cap but always follow manufacturer's instructions, especially when working with plastic tanks. Visually inspect cooling system for signs of external leaks.

P3-3 When testing a radiator cap, make note of the pressure setting on the cap.

P3-4 Install adaptor to test the pressure setting of the radiator cap.

P3-5 Use the pressure tester pump to raise the pressure exerted on the radiator cap to system pressure indicated on the cap. The cap should not leak air until the pressure setting of the cap has been surpassed.

P3-6 Relieve pressure from cooling system or cap before removing cooling system pressure tester. This is accomplished on the tester pictured by rotating the upper collar.

Water Pump Removal

If a water pump is thought to be defective, it should be removed from the vehicle and inspected as to the cause of the failure in order to avoid a repeat. Follow this sequence in order to remove the pump:

- Make sure the cooling system is at a safe temperature.
- Remove the radiator cap.
- Remove coolant by opening the radiator drain cock and allowing it to drain into a suitably sized container.
- Remove tension from the water pump drive belt.
- Remove water pump as per manufacturer's instructions.
- Inspect the water pump.

Water pumps are rebuildable by the technician but are most commonly replaced with a rebuilt unit. Rebuild centers are equipped with specialized tooling to perform the task of rebuilding the pumps quickly and efficiently. When a pump is rebuilt, inspect the components thoroughly; in many cases, especially where plastic impellers are used, only the housing and shaft are reused. Examine the shaft seal contact surface for wear, paying special attention to the drive gear teeth. The OEM instructions should be observed, and where ceramic seals are used, great care is required during installation to avoid cracking them. One critical measurement is the impeller to housing clearance, and failure to meet this will negatively affect the pump's efficiency.

Water Pump Installation

- Before installing the water pump, make sure the sealing areas of the pump and the mating housing are cleaned of any old gasket material or O-rings.
- Mount water pump with new gasket or O-ring and tighten mounting bolts to manufacturer's specifications.
- Inspect water pump drive belt condition, and adjust to manufacturer's specifications.
- Make sure the radiator drain cock is closed.
- Test condition of coolant and reuse if okay; otherwise recycle it or replace it.
- Pour coolant in through the filler neck of the radiator or expansion tank, allowing time for air to bubble up and out of the system.
- When coolant is filled to the correct level, run the engine to operating temperature so that the engine thermostat opens.
- Shut down and top up coolant level.

- Wash off any antifreeze in the engine compartment.
- Test for leaks.

THERMOSTATS

The thermostat is a valve in the cooling system that automatically adjusts the flow of engine coolant to maintain optimum engine operating temperature. A thermostat must perform the following functions:

- The thermostat must start to open at a specified temperature.
- The thermostat must be open fully at a set number of degrees above the start to open temperature.
- It must define a flow area through the thermostat in the fully open position.
- It must block coolant flow completely or define a small bypass quantity when the valve is fully closed.

Thermostat Operation

The thermostat is usually located in the coolant manifold or housing attached to the manifold. When the engine is cold, the thermostat will be closed, directing the coolant through the water pump and back into the engine (bypass circuit). This will allow the engine to warm up quickly. Once the engine has reached its normal operating temperature, the thermostat will begin to open, allowing coolant to flow past it and circulate throughout the cooling system. Due to the fact that the thermostat defines the flow area through which the coolant can pass, a system may use more than one thermostat.

The thermostat incorporates a heat-sensing element that actuates a piston attached to the seal cylinder. When the engine coolant is cold, it is bypassed from the radiator and routed directly to the water pump, where it is circulated through the engine. When the engine has sufficiently warmed to operating temperature, the seal cylinder blocks off the passage to the water pump and routes the coolant to the radiator. The heat-sensing element consists of a hydrocarbon or wax pellet in which the actuating shaft of the thermostat is immersed. As the coolant in contact with the heat-sensing element heats up, the medium inside the element heats up and expands. This forces the actuating shaft outward in the pellet, which in turn opens the thermostat.

Bypass Circuit

The **bypass circuit** describes the flow of coolant during startup before the thermostat opens. This allows the water pump to circulate coolant through the engine cylinder block and head. This flow of bypass coolant

Thermostat operation

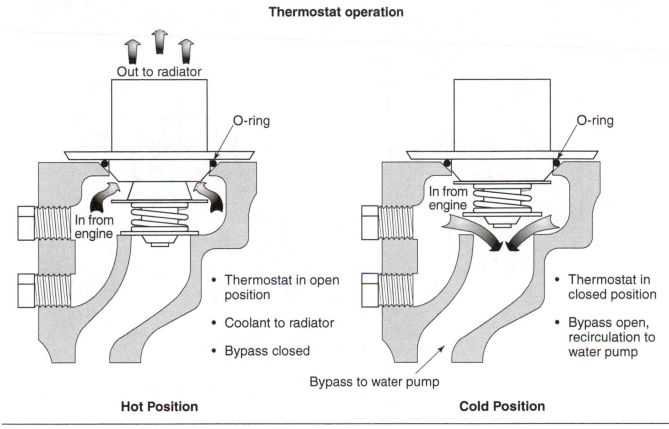

Hot Position   **Cold Position**

Figure 5-14 Thermostat closed in bypass mode and open at operating temperature. *(Courtesy of Navistar International Corp.)*

allows the engine to heat up faster to its proper operating temperature (see **Figure 5-14**).

Operating without a Thermostat

The practice of running a truck engine without a thermostat is not recommended by engine OEMs and may void warranty. Running in this condition also violates the EPA requirements with regard to tampering with emission control components. Removing the engine thermostat will result in the engine running too cool. This can cause water vapor in the crankcase to condense and cause corrosive acids and sludge in the crankcase. As well, when an engine is run at low temperatures, hydrocarbon emissions will increase, causing the engine to fail EPA requirements.

Thermostat Testing

Thermostats may be tested using a specialized tool that essentially consists of a tank, heating element, and an accurate thermometer. This apparatus can also be duplicated using a pot of water on a hot plate.

Place the thermostat in the container of water and heat the water, paying close attention to the thermometer. The temperature at which the thermostat starts to open

and the temperature when it is full open can be determined. If the test is satisfactory, the thermostat can be reinstalled. If the thermostat does not open at the correct temperature, it is defective and should be replaced. Always consult the OEM specifications for the thermostat you are testing. Also remember that there is a difference between start to open and fully open temperature values (see **Figure 5-15**).

HEATER CORE

The heater core is designed much like the truck's radiator although physically much smaller. The truck's heating system is built into the engine's cooling circuit. Hot engine coolant is pumped through the heater core while a fan circulates air from the cab through it. In this way, heat from the coolant is transferred to the cooler air by the tubes and fins of the core assembly. The warm air exiting the heater core is then circulated through the cab of the truck. The circulation of the engine coolant is performed by the truck's water pump. Control doors, sometimes referred to as vent doors, route the air to specific areas of the cab compartment to perform the heating and defrosting requirements of the truck. The vent doors may be actuated by vacuum, cable, air, or electric motor. The motors and cables of the system are controlled by either one or two control levers

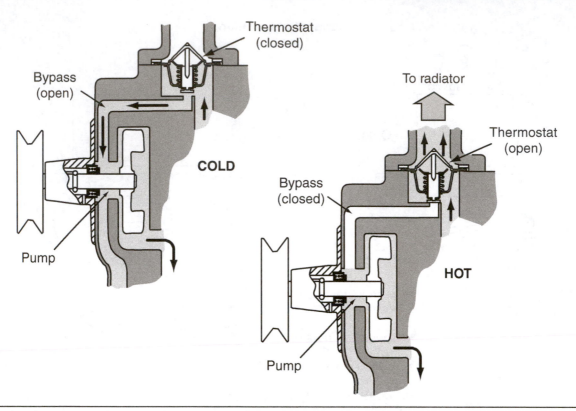

Figure 5-15 Thermostat closed and thermostat fully open.

or rotary dials, which vary the function and temperature of the system (see **Figure 5-16**).

Heater Core Maintenance and Inspection

Engine coolant kept in good condition is the best PM in prolonging the life of both the heater core and the radiator.

The heater core itself consists of an inlet tank connected by headers to a heat exchanger core. The heater core tank, tubes, and fins may become blocked in time by scale, rust, and mineral deposits circulated by the coolant.

If the heater core is suspected of being blocked or restricted, it can usually be determined by feeling the inlet and outlet lines leading to and from the heater core box. Both lines should feel hot or warm to the touch, depending on the position of the water valve. A leaking heater core will produce a pungent sweet smell in the truck cab and, if the leak is severe, will steam cover the windshield.

The heater core is located on the firewall of the truck and is usually buried deep within the dash. Replacement of the heater core is usually a time-consuming task, so it is well worth the effort to leak test the replacement core before installing it in position (see **Figure 5-17**).

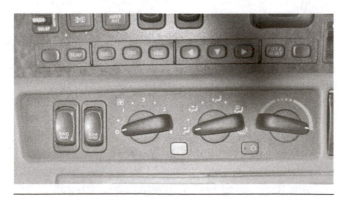

Figure 5-16 The dash controls for climate control of the truck cab. (*Courtesy of John Dixon*)

Figure 5-17 The case containing heater core, evaporator, blower fan, and mixing door and vents. (*Courtesy of John Dixon*)

HEATER CONTROL VALVE

The heater control valve may also be referred to as a coolant valve or water valve. This valve's function is to control the flow of engine coolant into the heater core. If the valve is in a closed position, no hot coolant is permitted to circulate through the heater core, allowing it to remain cool. If the valve is open fully, hot coolant will be allowed to circulate through the heater core, providing maximum heat. A variety of valve positions between fully open and fully closed allow the intensity of heat to be controlled. As previously stated, the engine thermostat that helps to control the engine coolant temperature plays a significant role in providing heat for the truck's cab. A malfunctioning thermostat can cause the engine to overheat or not allow the truck engine to reach its desired operating temperature at all. In both these instances, the performance of the truck's heating system will be compromised.

The climate control system for the cabin of the truck consists of the heater core and heater control valve for all the truck's heating and windshield defrosting needs and the evaporator for all vehicle cooling needs. The heater core and evaporator are compact units that are usually mounted below and to the right of the instrument panel. A variable-speed blower motor is mounted with the heater and evaporator core. It circulates air from the cab through both coils and supplies heated or cooled air to the ductwork behind the dash panel.

BUNK HEATER AND AIR CONDITIONING

Bunk or sleeper climate control systems use remotely mounted components, usually located in the storage compartment below the bunk (see **Figure 5-18**). These systems consist of a secondary heater core, evaporator core, and control valves. The bunk air-conditioning system is dependent on the vehicle's regular air-conditioning system, sharing the same compressor condenser and refrigerant. The bunk heater is plumbed directly to the engine and is usually dependent upon the truck's cab heater.

Bunk Climate Control Inspection

Visually check the air-conditioning refrigerant lines and heater hoses for signs of leakage, chafing, and proper mounting. Some systems require the servicing of an air filter for the ventilation system. Check the vehicle service manual for specific instructions.

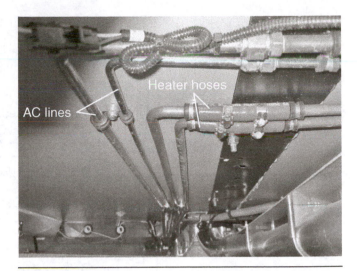

Figure 5-18 Heater hose lines and air-conditioning lines for the bunk climate control system. (*Courtesy of John Dixon*)

SHUTTERS

Shutters are located in front of the truck's radiator. Their purpose is to control the air flow through the radiator and engine compartment. Shutters are sets of louver-like slats that pivot on interconnected shafts and may be rotated in unison from fully closed to fully open positions, much like a window blind. The movement of the shutters is controlled by a **shutterstat** (see **Figure 5-19**). The shutterstat is a temperature-actuated control mechanism located in the coolant manifold. It receives a feed of system air pressure, which it will allow to pass through to the shutter cylinder until a predetermined temperature is reached.

When air from the shutterstat is delivered to the shutter cylinder, the plunger extends to actuate a lever and close the shutters. The shutters are spring loaded to open, so if there is a failure with the shutter system, they will be held open by spring force. During engine warmup, the shutters are normally held closed once there is enough air pressure delivered to the cylinder to overcome the opposing spring force. Shutters are not nearly as common today, with advances in engine technology to turbo-charged engines with air-to-air coolers requiring constant air circulation.

Shutter Inspection

To check the operation of the shutterstat, run the truck's engine up to operating temperature and ensure that the shutters open as the engine warms up.

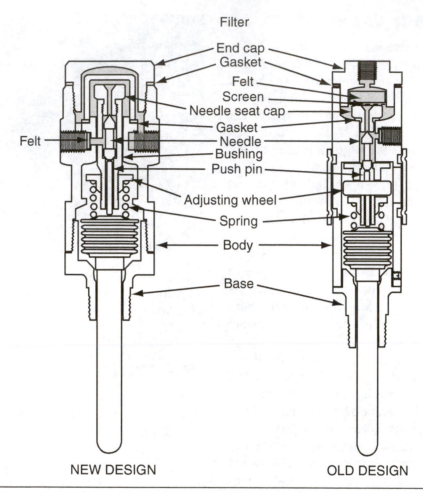

Filter

End cap
Gasket
Felt
Screen
Needle seat cap
Gasket
Felt — Needle
Bushing
Push pin
Adjusting wheel
Spring
Body
Base

NEW DESIGN   OLD DESIGN

Figure 5-19 The shutterstat, which controls the opening and closing of the shutters, if equipped.

COOLING FANS

OEMs generally used two basic types of engine cooling fans: suction type and pusher type. Suction-type fans are designed to pull outside air through the cooling system and engine compartment. Pusher-type fans, like those used on busses, push heated air out of the engine compartment while pulling in fresh air in from the outside. Highway trucks that use the ram air effect typically use a suction-type fan. Depending upon engine size, radiator design, and vehicle application, ram air is capable of performing the required cooling 95 percent of the time. The design of fan is an important consideration, as a typical modern fiberglass variable-pitch fan blade may draw around 6 BHP (brake horse power) from the engine while a large steel fan can require as much as 15 BHP. As a result, many OEMs have chosen the fiberglass blade over the steel or aluminum blade. Some of the more current blade designs use flexible pitch blades that alter the fan's efficiency proportionally to its driven speed, permitting efficient fan operation at low engine rpm (see **Figure 5-20**).

Cooling Fan Inspection

The cooling fan must be maintained regularly as part of the cooling system. This includes checking the condition of the blades (check for cracks or nicks), as fan blades

Figure 5-20 A fiberglass engine cooling fan. (*Courtesy of John Dixon*)

that are not balanced can set up harmonic vibrations that can destroy water pump bearings, create torsional stress failures of the crankshaft, and damage other related components, leading to total cooling system failure.

On/Off Fan Hubs

The fan assembly draws power from the engine while being driven, and most current truck engines use lightweight, temperature-controlled fans. Fanstats controlling air, electric, or oil pressure engaged clutch fans of the on/off type are used as well as thermo-modulated, viscous drive fans.

On/off fan hubs use clutches that are actuated electrically, by oil pressure, or air pressure. These clutches are controlled by a fanstat or the engine management system. As discussed in an earlier chapter, a vehicle's air-conditioning system may also control the fan hub cycles independently from the ECM using electric over pneumatic switching. The fanstat is usually located in the coolant manifold. It is a temperature-activated switch that will activate the clutch to lock up or to freewheel, using either an electrical signal or an air pressure signal as a means of engaging the clutch. Fan hubs are usually spring loaded to keep the clutch in the engagement mode, so the electrical or air pressure signal disengages the fan clutch, allowing it to freewheel. In this way, if there is a malfunction in the system, the fan will be engaged, preventing an overheating situation. On/off type fan hubs rob the least amount of power from the engine when the coolant temperatures are below the trigger value, which can be as much as 95 percent of the time. This percentage may be altered as ambient temperatures increase and the vehicle's air-conditioning system locks up the fan hub (see **Figure 5-21**).

Figure 5-21 Air-actuated fan hub. *(Courtesy of John Dixon)*

Engine oil pressure may also be used as the medium to couple the fan hub (some bus applications). The **fanstat** controls the flow of oil to the hub assembly. The fanstat sends oil pressure to disengage the fan hub, allowing it to freewheel. Once the fanstat has reached its set temperature limit, it will stop the flow of oil, allowing spring pressure to lock up the fan hub and engaging the fan. Oil pressure that is directed to the fan hub acts a fluid coupling, so there is always some slippage. Some electronically managed engines directly control the fan cycle and use it as a retarding mechanism as well as a cooling aid.

Fan Belts and Pulleys

Fan belts and pulleys should be checked on every PM service. Fan pulleys use bearings that may be roller, tapered roller, or the bushing type. The pulleys use external V or poly-V groove belts. Belt tension should be adjusted with the use of a belt tensioner in order to ensure that the belt is neither too loose nor too tight. If the belt is adjusted too tightly, it applies excessive loads on the pulley bearings, shortening both bearing life and belt life. If the belt is adjusted too loosely, the slippage will cause the belt to be destroyed even more quickly than if it were too tight.

Fan Operation Inspection

The fan operation can usually be tested for engagement by turning on the air-conditioning unit and allowing it to go through a cycle. This will allow the technician to verify hub lockup without actually running the truck engine up to a very high temperature. With regard to current fanstat operation, technicians should check for a fault code regarding to the cooling system. If no fault has been recorded, it may be assumed the fanstat is controlling the fan operation properly.

Check fan belt tension using a belt tensioning gauge according to manufacturer's specifications. Inspect the condition of the fan belt. Fan belts should be replaced when they show signs of being glazed, cracked, or nicked. It is much more cost effective to replace belts if they show any of these early signs of failure than it is for the truck to experience a roadside breakdown due to belt failure.

Thermatic Viscous Drive Fan Hubs/ Thermo-Modulated Fans

The **thermatic viscous drive** fan hub is an independent component and is not externally controlled. They use silicone fluid as the drive medium between the drive hub and the fan drive plate. The hub assembly may be broken down into three sections: the drive hub (input section), the driven fan drive plate (output section), and the control mechanism. This fan drive system does not use a mechanical connection between the drive and driven members. During minimum slip operation, torque is transmitted through the internal friction of the silicone drive fluid in the working chamber that couples the input and output sections. A wiper attached to the driven member continually wipes the fluid, and centrifugal force returns it to a supply chamber, where an open valve cycles it back to the working chamber. A bimetal strip located in the control mechanism senses temperature, and when the temperature within the engine compartment drops, the bimetal strip will contract, closing the valve that supplies the fluid medium to the working chamber, trapping it in the supply chamber. One of the advantages of a viscous drive fan hub is its ability to increase or decrease its efficiency through the variable amount of slippage allowed by the bimetal strip.

Fan Shrouds

Check the mounting of the fan shroud and ensure that it is in good condition. Fan shrouds usually consist of a molded fiber device bolted to the radiator assembly. The fan shroud may partially enclose the fan, providing some degree of safety for technicians working around running engines when the fan is engaged. The shroud increases air flow through the condenser and radiator by channeling the entire surface area of these components to the inlet side of the fan. The shroud also shapes the air flow through the engine compartment (see **Figure 5-22**). When weather conditions become hot, temperature

management problems can occur if the shroud is missing or damaged, as the shroud has an extreme effect on the efficiency or the cooling fan.

Cooling System Leaks

It is not uncommon for a coolant system to leak, and it should be inspected by the operator daily. Some leaks may be caused by the contraction of mated components at joints, especially around hose clamps. These leaks are referred to as cold leaks and often disappear as the engine comes up to operating temperature. The actual cause of a cold leak is that, as temperatures rise in the cooling system, the water necks that the coolant hoses connect to expand from the rising heat. The problem with this is that the hoses and clamps expand at different rates. The clamp does not allow the hose to expand as much as the water neck. The elastomer in the hose wall conforms to the size of the expanded neck. As the coolant temperature drops after engine shutdown, the water neck contracts and the seal between the neck and the hose diminishes. As the repetitive action occurs, the hose to neck seal is completely lost, resulting in water leakage at the hose connection as the system cools down.

Coolant hoses may be a rubber compound or silicone hose. Silicone hoses are more expensive than rubber compound hoses but have a longer service life. Silicone hoses require the use of special hose clamps that will not bite into the hose. These hoses are sensitive to being overtightened, so they should be torqued to specification (see **Figure 5-23**).

Most external leaks can be found with the use of a pressure tester. A typical cooling system pressure testing kit consists of a hand-actuated pump and a gauge assembly calibrated from zero to 25 psi (170 kPa) plus

Figure 5-22 The fan shroud increases air flow through the condenser and radiator by channeling their entire surface area to the inlet side of the fan. *(Courtesy of John Dixon)*

Figure 5-23 Hose clamp used with silicone hoses. *(Courtesy of John Dixon)*

Figure 5-24 Pressure gauge of a coolant pressure tester. *(Courtesy of John Dixon)*

various adaptors to allow installation to the different sizes of radiator fill necks and radiator caps. Some testers are capable of testing for vacuum as well (see **Figure 5-24**).

Testing for Leaks (see **Photo Sequence 4**)

The coolant pressure tester is installed onto the filler neck of the radiator in place of the radiator cap. The hand pump is operated until the pressure on the gauge indicates the maximum operating pressure as noted in the OEM manual. (Make sure that the radiator cap is also correct for its application.) The pressure on the gauge should not fall off. If the gauge does lose pressure, you may be able to locate the leak by the trail of coolant on the floor. If there is a loss of pressure but no external coolant, it may indicate the presence of an internal leak. Some external leaks, especially cold leaks, may be hard to locate and may be found by increasing the pressure by 10 percent, but never more.

Shop Talk: Never exceed the cooling system pressure by more than 10 percent, as internal sealing points may be damaged, causing internal leaks. Also, before pressure testing, make sure the radiator cap is the correct unit for the application.

CAUTION *The use of products that claim to stop leaks should be avoided, even in a situation that may be deemed an emergency. These products may work temporarily but in doing so may plug thermostats, radiator/heater cores, and oil cooler bundles. They may also insulate the components of the cooling system, reducing their ability to transmit heat. In short, they may cause more problems than they cure.*

COOLING SYSTEM MANAGEMENT

Most truck cooling systems are designed to manage the cooling system by first opening the thermostat. Next, the shutterstat will open the shutters. Finally the engine cooling fan will engage. The fan is the last in the chain, as it uses power from the engine (around 6 hp at rated speed). Therefore, most OEMs subscribe to the theory that the engine fan be run as little as possible.

Summary

- Most engine coolants are a mixture of water and antifreeze.
- Coolant should not be mixed in the cooling system but in an external container.
- Antifreeze lowers the freezing point and increases the boiling point of the coolant mixture while protecting the cooling system from corrosion, scaling, and foaming and inhibiting acid buildup.
- EG- or PG-based coolants must have the additive package routinely replenished.
- ELC can have a service life of up to 6 years with only one SCA recharge required.

- The most accurate instrument for measuring the freeze protection of a coolant mixture is a refractometer.
- The SCA use with EP and PG coolants must be tested and adjusted on a regular basis because the additives diminish over time.
- Radiators are usually positioned to take full advantage of the ram air effect for improved heat transfer.
- Radiator caps are designed to safely seal system pressure from the atmosphere and a vacuum valve allows for the expansion and contraction of the coolant.
- Coolant filters should be changed at OEM recommendations, as some contain SCAs.

PHOTO SEQUENCE 4 — Pressure Testing a Cooling System

P4-1 Inspect the radiator pressure cap and make note of the pressure rating. This cap has a 15 psi pressure rating

P4-2 Install the pressure tester onto the radiator by pushing downwards and twisting clockwise one quarter turn to lock into position.

P4-3 Pump the pressure tester until you have reached the limit noted from the radiator pressure cap. Do not exceed this pressure by more than 10 percent.

P4-4 Gauge reading 10 percent higher than the rated capacity of the cap.

P4-5 Inspect truck of dripping coolant. Check the floor for any signs of coolant leaks. Check all radiator hoses and heater hoses for signs of leakage.

P4-6 Twist the top fitting to relieve pressure within the cooling system before removing the pressure tester.

- The job of the thermostat is to maintain the engine temperature exactly to ensure the best performance, fuel economy, and minimum noxious emissions.
- When the engine is cold, the thermostat routes the coolant through the bypass circuit to quickly warm up the engine.
- When the engine warms sufficiently, the thermostat opens, directing coolant to flow through the radiator.

- The purpose of shutters is to control the air flow through the radiator and engine compartment.
- Shutters are controlled by the shutterstat, which directs system air pressure to open or close the shutters depending upon coolant temperature.
- Shutters are usually designed to fail in the open position.
- Most engine fans are temperature controlled by a fanstat located in the water manifold. Some

engine fans use flexible pitch blades that alter the fan's efficiency proportionally to its driven speed, permitting efficient fan operation at low engine rpm.

■ Viscous-type thermatic fans sense the temperature of the engine compartment and are driven by a fluid coupling designed to produce minimum slip at normal engine operating temperature.

■ Engine fans are driven by V or poly-V belts whose tension should be adjusted with a belt tensioning gauge to avoid premature bearing failure or belt slippage.

■ External leaks of cooling systems may be found using a handheld pressure tester. With the use of adapters found in the kit, the radiator cap may be tested as well.

Review Questions

1. Of the following coolant types, which is thought of as the most toxic?

   A. Pure water   C. PG

   B. EG   D. ELC

2. Of the following coolant types, which has the best ability to transfer heat?

   A. Pure water   C. PG

   B. EG   D. ELC

3. Which of the following components could cause cooling system hoses to collapse when the truck is left parked overnight?

   A. Defective thermostat   C. Defective radiator cap

   B. Improper coolant   D. This is normal.

4. What would be the most likely outcome if a radiator cap pressure valve fails to seal?

   A. Cooler operating temperatures

   B. Coolant boiling in the cooling system

   C. Increased hydrocarbon emissions

   D. Coolant from the overflow tank being drawn into the radiator

5. Which of the following best describes the type of water pumps typically found on truck diesel engines?

   A. Positive displacement   C. Constant volume

   B. Non-positive displacement centrifugal   D. Gear-type pump

6. Which of the following components control the drive efficiency of a thermatic, viscous drive fan hub?

   A. A bimetal strip   C. The fanstat

   B. The viscosity of the silicone drive medium   D. The solenoid

7. At what speed does a fiberglass flexible pitch fan blade produce its greatest efficiency?

   A. At high speed   C. At all speeds

   B. At low speed

8. If a diesel engine starts to overheat, at what location will the coolant begin to boil first?

   A. Within the thermostat housing

   B. Within the top radiator tank

   C. At the engine cooling jackets

   D. At the inlet to the water pump

9. From the following instrument list, which is the most accurate to test the degree of freeze protection of a diesel engine coolant?

   A. A spectrographic analyzer

   B. A color-coded test coupon

   C. A hydrometer

   D. A refractometer

10. Two technicians are having a discussion about ELC, and technician A says that some ELCs have a service life of up to 6 years, while technician B says that water should never be added to ELC even if the water is distilled. Which technician is correct?

    A. Technician A is correct.

    B. Technician B is correct.

    C. Both technicians A and B are correct.

    D. Neither technician A nor B is correct.

11. Two technicians are discussing the operation of the shutters on a truck, and technician A says that if the shutterstat fails, the shutters will be left in the open position. Technician B says that if the air pressure is not able to enter the shutter cylinder, the shutters will be left in the closed position.

    A. Technician A is correct.

    B. Technician B is correct.

    C. Both technicians A and B are correct.

    D. Neither technician A nor B is correct.

12. Two technicians are discussing the freeze protection of antifreeze, and technician A says the protection should be 10 degrees below the expected ambient temperature. Technician B says that due to the wind chill factor trucks experience while traveling down the road, the freeze protection must always be factored to the lowest expected wind chill factor. Which technician is correct?

    A. Technician A is correct.

    B. Technician B is correct.

    C. Both technicians A and B are correct.

    D. Neither technician A nor B is correct.

13. Two technicians are discussing the coolant mixtures. Technician A says that ELC should always be mixed at a ratio of 40 percent coolant with 60 percent distilled water. Technician B says that a 50/50 mixture of EG and water has better freeze protection than straight EG. Which technician is correct?

    A. Technician A is correct.

    B. Technician B is correct.

    C. Both technicians A and B are correct.

    D. Neither technician A nor B is correct.

14. During a PM service, a technician checks the coolant level in the expansion tank and finds that the level is very low. What action should the technician take?

    A. Check the coolant strength using a refractometer.

    B. Test the pH level of the antifreeze.

    C. Pressure test the cooling system and check for leaks.

    D. Test for TDS.

CHAPTER

6 Engine Service

Objectives

Upon completion and review of this chapter, the student should be able to:

- Choose the correct engine oil viscosity for the climatic conditions that the vehicle will be subjected to.
- Perform an engine oil change.
- Explain the function of an oil filter.
- Perform an oil filter change.
- Explain the function of a fuel filter.
- Perform a fuel filter change.
- Explain the function and need for a water separator.
- Explain how to service a water separator.
- Demonstrate the use of a hand primer pump.
- Explain how to prime a fuel system.
- Service an engine air filter and check restriction indicator.
- Explain how to check a harmonic balancer and why should be replaced at OEM intervals.
- List other checks that should be performed in the engine compartment while performing a PM engine service.

Key Terms

canister	dry positive filtration	positive filtration
cartridge	emulsified	semi-absorbed
centrifuge	free state	two-stage filtering
charging pump	harmonic balancer	vibration damper
coalesced	oil bath	

INTRODUCTION

This chapter is designed to guide a technician through the preventive maintenance procedures for engine service. Technicians need to familiarize themselves with the proper techniques and procedures to maintain today's sophisticated engines. Proper engine service can make the difference between many trouble-free miles and profits for the vehicle owner or very expensive repairs and denied warranty claims by the OEMs.

SYSTEM OVERVIEW

This chapter starts by explaining the properties that are engineered into engine oil to protect today's truck engines. It then goes on to list the SAE engine oil grades and the recommended temperature operating ranges.

The technician is then guided through the steps to change the engine oil, oil filter, fuel filter (water separator), and air filter. When fuel filters are serviced, often the fuel prime is lost, much like when the vehicle runs out of fuel. The technician must know how to prime the fuel system after this happens, and this is explained in detail. The chapter concludes by explaining how to inspect a viscous-type harmonic balancer and list items that should be inspected during a PMI.

ENGINE LUBRICATION OIL

Modern engine oil has been carefully developed by engineers and chemists to perform several important functions. The efficient operation of an engine depends on the oil doing the following:

- Permit easy starting
- Lubricate engine parts and prevent wear
- Reduce friction
- Protect again rust and corrosion
- Keep engine parts clean
- Reduce combustion chamber deposits
- Fight soot
- Cool engine parts
- Seal combustion pressures
- Be non-foaming

Shop Talk: Engine oil and filters should be changed at regular intervals according to the manufacturer's recommended intervals.

SAE Viscosity Grades

The following lists the SAE engine oil grades and the recommended temperature operating ranges. The W stands for a winter grade lubricant.

MULTI-GRADE ENGINE OILS

- 0W-30 is recommended for use in arctic and sub-arctic winter conditions.
- 5W-30 is recommended for winter use where temperatures frequently fall below 0°F (−18°C) and seldom exceed 60°F (15°C).

- 5W-40 is recommended for severe-duty winter use where temperatures frequently fall below 0°F (−18°C).
- 5W-50 is recommended for severe-duty winter use in arctic and sub-arctic conditions where temperatures frequently fall below 0°F (−18°C). A synthetic oil viscosity grade.
- 10S-30 is recommended for winter use where temperatures never fall below 0°F (−18°C).
- 10W-40 is recommended for severe-duty winter use where temperatures never fall below 0°F (−18°C).
- 15W-40 is recommended for use in climates where temperatures never fall below 15°F (−9°C). Despite this and diesel engine OEM recommendations that support the use of lighter multi-grades in winter conditions, this is by far the most commonly used viscosity grade for truck and bus engines year round, often in climates that have severe winters. A true 14W-40 engine oil will freeze to a grease-like consistency in subzero conditions, making the engine impossible to crank. Additionally, engine wear is accelerated during the warmup phase of operation. However, many oils with the normal 15W-60 grading are actually formulated to perform effectively at temperatures of 0°F (−18°C).
- 20W-40 is recommended for use in high performance engines in climates where temperatures never fall below 20°F (−6°C).
- 20W-50 is recommended for use in high performance engines in climates where temperatures never fall below 20°F (−6°C).

STRAIGHT GRADES

- 10W is recommended for winter use in climates where temperatures never fall below 0°F (−18°C) and never exceed 60°F (15°C).
- 20W-20 is recommended for use in climates where temperatures never fall below 20°F (−6°C).
- 30 is recommended for use in climates where temperatures never fall below 32°F (0°C).
- 40 is recommended for severe-duty use in climates where temperatures never fall below 40°F (4°C).
- 50 is recommended for severe-duty use in climates where temperatures never fall below 60°F (15°C). Often used to disguise engine problems, especially oil burning and leakage.

Engine Oil Change

Before the engine lubrication is changed, the engine should first be warmed up to operating temperature.

Note: Draining lubricants when warm ensures that contaminants are still suspended and also reduces drain time, and the oil is drained more completely.

Once the engine is warm, shut the engine off and remove the drain plug from the oil pan. Have a container large enough to capture all the oil in the engine. (Consult the engine manufacturer's literature for oil quantities.) The truck must be kept in a level position to ensure all the oil is allowed to drain from the oil pan. It is important to get as much of the oil out as possible, as most of the dirt particles are in the last few quarts (liters) of oil to drain from the oil pan. While the oil is draining, check the drain plug for metal cuttings. Metal debris on the magnetic tip of the drain plug can be an indication of damage or imminent failure, so let the service manager or an experienced technician know of the problem before proceeding so that they may attempt to identify the problem. A few small particles should not be a problem. Clean the drain plug and replace it after the lube has been completely drained. Engine oil must be disposed of in accordance with federal and local regulations. Fill the engine oil with the recommended quantity of oil.

Note: Always follow the manufacturer's recommendations for oil viscosity ratings for the ambient conditions the piece of equipment will be subjected to.

Dipsticks

The dipstick is a rigid band of hardened steel that is inserted into a round tube to extend into the oil sump. Checking the engine oil level is performed daily by the vehicle operator, so its location is always accessible. In cab over engine (COE) chassis, the dipstick must be accessible without raising the cab, so it may be of considerable length. It is crucial that the correct dipstick be used for an engine.

OIL FILTERS

The purpose of a diesel engine oil filter is to remove and trap contaminants while providing the least amount of flow restriction in the lubrication circuit. Filters use several different principles to accomplish this objective. The term **positive filtration** is used to describe a filter that operates by forcing all the fluid to be filtered through the filtering medium. Most engine oil filters use a positive filtration principle. It should be noted that filters function at higher efficiencies when the engine oil is at operating temperature.

Bypass Oil Filters

Bypass oil filters are used to complement the full flow filter on current highway diesel engines. These are plumbed in parallel in the lubrication circuit, usually by porting them into the main engine oil gallery. They filter more slowly, but are rated to entrap particles down to 10 μ (microns, one millionth of a meter) in size.

Replacing Oil Filters

Oil filters are removed using a band strap or socket wrench. Make sure that the filter gasket and seal (if used) are removed with the filter. Precautions should be taken to capture oil that spills when the filter is removed. Filters mounted vertically will usually spill the contents of the oil in the mounting pad assembly. Disposable filters and elements are loaded with toxins and must be disposed of in accordance with federal and local regulations. Most OEMs require that new oil filters be primed. If this is not done, in some cases, the lag required to charge the oil filters is sufficient to generate a fault code. Priming an oil filter requires that it be filled with new engine oil on the inlet side of the filter until it is just short of the top of the filter; this will take a little time, as the oil must pass through the filtering media to fill the outlet area inside the element. The sealing gasket should be lightly coated with engine oil. OEMs are usually very specific about how much the filter should be tightened and caution against overtightening. In most cases, the filter should be tightened by rotating it one-half to a full turn after the gasket and filter pad mounting face make contact.

FUEL FILTERS

Fuel injection systems are manufactured with minute clearances. Because of this, any impurities in the fuel that are not removed by the fuel filters can cause premature failures. Most dirt found in fuel is a result of conditions in stationary fuel storage tanks, refueling practices, and improper fuel filter priming techniques by service technicians. The function of a fuel filter is to capture particulate (fine sediment) in the diesel fuel and, while some water in its free state will not pass through the filtering media, a water separator is often used to remove H_2O.

A typical fuel subsystem with a suction circuit and a charge circuit will, in most cases, use a two-filter system, one in each of the suction and charge circuits. Two basic type of filters are used: the currently more common spin-on, disposable **cartridge** type and the **canister** and disposable element type. Spin-on filters are obviously easier to service and are the filter design of choice by most manufacturers. (See **Figure 6-1**).

Primary Filters

Primary filters represent the first filtration stage in a typical **two-stage filtering** fuel subsystem. Primary filters are therefore usually under suction, plumbed in series between the fuel tank and the fuel transfer pump. They are designed to entrap particles sized larger than 20 to 30 μ (1 μ = one millionth of a meter) depending on the fuel system, and achieve this using media ranging from cotton-threaded fibers and synthetic fiber threads to resin-impregnated paper.

Secondary Filters

As the name implies, the secondary filters are the second filtration system in the two-stage filtering system. In a typical fuel subsystem, the secondary filter is charged by the transfer pump, and this enables use of more restrictive filtering media. The secondary filter would therefore normally be located in series between the transfer or **charging pump** (the pump responsible for pulling fuel from the fuel tank and charging the fuel injection components) and the fuel injection apparatus. In some diesel fuel subsystems using two-stage filtering, a primary and secondary filter may be both located on the same circuit (usually the charge circuit). In such cases both filters are mounted in the same base pad with the primary filter feeding the secondary filter. This type of arrangement is more likely to be found on diesel engines in off-highway applications. Current secondary filters may entrap particulate that is as small as 1 μ (micron = one millionth of a meter) but filtering efficiencies of 2 to 4 μ are more common.

Water in its free or emulsified state will not be pumped through many of the current secondary fuel filters. This results in the filter plugging on water and shutting down the engine, starving it for fuel. Secondary filters use a variety of media, including chemically treated pleated paper and cotton fibers.

Fuel Filter
Two-stage box-type filter

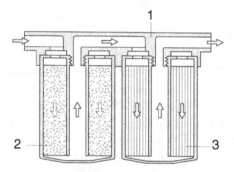

Multistage Filter
With spiral V-form filter element

1. Filter cover with mounting
2. Coarse filter
3. Fine filter

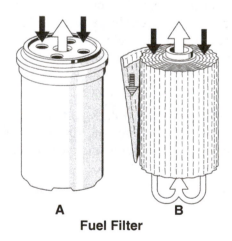

Fuel Filter

A. Easy-charge filter
B. Spiral V-form filter element

Figure 6-1 Types of fuel filters. *(Courtesy of Robert Bosch Corporation)*

In a fuel subsystem that is entirely under suction such as some Cummins systems, the terms *primary* and *secondary* are not used to describe multiple filters when fitted to the circuit. Because every filtering device used in the fuel subsystem is under suction, an inlet restriction specification is critical, and if it exceeded the maximum, it would result in a loss of power caused by fuel starvation.

Filter Servicing

Most fuel filters are routinely changed on PM services that are governed by highway miles, engine hours, or calendar months. They are seldom tested to determine serviceability. When filters are tested, it is usually to determine if they are restricted (plugged) to the point where engine power is reduced due to fuel starvation.

Servicing Spin-On Filter Cartridges

Improper techniques used by service technicians are responsible for much of the dirt that finds its way into diesel fuel systems. Most diesel service technicians realize that sets of replacement filters should be primed, that is, filled with fuel before installation, but few concern themselves about the source of the fuel. Fuel filters should be primed with filtered fuel. Shops performing regular engine service should have a reservoir of clean fuel; any process that requires the technician to remove fuel from a vehicle fuel tank will probably result in its becoming contaminated, at least to some extent, no matter how careful the technician is. The container used to transport the fuel from the tank to the filter should be cleaned immediately before it is filled with fuel. Paint filters (the paper cone-shaped type) can be used to filter fuel. The inlet and outlet sections of the filter cartridge should be identified. The filter being primed should be filled only through the inlet ports, usually located in the outer annulus (ring) of the cartridge, and never directly into the outlet port, usually located at the center. Some manufacturers prefer that only the primary filters be primed before installation during servicing. After the primary filter has been primed and installed, the secondary filter should be installed dry and primed with a hand primer pump.

Replacement Procedure

1. Remove the old filter cartridge from the filter base pad using an appropriately sized filter wrench. Drain the fuel to an oil disposal container.
2. Ensure that the old filter cartridge gasket(s) have been removed from the filter base. Wipe the filter pad gasket face clean with a lint-free wiper.

3. Remove the new filter cartridge from the shipping wrapping. Fill the filter cartridge with clean, filtered fuel poured carefully into the inlet section. The inlet ports are usually located in the outer annulus of the cartridge. Fuel poured into the filter inlet ports will pass through the filtering media and fill the center or outlet section of the filter; this method will take a little longer because it will require some time for the fuel to seep through the filtering medium.
4. The fuel oil itself should provide the gasket and/or O-ring and mounting threads with adequate lubricant; it is not necessary or good practice to use grease or white lube on filter gaskets.
5. Screw the filter cartridge clockwise (right-hand threads are used) onto the mounting pad; after the gasket contacts the pad face, a further rotation of cartridge is usually required. In most cases, hand tightening is sufficient, but each filter manufacturer has its own specific recommendations on the tightening procedure, and these should be referenced.

Shop Talk: When a hand primer pump is fitted to a fuel subsystem, externally prime only the primary filter, ensuring that all the fuel is poured through the inlet side only. Install the secondary filter dry and prime using the hand primer pump.

CAUTION *When removing filter cartridges, ensure that the gasket is removed with the old filter. A common source of air in the fuel subsystem is double gasketing of the primary filter.*

WATER SEPARATORS

Currently, most manufacturers of diesel-powered vehicles use fuel subsystems with fairly sophisticated water removal devices. See **Figure 6-2.** Water appears in diesel fuel in three forms: **free state, emulsified,** and **semi-absorbed.** Water in its free state will appear in large globules and, because its weight is greater than that of diesel fuel, will readily collect in puddles at the bottom of fuel tanks or storage containers.

Water emulsified in fuel appears in small droplets; because these droplets are minutely sized, they may be suspended for some time in the fuel before gravity takes

120   360   490   6120   6401

Figure 6-2 Racor filter/separators. *(Courtesy of Parker Hannefin)*

them to the bottom of the fuel tank. Semi-absorbed water is usually water in solution with alcohol, a direct result of the methyl hydrate (type of alcohol added to fuel tanks as deicer or in fuel conditioner) added to fuel tanks to prevent winter freeze-up. Water that is semi-absorbed in diesel fuel is in its most dangerous form because it may emulsify (the fine dispersion of one liquid into another) in the fuel injection system, where it can seriously damage components.

Generally, water damages fuel systems for three reasons. Water possesses lower lubricity than diesel fuel, has a tendency to promote corrosion, and its differing physical properties affect the pumping dynamics. Diesel fuels are compressible at approximately 0.05 percent per 1,000 psi; water is less compressible, at approximately 0.35 percent per

1,000 psi. Fuel injection apparatus is engineered to pump diesel fuel, and if water with its lower lubricity and compressibility is pumped through the system, the increased pressure that results can cause structural failures, especially at the nozzle area of the fuel injector.

Water separators have been used in diesel fuel systems for many years. These were most often rather crude devices that used gravity to separate the heavier water from the fuel. However, over the past two decades, as both injection pumping pressures have steadily risen and the consumers' expectation of engine longevity greatly increased, water separators have developed accordingly. Often a water separator will combine a primary filter and water-separating mechanism into a single canister. Many of these combination primary filter/water separators are manufactured by aftermarket suppliers. These use a variety of means to separate and remove water in free and emulsified states; they will not remove water from fuel in its semi-absorbed state. Water separators use combinations of several principles to separate and remove water from fuel. The first is gravity. Water in its free state or emulsified water that has been **coalesced** (where small droplets come together to combine into larger droplets) will, because of its heavier weight, be pulled by gravity to the bottom of a reservoir or sump. Some water separators use a **centrifuge** to help separate both larger globules of water and emulsified water from fuel, the centrifuge subject's fuel passing through it to centrifugal force, throwing the heavier water to the sump walls where gravity can pull it into the sump drain. A centrifuge will act to separate particles from the fuel in the same manner. Fuel directed through a fine resin-coated, pleated paper medium will pass through the medium with greater ease than water. Water entrapped by the filtering medium can collect and coalesce in large enough droplets to permit gravity to pull it down into the sump drain. In many cases, aftermarket water separator/fuel filters are designed to replace the fuel system OEM's primary filter; in others, this unit may work in conjunction with the primary filter. When installing an aftermarket water separator/filter unit on the suction side of a fuel subsystem, it is good practice to locate the manufacturer's maximum restriction specification and test that it is not exceeded.

Servicing water separator units is a simple process but one that should be undertaken with a certain amount of care, because it is easy to contaminate the fuel in the separator canister, either by priming it with unfiltered fuel or by permitting dirt to enter when the canister lid is removed. Most aftermarket water separators have a clear sump, making it easier to see the presence of water. All water separators are equipped with a drain valve so the water can be drained from the sump of the separator, and this should be performed on a regular basis. The filter elements used in combination water separator/primary filter units should be replaced in most instances with the other engine fuel filters at each full service. However, some manufacturers claim that their filter elements have an in-service life that may exceed the oil change interval by two or more times. Whenever a water separator is fully drained, it should be primed before attempting to start the engine.

HAND PRIMER PUMPS

Many manufacturers of fuel subsystems will incorporate a hand primer pump in the system, located on the fuel transfer pump body or a filter mounting pad. See **Figures 6-3** and **6-4**. A hand primer pump can be a useful addition to the technician's toolkit, and it can be fitted to a fuel subsystem when priming is required. The function of a hand primer pump is to prime the fuel system whenever prime is lost. Typically, they consist of a hand-actuated plunger and use a singe-acting pumping principle. On the outward stroke, the plunger exerts suction on the inlet side, drawing in a charge of fuel to the pump chamber; on the downward stroke, the inlet valve closes and fuel is discharged to the outlet. When using a hand primer pump, it is important to purge air downstream from the pump unit's charge side.

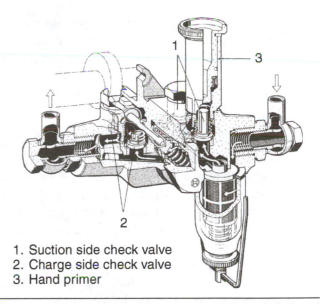

1. Suction side check valve
2. Charge side check valve
3. Hand primer

Figure 6-4 Bosch charging pump with integral hand primer and primary filter. *(Courtesy of Robert Bosch Corporation)*

Some fuel subsystems mount a hand primer to the transfer pump housing. Some newer fuel systems have self-contained, electric priming pumps whose function is to prime the system after servicing.

Priming a Fuel System

Priming of a fuel system is usually a simple task, but it is advisable to check the service manual for the vehicle you are working on.

Caterpillar suggests pressurizing the fuel tank(s) with shop air regulated to 5 psi (35 kPa) and cranking the engine.

CAUTION *Never exceed 8 psi. (55 kPa), or the tank may be damaged. Most OEMs prefer that the technician avoid pressurizing air tanks. It should be remembered that diesel fuel contains volatile fractions and the act of pressurizing a fuel tank with air pressure will vaporize some fuel, while air exiting an air nozzle creates friction and the potential for ignition. Avoid the practice in extremely hot weather conditions.*

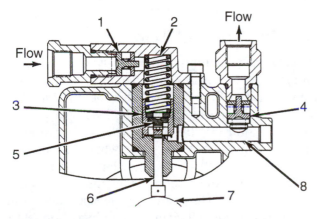

Fuel transfer pump

1. Inlet check valve
2. Spring
3. Piston assembly
4. Outlet
5. Piston check valve
6. Tappet assembly
7. Cam
8. Passage

Figure 6-3 Caterpillar transfer pump. *(Courtesy of Caterpillar)*

Priming Procedure

1. When a vehicle runs out of fuel and it is determined that the fuel system requires priming, remove the filters and fill with filtered fuel.

2. Locate a bleed point in the system, often on an inline injection pump system, this will be at the exit of the charging gallery, and crack open the coupling. A fuel manifold outlet coupling in the cylinder head should be opened in most other systems.

3. Next, if the system is equipped with a hand prime pump, actuate it until air bubbles cease to exit and a clear stream of fuel exits from the cracked open coupling. If the system is not equipped with a hand primer pump, fit one upstream from the transfer pump and actuate until air bubbles cease to exit and a clear stream of fuel exits from the cracked open coupling.

4. Retorque the coupling. Crank the engine for 30-second segments with at least 2-minute intervals between cranking until the engine starts; this will avoid damage to the starter motor due to overheating. In most diesel engine systems, the high-pressure circuit will self-prime once the subsystem is primed.

CAUTION *When refueling tanks, many drivers and technicians overlook the fact that diesel fuel vaporizes and combines with air to form combustible mixtures that only require an ignition source to cause an explosion. Diesel fuel is less volatile than gasoline, but it should always be handled with care, especially in the heat of summer.*

FUEL SYSTEM OUT-OF-SERVICE CRITERIA

(Taken from the North American Standard Out-of-Service Criteria by the Commercial Vehicle Safety Alliance, April 1, 2006)

Liquid Fuel Systems

1. Any fuel system with a dripping leak at any point (including refrigeration or heater fuel systems)
2. A fuel tank not securely attached to the vehicle

Note: Some fuel tanks use springs or rubber bushings to permit movement.

AIR FILTER SERVICE/ REPLACEMENT

Air cleaners filter all air entering the engine for the combustion process. In time, the air cleaner will become dirty, causing the air to be restricted from entering the intake of the engine, resulting in lost horsepower, increased fuel consumption, and shortened engine life. The speed with which the air cleaner becomes fouled is proportionate to the conditions the unit is operating in. For example, the air cleaner will become dirty faster if the truck is operated in dusty areas or on secondary or gravel roads. Air cleaners should be serviced or replaced at every PM service. There are two styles of air cleaners used in trucking industry, oil bath type and dry type. Older model trucks used air cleaners of the **oil bath** type while current units use a **dry positive filtration** type air filter.

Dry Positive Filters

Dry positive filtration air cleaners are used in all contemporary North American commercial vehicle diesel engines. Because they use a positive filtration principle, all of the air entering the intake system must pass through the filtering media, usually resin-impregnated pleated elements. Filtering efficiencies are high throughout the speed and load range of the engine, usually better than 99.5 percent, and highest just before replacement is required. Dry paper elements filters are designed to last for as long as 12 months in a linehaul application and should not be serviced unless the inlet restriction specification exceeds the OEM maximum.

Servicing Dry Air Filters

Many manufacturers using dry-type air cleaners will install an air restriction indicator in the air intake elbow. Refer to **Figure 6-5** to see the air restriction indicator. This indicator should be inspected periodically to confirm that the air cleaner is not restricted. Replace the dry air filter cartridge when the red signal remains in view with the engine shut off. To replace the dry-type air filter, stop the engine and remove necessary clamps. Discard old air cleaner and install new unit. Be certain all clamps are tight and fitment is correct. Any leaks in the air intake will allow foreign material directly into the engine, causing damage and premature wear. **Figure 6-6** shows a typical dry-type air cleaner and intake system. Depress the reset button on top of the restriction indicator after replacement has been performed.

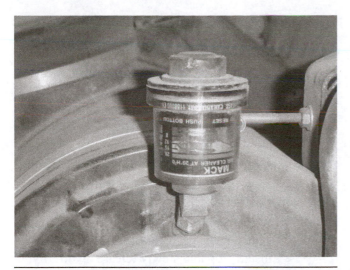

Figure 6-5 Air restriction indicator. *(Courtesy of John Dixon)*

Figure 6-6 A typical dry-type air cleaner and intake system. *(Courtesy of John Dixon)*

Oil Bath Filters

Low operating efficiencies have made these types of non-positive filters a thing of the past, but they are still in service on some older model vehicles. They consist of a mesh-filled canister with a sump filled with engine oil. Air flow is cycloned through the canister and acts to wet down a cylindrical mesh with engine oil to which dirt particles will attach themselves. The principle works effectively to remove large particles from the air stream and has higher efficiencies when induced air flow is highest, that is, at rated speed. They possess low filtering efficiencies at low-load, low-speed operation.

Oil Bath Type Filter Service

The oil bath air cleaner is a serviceable unit. Simply all that needs to be done is to:

- Remove the lower cup of the assembly and discard the old dirty oil.
- Wash the cup in solvent to remove sludge that has adhered to the cup assembly.
- Dry the cup of any solvent residue.
- Fill the cup to the oil level mark with clean oil of the same weight as that in the engine's crankcase.
- DO NOT OVERFILL THE CUP.

HARMONIC BALANCERS

A **harmonic balancer,** otherwise known as a **vibration damper,** is mounted on the free end of the crankshaft, usually at the front of the engine. Its primary function is to reduce crankshaft tensional vibration. They consist of a damper drive or housing and inertia ring. The housing is coupled to the crankshaft and, using springs, rubber, or viscous medium, drives the inertia ring. Viscous-type harmonic balancers have become almost universal in truck and bus diesels.

Most OEMs recommend the replacement of the harmonic balancer at each major overhaul, but this is seldom done due to the expense and the fact that these components usually exceed the OEM-projected lifespan. The consequences of not replacing the damper when scheduled are economic, as they may result in a failed crankshaft. The shearing action of the silicone gel produces friction, which is released as heat. This leads to eventual breakdown of the silicone gel, a result of prolonged service life or old age.

Inspecting a Viscous-Type Damper

The critical functional element in a viscous-type harmonic balancer is the silicone gel damping/drive medium that is sealed inside the drive ring. The typical service facility lacks the equipment to check the operating or dynamic effectiveness of these devices. However, if an engine balance irregularity is suspected and the OEM projected service life is known to have been exceeded, the damper should be replaced. If the harmonic balancer service life has not been exceeded, there are some external checks that you can perform in your shop to help diagnose the condition:

1. Visually inspect the damper housing, noting any dents or signs of warping; evidence of either is reason to reject the component.

2. Using a dial indicator, rotate the engine manually and check for damper housing radial and axial runout against the OEM specification. This is a low tolerance specification, usually 0.005 inch (0.127 mm) or less.

3. Check for indications of fluid leakage, initially with the damper in place. Trace evidence of leakage justifies replacement of the damper.

4. Should the damper not be condemned using the tests above, remove it from the engine. By hand, shake the damper; any clunking or rattle is reason to replace it.

5. Next, using a gear hotplate or component heating oven, heat the damper to its operating temperature, usually close to the engine operating temperature, around 180°F (90°C). This may produce evidence of a leak.

6. A final strategy, if the shop is so equipped, is to mount the damper in a lathe using a suitable mandrel. It should be run up through the engine operating range and monitored using a balance sensor and strobe light.

ENGINE COMPARTMENT CHECKS

While performing the required preventive maintenance on the engine, there are many other components that will require your attention. Perform these inspections while waiting for the oil to drain completely.

- Check the alternator mounting for wear or any signs of damage.
- Check the electrical connection for the alternator, starter motor, and preheater for dirt, corrosion, and damage.

Next Check the Drive Belts

1. Using a belt tension gauge, check fan belts for correct tension.

....

Note: This is for manual tensioning systems only.

....

2. Inspect the belts for signs of wear, fraying, or cracking.
3. Look along the belts for proper alignment.
4. Visually inspect hubs and pulleys for signs of wear or damage.
5. For systems with tensioner units:

- Check if the tensioner is against the install stop or free-arm stop. Replace the belt if this is the case.
- Inspect if the belt is tracking all the way to one edge of the tensioner pulley. A witness mark, considerably wider than the belt, may be seen on the pulley under this condition, and the tensioner should be replaced.

6. With a breaker bar, pull the tensioner back so that the belt can be removed. Slowly return the tensioner to its free-arm stop. Perform the following:

- With the breaker bar, slowly pull the tensioner back from the free-arm stop to the install stop and slowly release again. If you feel excessive roughness or binding during this maneuver, replace the tensioner.
- Check for metal-to-metal contact between the arm and spring case. Replace if metal-to-metal contact is seen.
- Check to see if there are any cracks in the tensioner or if the stops on the springcase are sheared off. Replace the tensioner if either of these conditions exists.

7. Install the belt, making sure the belt is properly seated in the groves of all the pulleys.

Check Engine Mounting

1. Check all brackets, bolts, and other engine mounting parts for dirt, wear, and damage.

Check Radiator Mounting

1. Check all brackets, bolts, and other radiator mounting parts for dirt, wear, and damage.

Check Radiator Fan, Bearing Tolerance, Bolt Unions, Fan Shroud, and Fan Ring with Rubber Seal

1. Inspect the fan blades for any signs of damage.
2. Inspect that there is enough clearance between the tips of the fan blades and the ring of the shroud.

....

Note: The minimum clearance is 0.12 inch (3 mm).

....

3. Loosen the belts and rotate the fan hub assembly. Check for binding or roughness in the bearings.

Check that the end play does not exceed 1/16 inch (1.5 mm).

4. Check fan clutch for lining wear or air leaks. Lubricate fan hub, if applicable.
5. Inspect the idler pulley for cracks.
6. Check for bearing of idler pulley for roughness or binding.
7. Inspect the belt-driven water pump for bearing wobble and any evidence of leakage.
8. Inspect fan shroud rubber molding and all fasteners for wear and tightness.

Check Intercooler Pipes, Hoses, and Air Flow

1. Inspect the charge-air cooler for cracks.

Note: Do not operate a vehicle with a damaged or broken charge-air cooler. This can void the engine manufacturer's warranty, and the engine will not meet emission regulation requirements.

Check Tightness of Engine and PTO

1. Check that the engine does not leak oil or coolant.
2. Check that the fittings, flanges, and hose connections are tight.
3. Check that hoses or pipes do not chaff or are cracked.
4. Check that the PTO or lines do not leak oil.

Check Fuel Lines on Engine

1. Check that the hose and pipe fittings for the fuel lines do not leak.
2. Check that the fuel lines do not leak or chaff.

Check for Exhaust Leakage

1. Check the exhaust flex line for proper routing.
2. Ensure that nothing that is affected by heat is near the exhaust flex line.
3. Check for signs of exhaust leaks.
4. Check for leaks on the exhaust ports on the cylinder heads, at the intake exhaust manifolds and flanges.
5. Check for leaks on the exhaust pressure governor and joining clamp to the exhaust pipe.

EXHAUST SYSTEM OUT-OF-SERVICE CRITERIA

(Taken from the North American Standard Out-of-Service Criteria by the Commercial Vehicle Safety Alliance, April 1, 2006)

Exhaust System

- Any exhaust system, other than that of a diesel engine, leaking at a point forward of or directly below the drive/sleeper compartment and when the floor pan is in such condition as to permit entry of exhaust fumes
- Any bus exhaust system leaking or discharging under the chassis more than 6 inches (152 mm) forward of the rearmost part of the bus when powered by a gasoline engine, or more than 15 inches (381 mm) forward of the rearmost part of the bus when powered by other than a gasoline or diesel engine
- No part of the exhaust system of any motor vehicle shall be so located as to be likely to result in burning, charring, or damaging the electrical wiring, the fuel supply, or any combustible part of the motor vehicle.

Check Air Pipe between Air Intake and Turbo

1. Check the air pipe between the air intake and turbo for wear, damage, and cracks.

Check Turbocharger and Regulator

1. Check the turbocharger, attached oil pipes, clamps, nuts, and air lines for wear, damage, dirt, and cracks.
2. Check the regulator for wear, damage, dirt, and cracks.

Check Tightness of Power Steering Pump, Oil Lines, and Steering Gear

1. Check that the power steering pump is secured tightly to the transmission. Also check for excessive dirt, wear, or damage.
2. Check that the oil lines are secure and free from excessive dirt, wear, or damage.
3. Ensure that the steering gear is secure and free from excessive play.
4. Ensure that the power steering hoses are secured and free of leaks.

Summary

- Engine oil and filters should be changed at regular intervals according to the manufacturer's recommended intervals.

- Before the engine lubrication is changed, the engine should first be warmed up to operating temperature.

- The engine must be kept in a level position to ensure all the oil is allowed to drain from the oil pan.

- Always follow the manufacturer's recommendations for oil viscosity ratings for the ambient conditions the piece of equipment will be subjected to.

- Oil filters remove and trap contaminants while providing the least amount of flow restriction in the lubrication circuit.

- The function of a fuel filter is to capture particulate (fine sediment) in the diesel fuel to protect the fuel injection system.

- Most fuel filters are routinely changed on PM services.

- Fuel filters should be primed with filtered fuel.

- If water with its lower lubricity and compressibility is pumped through the system, the increased pressure that results can cause structural failures, especially at the nozzle area of the fuel injector.

- Diesel fuel is less volatile than gasoline but it should always be handled with care, especially in the heat of summer.

- Air cleaners filter all air entering the engine for the combustion process.

- Air cleaners should be serviced or replaced at every PM service.

- The harmonic balancer's function is to reduce crankshaft tensional vibration.

- The consequence of not replacing the harmonic balancer is a failed crankshaft.

Review Questions

1. Which of the following SAE multi-grade oils would likely be recommended by most highway diesel engine OEMs for North American summer conditions?

   A. 5W-30   C. 20W-20

   B. 15W-40   D. 20W-50

2. Which of the following SAE multi-grade oils would likely be recommended by most highway diesel engine OEMs for North American midwinter conditions?

   A. 5W-30   C. 20W-20

   B. 15W-40   D. 20W-50

3. Which of the following correctly describes the unit of measurement known as micron?

   A. A thousandth of an inch   C. A thousandth of a meter

   B. A millionth of an inch   D. A millionth of a meter

4. Which of the following air filter types has the highest filtering efficiencies?

   A. Centrifugal pre-cleaners   C. Dry, positive

   B. Oil bath

5. What could happen if harmonic balancer is not replaced at its scheduled replacement as recommended by the OEM?

   A. The driver will complain that the engine is running rough.

   B. Accessory drive belt will wear out prematurely, due to improper belt alignment.

   C. The crankshaft can fracture.

   D. None of the above

6. Technician A states that the most commonly used type of vibration damper used on truck diesel engines is the viscous type. Technician B states that the inertia ring drive medium on a viscous-type vibration damper is a rubber compound. Who is right?

   A. Technician A only

   B. Technician B only

   C. Both Technicians A and B are correct.

   D. Neither Technician A nor B is correct.

7. A technician is priming a fuel system by pressurizing the fuel tank with air pressure. What is the recommended pressure and the maximum pressure that can be used but not exceeded?

   A. 3 psi (21 kPa) and 6 psi (41 kPa) maximum

   B. 4 psi (28 kPa) and 7 psi 48 kPa) maximum

   C. 5 psi (35 kPa) and 8 psi (55 kPa) maximum

   D. 6 psi (41 kPa)and 9 psi (62 kPa) maximum

8. In a fuel system, which filter will catch the smallest particles of debris?

   A. The fuel water separator filter

   B. The primary fuel filter

   C. The secondary fuel filter

   D. None of the above

9. Harmonic balancers are seldom replaced because:

   A. of the time involved in changing it.

   B. they should last the life of the vehicle.

   C. they are expensive.

   D. they have no effect on vibration.

10. The majority of new trucks use oil-bath style air filters because of their superior filtering efficiency over dry positive type air filters.

    A. True

    B. False

Drive Train Service and Inspection

Objectives

Upon completion and review of this chapter, the student should be able to:

- Perform a drive axle lube service.
- Explain the effects of mixing lubricants.
- Perform a check of the fluid level on a drive axle.
- Explain the function of the drive shaft and U-joints.
- Service the drive shaft and U-joints.
- Check the oil level in a manual transmission.
- Perform a service and inspection on a manual transmission.
- Perform adjustments on various styles of clutches.
- Perform service and inspection on various types of clutch linkages.
- Perform service and inspection on an automatic transmission.

Key Terms

American Petroleum Institute	interaxle differential	TransSynd
breather	lockstrap	trunnions
center support bearings	pull-type clutch	U-joint
clutch	push-type clutch	wear compensator
clutch break	release bearing	yokes
differential carrier	slip splines	zerk-type
driveshaft	spider	
hanger bearings	throw-out bearing	

INTRODUCTION

A technician must know how to perform a proper inspection and service on the complete driveline. Preventive maintenance of this area will mean the difference between a successful trip or the vehicle being stranded and waiting for emergency service by the side of the road.

SYSTEM OVERVIEW

This chapter starts by explaining how to inspect and service a drive axle and driveshafts, including universal joint service and inspection. Servicing of manual transmission and automatic transmission are also discussed.

Clutch maintenance is also explained in this chapter, as well as how to make proper clutch adjustment and linkage adjustment and perform a proper inspection.

DRIVE AXLE LUBE SERVICE

The drive axle is completely dependent upon lubrication to:

- Provide a lubricating film between the moving parts to reduce friction.
- Help cool components subject to friction.
- Keep dirt and wear particles away from mating components.

Proper maintenance of the drive axle depends upon using the correct lubricant, changing it at the proper intervals, and consistently maintaining the correct fill level in the axle. Always follow the OEM service literature for specific maintenance and drain intervals.

Synthetic lubricants have become common in recent years, causing manufacturers to increase their drain intervals. Synthetic oils provide so many advantages that most component manufacturers endorse their use. The upfront cost of purchasing synthetic lubricants is greater when compared to mineral-based lube, but if the drain interval is increased by three times, the synthetic becomes the cheaper option. Generally, synthetic oils can be said to possess:

- Higher lubricity
- Much higher boil points than mineral-based oils (600°F vs. 350°F)
- Better filming properties (boundary lubrication)
- Better cold weather performance

Some OEMs suggest that the initial drain and flush of the factory-fill axle lubricant on a new drive axle is unnecessary when using synthetic lubricants. This extension in drain intervals reduces labor costs and is embraced by most of the major fleets.

Mixing Lubricants

Many trucks are serviced at multiple truck shops, which makes it inevitable that the rear axle lubricant will get mixed. Despite the fact that some manufacturers of synthetic oils state that their product is compatible in a mix with other great lubes, most service experts in the trucking industry suggest that mixing rear axle lubricants accelerates breakdown. All efforts should be made to avoid mixing different gear lubes in the axle housing, because some of the additives will conflict. In some cases, when petroleum-based stock and synthetic lubes are mixed, thickening and foaming can result, producing premature failure of the drive axle.

Approved Lubricants

All lubricants used in a **differential carrier** (rear end) assembly must meet the **American Petroleum Institute** (API) and Society of Automotive Engineers (SAE) GL standards. The best practice is to follow OEM recommendations. Currently all OEMs approve the use of synthetic lubricants meeting the GL-5 performance classification.

API-GL-5 and synthetics are available in several viscosities. The viscosity used in an application depends mostly on the expected operating temperatures. **Table 7-1** shows the appropriate gear lube grades to use for the operating temperatures.

Lube Change Interval

If a mineral-based gear lube is used, the initial lube change should be made at 1,000 to 3,000 miles, with subsequent lube changes at 100,000 mile intervals; for linehaul operation, that is terminal to terminal highway operation. Other types of operations will require more frequent changes. If the truck does not accumulate enough mileage to require a lube change on the basis

TABLE 7-1: AXLE GEAR LUBE VISCOSITY	
Ambient Temperature Range	**Proper Grade**
−40°F to −15°F (−40°C to −26°C)	75W
−15°F to 100°F (−26°C to 38°C)	80W–90
−15°F and above (−26°C and above)	80W–140
10°F and above (−12°C and above)	85W–140

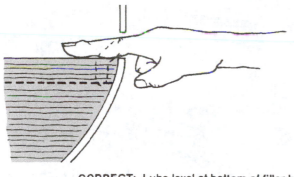

_____ CORRECT: Lube level at bottom of filler hole.

- - - - - INCORRECT: Lube level below filler hole.

Figure 7-1 Correct and incorrect procedure for checking the drive axle lubricant level. *(Courtesy of Roadranger Marketing. One great drive train, two great companies—Eaton and Dana Corporations.)*

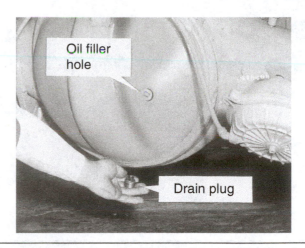

Figure 7-2 Drive axle filler hold and drain plug. *(Courtesy of Roadranger Marketing. One great drive train, two great companies—Eaton and Dana Corporations.)*

of mileage completed, it is good practice to change the lubricant once a year. When using synthetics, use the OEM recommendation or fleet practice.

If the level of the lubricant falls below its proper level between changes, it should be replenished as needed. If loss is excessive, troubleshoot the problem. Use an API-Gl-5 gear lube to maintain proper viscosity levels. However, do not mix lube grades when adding to an existing supply.

Checking Lube Level

Remove the fill hole plug located in either the banjo housing or differential carrier housing. The lube should be level with the bottom of the fill hole, as shown in **Figure 7-1.** To be seen or touched is not sufficient; it must be exactly level with the fill hole. Do not overfill the housing because that can cause problems with aerating the oil. When checking the lube level, also check and clean the housing breather. The breather is usually located on the top of the banjo housing offset from the banjo. If this plugs, the wheel seals can be blown out.

Draining Axle Lube

Draining both the differential carrier housing and the **interaxle differential** should be performed while still warm from the road. Keep in mind that rear axle lubricant runs at high temperatures, so be careful not to burn yourself.

....

Shop Talk: Draining lubricants when warm ensures that contaminants are still suspended and also reduces drain time.

....

The location of the drain plug can be seen in **Figure 7-2.** Check the drain plug for metal cuttings. Metal debris on the magnetic tip of the drain plug can be an indication of damage or imminent failure, so let the service manager or an experienced technician know of the problem before proceeding so that they may attempt to identify the problem. A few small particles should not be a problem. Clean the drain plug and replace it after the lube has been completely drained. The oil should be disposed of in accordance to the regulations in your area.

Refill the Rear Axle

Before you proceed to refill the rear axle, make sure you have the correct lubricant on hand. The specific lubricant used for the refill should be recorded in the vehicle file to prevent mixing of fluids during subsequent top-ups. Pump oil into the fill hole. This may be located in either the differential carrier or in the banjo housing. With the truck parked on a level surface, the oil level should be exactly equal to the bottom of the fill plug hole.

If the drive axle is equipped with an interaxle differential, this also should be filled. Some OEMs recommend that this be filled first with a couple of quarts of lube, followed by filling the axle housing to the correct level. The interaxle differential is filled using the filler plug hole as shown in **Figure 7-3.**

The angle of the differential carrier pinion usually determines which oil fill is to be used to fill and set the oil level in a rear axle. Measure the differential carrier pinion angle using a protractor or inclinometer, as shown in **Figure 7-4.** If the angle is less than 7 degrees (above horizontal), use the fill hole located in the side of the

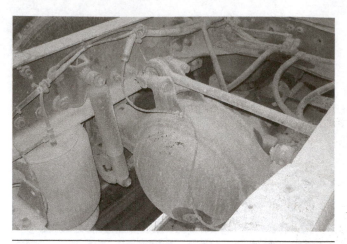

Figure 7-3 The drive axle oil fill plug at the lower portion of the banjo housing. *(Courtesy of John Dixon)*

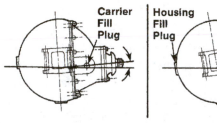

Pinion Angle Less Than 7°— Fill to carrier fill plug hole.

Pinion Angle More Than 7°— Fill to housing fill plug hole.

Figure 7-4 Measuring the differential carrier pinion angle. *(Courtesy of Arvin Meritor)*

carrier. If the angle is more than 7 degrees (above the horizontal), use the hole located in the banjo housing.

There are some manufacturers that only use one lube fill hole located in the banjo housing. When this is the case, you have no choice but to use this lube fill hole, regardless of differential carrier pinion. Rear differential carrier axles sometimes have a smaller plug located nearby, usually just below the lubricant level plug. If there is a plug in this hole, it is for a lubricant temperature sensor that is not installed. It should not be used as a fill hole for determining the correct axle oil level.

When you have completely drained and refilled the rear axle housing lubricant, drive the truck for a few miles to circulate the lubricant throughout the axle and differential carrier and allow vehicle to rest for 5 minutes to allow lubricant to settle. It should be exactly level with the bottom of the fill hole, as shown in **Figure 7-1.** If not, adjust level as necessary.

Drive Axles with Lube Pump

When draining an axle that uses a lube pump, remove the magnetic strainer from the power divider cover and inspect for wear material in the same way

Figure 7-5 Remove the magnetic strainer from the power divider. *(Courtesy of Roadranger Marketing. One great drive train, two great companies—Eaton and Dana Corporations.)*

you would when inspecting a magnetic drain plug. The location of the strainer is shown in **Figure 7-5.** Wash the strainer in solvent and remove any residue by blowing off with compressed air.

WHEEL BEARING LUBRICATION

The wheel bearings are lubricated by the same oil that is in the differential carrier housing. This is one of the reasons why the correct fluid level is so important in the rear end. There is no external visual means of checking the lubricant level of the wheel bearing (unlike a non–drive axle) so the importance of making sure the drive axle lubricant lever is correct cannot be overemphasized. Wheel bearings are entirely lubricated by the gear oil in the banjo housing. There should always be some lube in the cavity of each wheel end (See **Figure 7-6**).

When the drive wheels are installed, the hub cavities should be pre-lubed with the same lubricant used in the differential carrier. Otherwise, they can be severely damaged before gravity and normal action of the differential gearing and axle shafts can distribute lube to the wheel hubs.

Procedure

If the wheel assemblies have not been pulled, you should use this procedure to ensure the wheel bearing cavities have a supply of lubricant:

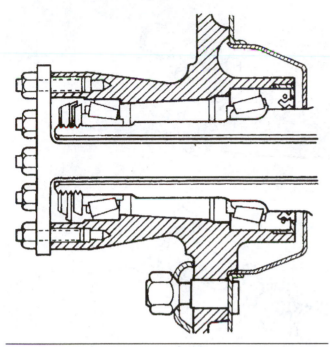

Figure 7-6 Location of wheel hub lube cavity. *(Courtesy of Roadranger Marketing. One great drive train, two great companies—Eaton and Dana Corporations)*

1. Fill the drive axle with lube to the correct level using the fill hole in either the differential carrier or banjo housing.
2. Jack up the left side of the axle, and maintain the position for 1 minute to allow oil from the differential to flow into the right wheel end.
3. Jack up the right side of the axle, and maintain the position for 1 minute to allow oil from the differential to flow into the left wheel end.
4. Lower the axle to the shop floor in a level position, and then check the lube level at the differential carrier fill hole. Add as necessary.

CAUTION *On most drive axles, there is no external visual means of checking lubricant level in the wheel end, so the importance of making sure the drive axle lubricant level is correct cannot be overemphasized. Raising each side of an axle with a jack ensures oil fills the wheel-end hub cavity. Make a final check of the differential carrier oil level after tilting the axle from both sides.*

DRIVESHAFT ASSEMBLIES

The function of the **driveshaft** is to transmit drive torque from one driveline component to another. This should be accomplished in a smooth, vibration-free manner. In a heavy duty truck, that means transmitting engine torque from the output shaft of the transmission to a rear axle or to an auxiliary transmission.

In most cases, a driveshaft is required to transfer torque at an angle to the centerlines of the driveline components it connects to. Because the rear drive axle is part of the suspension and not connected to the ridged frame rails of the truck, the driveshaft must be capable of constantly changing angles as the rear suspension reacts to road profile and load effect. In addition to being able to sustain constant changing angles, a driveshaft must also be able to change in length when transmitting torque. When the rear axle reacts to road surface changes, torque reactions, and braking force, it pivots both forward or backward, requiring a corresponding change in the length of the driveshaft.

Driveshaft Construction

The driveshaft assembly is made of a **U-joint** (universal joint), **yokes, slip splines,** and driveshafts. Driveshafts have a tubular construction designed to sustain high torque loads and be light in weight. **Figure 7-7** shows the components of a typical heavy-duty truck driveshaft. In the next section we will study the role each of these components play.

Driveshafts

To transmit engine torque from the transmission to the rear drive axles, driveshafts have to be durable and strong. If an engine produces 1,000 pound-feet of torque, when this is multiplied by a 12:1 gear ratio in the transmission, it produces 12,000 pound feet of torque at the driveshaft. The shaft has to be tough enough to deliver this twisting force to a fully loaded axle without deforming.

In a simple driveshaft, a yoke at either end connects the shaft to other driveline components. A yoke is usually welded to the shaft tube at one end. At the other end, a slip spline connects to a slip yoke or a second section of driveshaft. The function of a slip spline is to accommodate the variations in driveshaft length required to connect driveshafts between transmissions and drive axle assemblies (the suspension allows the drive axle to move up and down) assemblies. The yokes at either end of a driveshaft are connected by means of U-joints to end yokes on the output and input shafts to the transmission and drive axle(s). A detail of a driveshaft assembly highlighting the slip splines is shown in **Figure 7-8**.

A slip joint is made up of a hardened splined shaft stub welded to the end of the driveshaft tube. The male end of the slip yoke with external splines is inserted

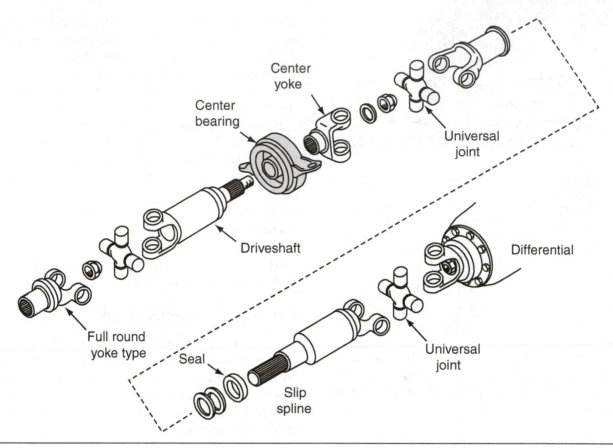

Figure 7-7 Exploded view of a heavy-duty truck driveshaft assembly. *(Courtesy of Volvo Trucks North America, Inc.)*

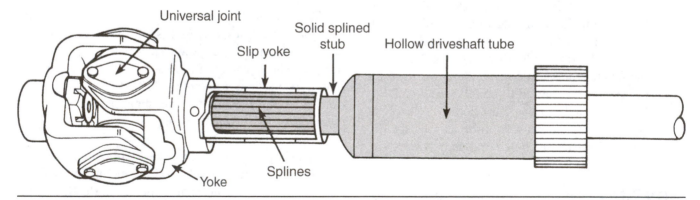

Figure 7-8 Driveshaft slip splines. *(Courtesy of Chicago Rawhide)*

into a female slip yoke with mating internal splines. The splines allow the driveshaft to change in length at the same time as it transmits full torque loads. The slip splines are lubricated with grease and, additionally, are polymer nylon coated to permit slippage while reducing wear. A threaded dust cap contains a washer and seal (usually synthetic rubber or felt) that exclude contaminants and contain the grease.

U-JOINTS

The universal joint, also known in the industry as a U-joint, is used to connect the driveline components while permitting them to operate at different and constantly changing angles. The hub of a U-joint consists of a forged journal cross, also known as a **spider.** The forged cross is machined for grease fitting and four sets of needle bearings. The ends of the U-joints are called **trunnions.** The trunnions are case-hardened ground surfaces on which the needle bearings ride. Bearing caps contain the needle bearings, and these fit tightly to yoke bores to retain the U-joint within a pair of yokes offset from each other. The needle bearings and trunnions are lubricated with chassis grease. A complete U-joint assembly with one

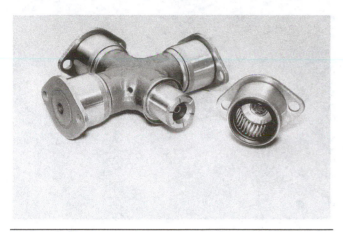

Figure 7-9 Universal joint assembly. (*Courtesy of Chicago Rawhide*)

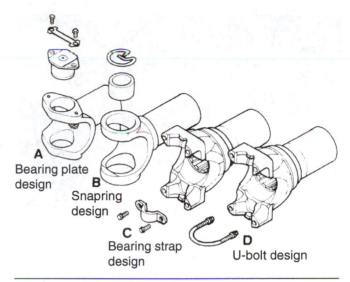

Figure 7-11 Four methods of fastening a universal joint in a yoke: (A) bearing plate; (B) snapring; (C) strap; and (D) U-bolts. (*Courtesy of Spicer Universal Joint Division/Dana Corporation*)

bearing cap removed from the trunnion is shown in **Figure 7-9.**

U-JOINT GREASE FITTINGS

The U-joints of heavy duty trucks have cross-drilled grease passages and grooves on the ends of the trunnions to permit the needle bearings to be lubricated. Either one or two **zerk-type** grease fitting are used to charge grease into the U-joint.

In the center of each trunnion on some U-joints is a stand pipe (a type of check valve), which prevents reverse flow of the hot liquid lubricant generated during operation. When the U-joint is stationary, one or more of the trunnions has to be upright See **Figure 7-10**). Without the stand pipe, lubricant would flow out of the upper grease passage and trunnion, resulting in a partially dry startup that could cause wear. The stand pipe helps ensure adequate lubrication of the trunnions and needle bearings at each startup. Other U-joints have rubber check valves in each cross that perform the same function. A lubed-for-life U-joint

should not require servicing: These have an extensive service life providing the driveshafts are not frequently separated.

BEARING ASSEMBLIES

Each cross consists of four bearing assemblies, one for each trunnion race. These assemblies consist of bearings in a bearing cup and a rubber seal around the open end of the cup. Needle bearings are used because of their strength and durability and are capable of sustaining high loads that result from the oscillation action of rotating driveshafts.

MOUNTING HARDWARE

U-joints can be connected to yokes in different ways. **Figure 7-11** shows four common methods of securing U-joints to yokes. Half-round end yokes are clamped to the U-joint using bearing straps or U-bolts. Full-round end yokes use snaprings or bearing lock plates to secure the joint in the yoke bore.

CENTER SUPPORT BEARINGS

Center support bearings are used when the distance between the transmission (or auxiliary transmission) and the rear axle is too great to span with a single driveshaft. These are commonly known as **hanger bearings** (**Figure 7-12**). The center support bearing is fastenened to the frame and aligns a pair of connecting driveshafts. Hanger bearings also buffer driveline and frame vibrations by surrounding the bearings with a rubber insulator.

A center support bearing is housed in a stamped steel bracket that aligns and fastens the bearings assembly to

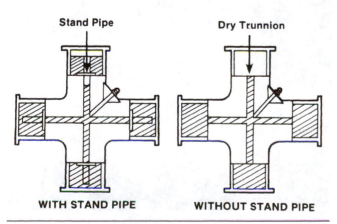

Figure 7-10 A stand pipe at the end of each trunnion ensures constant lubrication and prevents dry start-up. (*Courtesy of Chicago Rawhide*)

Figure 7-12 Typical hanger bearing assembly. *(Courtesy of John Dixon)*

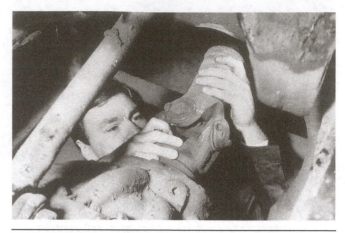

Figure 7-14 Check the driveshaft yokes for excessive radial play. *(Courtesy of Chicago Rawhide)*

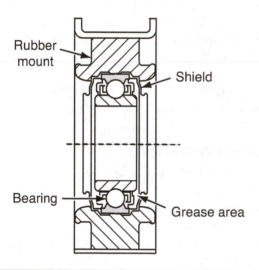

Figure 7-13 Cross-section of a typical center bearing.

Labels: Rubber mount, Shield, Bearing, Grease area

1. Check the yokes on both the transmission and drive axle(s), for looseness as shown in **Figure 7-14.** If loose, disconnect the driveshaft and retorque the end yoke retaining nut to the OEM specification. If this does not correct the problem, yoke replacement may be necessary. If you do have to replace a yoke, check for the OEM recommendation regarding replacement frequency of the end yoke retaining nut.
2. If the end yokes are tight, check for excessive radial looseness of the transmission output shaft and drive axle input and output shafts in their respective bearings, as shown in **Figure 7-15.** Consult transmission and axle OEM specifications for acceptable radial looseness limits and method of checking. If the radial play exceeds the specifications, the bearings should be replaced.

the frame cross-member. The rubber insulator supports the bearings with a small margin of forgiveness. Most hanger bearings are sealed. A center support bearing is usually required on driveshafts that exceed 70 inches (178 cm) in length. **Figure 7-13** illustrates a cross-section of a typical center support bearing.

DRIVESHAFT INSPECTION

Whenever a truck is moving, its driveshafts are working. Driveshafts should be routinely inspected and lubricated. Driveline vibration, U-joint failures, and hanger bearing problems are caused by such things as loose end yokes, excessive radial play (side to side), slip spline play, bent driveshaft tubes, and missing lube plugs in slip joint assemblies. A simple inspection during PM lubricating schedules can prevent many of the previously mentioned problems from starting or becoming worse:

Figure 7-15 Check the transmission output yoke and the rear axle input yoke for excessive radial play. *(Courtesy of Chicago Rawhide)*

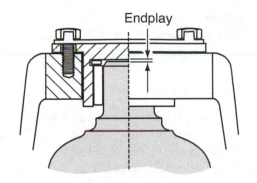

Figure 7-16 Check for excessive play in the U-joint bearings.

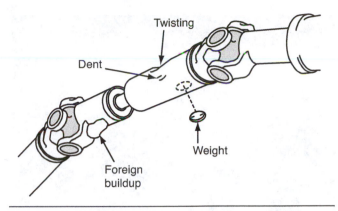

Figure 7-17 Points to inspect when checking a driveshaft for twisting, dents, missing weights, and buildup of foreign material.

3. Check for looseness across the U-joint bearing caps and trunnions (See **Figure 7-16**). This looseness should not exceed the OEM specification, typically as little as 0.006 inch (0.152 mm).

4. Check the slip splines for radial movement. Radial looseness between the slip yokes and the driveshaft stub should not exceed the OEM specification, typically as little as 0.007 inch (0.178 mm).

5. Inspect the driveshaft for damage, bent tubing **(see Figure 7-17)** for missing balance weights. Ensure there is no buildup of foreign material on the driveshaft, such as asphalt or concrete. Anything adhering to the driveshaft has the potential to unbalance it, causing vibration.

6. Check the hanger bearing visually; make sure it is mounted securely. Check for leaking lubricant from the bearing, damaged seals, or rubber insulator failure. Replace the hanger bearing if there is evidence of damage. Do not attempt to repair or lubricate it.

Lubrication

One of the most common causes of U-joint and slip joint problems is the lack of proper lubrication. If U-joints are properly lubricated at the recommended intervals, they will actually last longer than the manufacturer's intended life span. Regular lubricating ensures

that the bearings have adequate grease and, additionally, that the trunnion races are flushed, which removes contaminants from the critical surface contact areas.

It is important to remove the accumulated dirt from the zerk fitting before greasing to prevent abrasive material from being forced through the nipple into the bearing during lubrication.

Heavy-duty driveshafts typically use a lithium soap-based, extreme pressure (EP) grease meeting National Lubricating Grease Institute (NLGI) classification grades 1 or 2 specifications. Grades 3 and 4 are not recommended because of their greater thickness, meaning that they function less effectively when cold. Most lubed-for-life bearings use synthetic greases.

Lubrication cycles vary depending on the service requirements and operating conditions of the vehicle. A typical recommended lube cycle for U-joints is shown in **Table 7-2**.

On-highway operation is generally defined as an application that operates the vehicle running less than 10 percent of total operating time on gravel, dirt, or unimproved roads. Vehicles running more than 10 percent operation time on poor road surfaces are classified as off-highway in terms of preventive maintenance.

TABLE 7-2: RECOMMENDED LUBE CYCLE FOR U JOINTS		
Type of Service	Miles/Kms	Or Time
City	5,000 – 8,000 / 8,000 – 13,000	3 months
On highway	10,000 – 15,000 / 16,000 – 24,000	1 month
On/off highway	5,000 – 8,000 / 8,000 – 13,000	3 months
Extended (linehaul)	50,000 / 80,000	3 months
Severe usage off highway (4 x 4)	2,000 – 3,000 / 3,000 – 5000	1 month

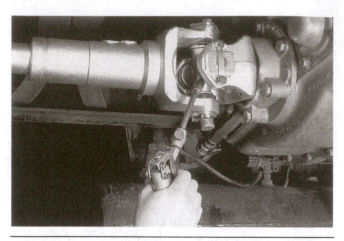

Figure 7-18 Apply grease until grease exits the trunnion seals. *(Courtesy of Dana Corporation)*

LUBRICATING A U-JOINT

Adhere to the following procedure when lubricating U-joints:

1. Apply chassis grease through either one of the two zerk fittings on the U-joint cross. Wipe dirt and contaminants off the grease fitting with a clean shop rag. Fit the grease nozzle over the zerk nipple as shown in **Figure 7-18.** Pump grease slowly into the zerk fitting until each of the four trunnion seals pops. Allow a small quantity of the old grease to be forced out of the bearings. This flushes contaminants out of the bearings and helps ensure that all four have taken grease. The U-joint is properly greased when evidence of purged grease is seen at all four bearing trunnion seals.

2. If grease does not exit from a U-joint trunnion seal, try forcing the driveshaft from side to side when applying gun pressure. This allows greater clearance on the thrust end of the bearing assembly that is not purging. If the U-joint has two grease fittings, try greasing from the opposite fitting. If this does not work, proceed to the next step.

3. Back off the bearing cap bolts on the trunnion that is not taking grease and pop the cap out about 1/8 inch (3 mm). Apply grease. This usually will cure the problem, but if it does not, you should remove the U-joint to investigate the cause.

4. After working on a U-joint that fails to take grease, make sure you torque the bearing cap bolts to OEM specification.

Shop Talk: Half-round end yoke self-locking retaining bolts should not be reused more

than five times. If in doubt as to how many times bolts have been removed, replace with new bolts.

LUBRICATING SLIP SPLINES

The slip splines can be lubricated with the same grease that is used on the U-joints. An EP grease meeting NLGI grade 1 or 2 specifications is required. Slip splines should be lubricated in the same service schedule as the U-joints at the intervals outlined in **Table 7-2.**

Use the following procedure:

1. Wipe dirt and contaminants off the grease fitting with a clean shop rag. Apply grease to the zerk fitting until lubricant appears at the relief hole at the slip yoke end of the slip spline assembly.

2. Seal the pressure relief hole with a finger and continue to apply grease until it starts to exit at the slip yoke seal, as shown in **Figure 7-19.** Sometimes it is possible to purge the slip yoke by removing the dust cap and reinstalling it after grease appears.

CAUTION *In cold temperatures, you should drive the vehicle immediately after lubricating drive shafts. This activates the slip spline assembly and removes excessive lubricant. Excess lubricant in slip splines can freeze in cold weather to a wax consistency and force the breather plug out. This would expose the slip joint to contaminants and eventually result in wear and seizure.*

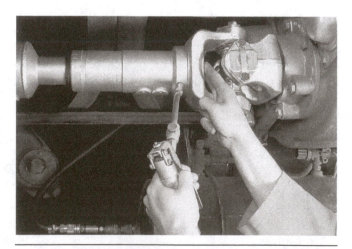

Figure 7-19 Apply grease to the slip joint until grease begins to exit the relief hole. Then cover the hole and continue to apply lubricant until grease begins to ooze out around the seal. *(Courtesy of Dana Corporation)*

LUBRICATING HANGER BEARINGS

Hanger bearings are usually lubricated for life by the manufacturer and are not serviceable. However, when replacing a support bearing assembly, fill the entire cavity around the bearing with chassis grease to shield the bearings from water, salt, and other contaminants.

You should put in enough grease to fill the cavity to the edge of the slinger surrounding the bearing.

When replacing a hanger bearing, make sure you look for and do not lose track of the shim pack that is usually located between the bearing mount and cross-member. The shims set the driveshaft angles, and omitting them will result in a driveline vibration.

STANDARD TRANSMISSION SERVICE

Proper lubrication is one of the keys to a good preventive maintenance program. Maintaining the correct transmission oil level and OEM-recommended type are both critical to ensuring the transmission performs for its expected service life.

Recommended Lubricants

Only the lubricants that the manufacturer recommends should be used in the transmission. Most transmission manufacturers today prefer synthetic lubricants formulated for use in transmissions; they also suggest a specific grade and type of transmission oil. In the past, heavy-duty engine oils and straight mineral oils tended to be more commonly used in truck transmissions. But the demand for extended service intervals has driven the industry to use synthetic oil almost exclusively.

Today Eaton recommends E-500 lubricant for its transmissions. E-500 lube is designed to run 500,000 linehaul miles with no initial drain interval required. E-250 lubricant is rated for 250,000 linehaul miles before a change is required. Most synthetic transmission lubricants exceed the stated viscosity grade ratings, so they can be expected to perform effectively through various geographic and seasonal temperature conditions.

Although most transmission manufacturers recommend the use of synthetic lubes in their transmissions, mineral oil gear lubes are still available and used by some operators. These mineral-based lubricants are slightly cheaper than synthetic lubricants to purchase, but because their projected service life is a fraction of synthetic lubricants, it really doesn't make sense to use them. The service requirements of both mineral-based and synthetic lubricants are covered here.

Mineral-Based Lubes

Most manufacturers will recommend an early initial oil change after the transmission is placed in service. Usually the first oil change will be made between 3,000 and 5,000 miles in linehaul service. In off-highway use, the first transmission oil change should be made after 24 and before 100 hours of service. Although transmission oil is never required to be changed as frequently as engine oil, this first oil change is important because it will flush out any cutting debris created by virgin gears meshing.

There are a number of factors that affect the service interval of mineral-based transmission oil change periods. A key factor is the application: Linehaul operation tends to be gentle to transmission oil, while operating on a construction site can reduce performance life. Manufacturer suggestions for mineral-based transmission oil change intervals vary between 50,000 and 100,000 miles for linehaul applications. Off-highway operation usually requires oil change intervals ranging from 1,000 hours to a maximum of 2,000 hours of service.

Synthetic Lubes

The fact that transmission manufacturers almost exclusively recommend the use of synthetic gear lubes really makes mineral-based lubes obsolete. As previously stated, Eaton recommends E-500 lube for its transmissions, with no initial drain interval. Modern production machining accuracy has all but eliminated the tendency of a gearset to produce break-in cuttings when new. As a result, the initial drain interval, once considered so important to maximize transmission service life, can be eliminated.

It is important that synthetic lubricants not be mixed with mineral-based lubes in transmissions. Although mixing of dissimilar lubricants may not produce an immediate failure, the service of the lube can be greatly reduced. Mixing dissimilar lubricants can sometimes thicken oil and produce foaming.

CAUTION *Do not add transmission lubricant without first checking what lubricant the transmission is using. Mineral-based gear oils, mineral-based engine oils, and synthetic gear lubes are all approved for use in transmissions and none of them is particularly compatible. Mixing transmission oils causes accelerated lube breakdown, resulting in lubrication failures.*

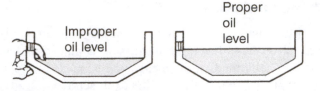

Figure 7-20 The oil level must be level with the bottom of the plug hole. *(Courtesy of Roadranger Marketing. One great drive train, two great companies—Eaton and Dana Corporations.)*

Checking Oil Level

The transmission oil level should be routinely checked at each A-type service, typically at intervals of 5,000 or 10,000 highway miles. When adding oil to transmissions, care should be taken to avoid mixing brands, weights, and types of oil. When you top up or fill the transmission oil level, it should be exactly even with the filler plug opening, as shown in **Figure 7-20**. Overfilling can cause oil aeration, and underfilling results in oil starvation to critical components.

Draining Oil

Drain the transmission oil while it is still warm. Remember that it can be hot enough to severely burn your hand. To drain the oil, remove the drain plug at the bottom of the housing. Allow it to drain for at least 10 minutes after removing the plug. Check the drain plug for cuttings and thoroughly clean it before reinstalling.

Refill

Clean away any dirt or contaminants around the filler plug, then remove the plug from the die of the transmission and refill with the appropriate grade of the new oil. Fill until the oil just begins to spill from the filler plug opening. If the transmission housing has two filler plugs, fill both until oil is level with each filler plug hole.

CAUTION *Do not overfill the transmission. Overfilling usually results in oil breakdown due to aeration caused by the churning action of the gears. Premature breakdown of the oil will result in varnish and sludge deposits that plug up oil ports and build up on splines and bearings.*

PM INSPECTIONS

A good PM program can help avoid failures, minimize vehicle downtime, and reduce the cost of repairs. Often, transmission failure can be traced

directly or indirectly to poor maintenance. **Figure 7-21** identifies key areas of a transmission that should be routinely checked.

Daily Maintenance

Some of the practices listed here are part of the driver's pre-trip inspection.

- **Air tanks:** Drain air tanks to remove water or oil. To be sure of removing all liquid contaminants from an air tank, the drain cock must be fully opened and all air discharged. Most of the liquid will drain from the tank after the air has been removed.
- **Oil leaks:** Visually check for oil leaks around bearing covers, PTO covers, and other machined surfaces. Check for oil leakage on the ground before starting the truck each morning.
- **Shifting performance:** Report any shifting performance problems such as hard shift or jumping out of gear.

"A" Inspection

The following should be checked at each A or lube inspection, approximately every 10,000 miles:

- **Fluid level:** Remove filler plug(s) and check lubricant lever. Top off if necessary and tighten plugs securely.
- **Fasteners and gaskets:** Use a wrench to check to torque on bolts and plugs, paying special attention to those on PTO covers/flanges and the rear bearing cover assembly. Look for oil leakage at all gasket-mating surfaces.
- **Output yoke seal:** Check for leaks around the seal, especially if the transmission has recently been serviced or rebuilt.

"B" Inspection

Numbers refer to **Figure 7-21**.

- **Air control system:** (1) Check for leaks, worn hoses and air lines, loose connections, and loose fasteners.
- **Bell/clutch housing mounting flange:** (2) Check fastener torque.
- **Clutch shaft yoke bushings:** (4) If the clutch shaft bushings are equipped with zerk fittings, grease them lightly. Pry upward on the shaft to remove the clutch release mechanism and check for worn bushings.
- **COE (cab over engine) remote shift linkage:** Check the linkage U-joints for wear and

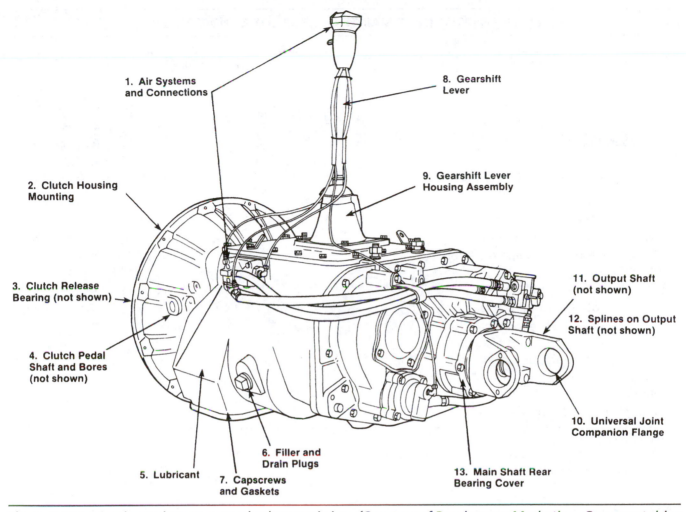

1. Air Systems and Connections
2. Clutch Housing Mounting
3. Clutch Release Bearing (not shown)
4. Clutch Pedal Shaft and Bores (not shown)
5. Lubricant
6. Filler and Drain Plugs
7. Capscrews and Gaskets
8. Gearshift Lever
9. Gearshift Lever Housing Assembly
10. Universal Joint Companion Flange
11. Output Shaft (not shown)
12. Splines on Output Shaft (not shown)
13. Main Shaft Rear Bearing Cover

Figure 7-21 Inspection points on a standard transmission. *(Courtesy of Roadranger Marketing. One great drive train, two great companies—Eaton and Dana Corporations.)*

binding. Lubricate the U-joints. Check any bushings in the linkage for wear.

- **Air filter:** Check and clean or replace the air filter element.
- **Transmission output yoke:** (10) Uncouple the U-joint and check the flange nut for proper torque. Tighten if necessary.
- **Output shaft assembly:** (11) Pry upward on the output shaft to check radial play in the mainshaft rear bearing. Check the splines on the output shaft (12) for wear from movement and chucking action of the U-joint yokes.

"C" Inspection

- **Check lubricant change interval:** This means checking the type of lubricant used in the transmission. Remember that many synthetic lubes are performance rated for oil change intervals up to 500,000 linehaul miles, so it is wasteful to change them more frequently. If an oil change is required, drain and refill the transmission with the specified oil. Transmission oil analysis can be used to establish more precise oil change intervals that are better suited to the actual operating condition of the truck.
- **Gearshift lever:** (8) Check for bending and free play in the tower housing. A lever that is excessively loose indicates wear.
- **Shift tower assembly:** (9) Remove the air lines at the slave valves and remove the shift tower from the transmission. Check the tension spring and washer for wear and loss of tension. Check the gearshift lever spade pin/shift finger for wear. Also take a look at the yokes and blocks in the shift bar housing, checking for wear at all critical points.

Preventive maintenance practices vary by manufacturer and by fleets, many of which have tailored PM to exactly suit their needs. **Table 7-3** outlines a preventive maintenance schedule recommended by Eaton Roadranger, but you should note that this assumes the use of mineral-based oil.

TABLE 7-3: PREVENTIVE MAINTENANCE RECOMMENDATIONS**

PM Operation	Daily	5,000	10,000	20,000	30,000	40,000	50,000	60,000	70,000	80,000	90,000	100,000**
Bleed air tanks and listen for leaks	X											
Inspect for oil leaks	X											
Check oil level			X	X	X	X	X	X	X	X	X	X
Inspect air system connections				X		X		X		X		X
Check clutch housing capscrews for looseness				X		X		X		X		X
Lube clutch pedal shafts				X		X		X		X		X
Check remote control linkage				X		X		X		X		X
Check and clean or replace air filter element				X		X		X		X		X
Check output shaft for looseness				X		X		X		X		X
Check clutch operation and adjustment						X				X		
Change transmission oil	X*						X					X

*Initial fill on new units.
**REPEAT SCHEDULE AFTER 100,000 MILES.

Courtesy of Eaton Corporation

CLUTCH

The function of a **clutch** is to transfer torque from the engine flywheel to the transmission. At the moment of clutch engagement, the transmission input shaft may either be stationary, as when the truck is stationary, or rotating at a different speed than the flywheel, as in the case of upshifting or downshifting. At the moment the clutch fully engages, however, the flywheel and the transmission input shaft must rotate at the same speed.

Torque transmission through a clutch is accomplished by bringing a rotating drive member connected to the engine flywheel into contact with one or more driven members splined to the transmission input shaft. Contact between the driving and driven members is established and maintained by both spring pressure and friction surfaces. Pressure exerted by springs on the driven members is unloaded by the driver by depressing the clutch pedal: This "releases" the clutch.

A clutch is equipped with one or two discs that have friction surfaces known as facing (See **Figure 7-22**). When the operator pushes down on the clutch pedal, the pressure plate is moved away from the flywheel, compressing the springs and freeing the friction disc(s) from contact with the flywheel friction surface. The

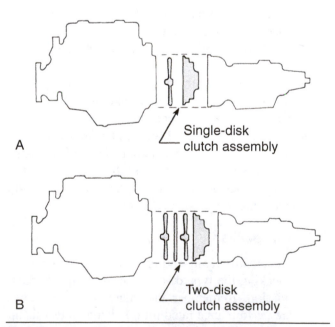

A — Single-disk clutch assembly

B — Two-disk clutch assembly

Figure 7-22 Different clutch styles. Example A uses one friction disc while example B uses two friction discs to couple the engine to the transmission. *(Courtesy of Eaton Corp.—Eaton Clutch Div.)*

clutch is now disengaged, and torque transfer from the engine to the transmission is interrupted.

When the clutch pedal is released by the driver, the pressure plate moves toward the flywheel, allowing the springs to clamp the disc(s) between the flywheel and the pressure plate. The discs are designed for moderate slippage as they come into contact with the rotating flywheel. This minimizes the torsional or twisting shock from being transmitted through to driveline components. As clutch clamping pressure increases, the discs accept the full torque from the flywheel. At this point engagement is complete, and engine torque is transferred to the transmission. Once engaged, a properly functioning clutch transmits engine torque to the transmission without slippage.

Clutch Adjustment Mechanisms

As the friction facings wear, there must be a means of compensating for the loss of friction material. There are two styles of clutch adjustment mechanism: manually adjusted and self-adjusting clutches.

MANUALLY ADJUSTED CLUTCHES

These clutches have a manual adjusting ring that permits the clutch to be adjusted to compensate for friction facing wear. The ring is positioned behind the pressure plate and is threaded into the clutch cover. A **lockstrap** or lock plate secures the ring so that it cannot turn during normal operation. When the lockstrap is removed, the adjusting ring can be rotated in the cover to adjust for wear. This forces the pivot points of the levers to advance, pushing the pressure plate forward and compensating for the wear. A manual adjusting ring is shown in **Figure 7-23.**

SELF-ADJUSTING CLUTCHES

Self-adjusting clutches automatically take up the slack between the pressure plate and clutch disc as wear occurs. The adjusting ring has teeth that mesh with a worm gear in a **wear compensator** (See **Figure 7-24**). The wear compensator is mounted in the clutch cover and has an actuator arm that fits into a hole in the release sleeve retainer (See **Figure 7-25**.) As the retainer moves forward each time the clutch is engaged, the actuator

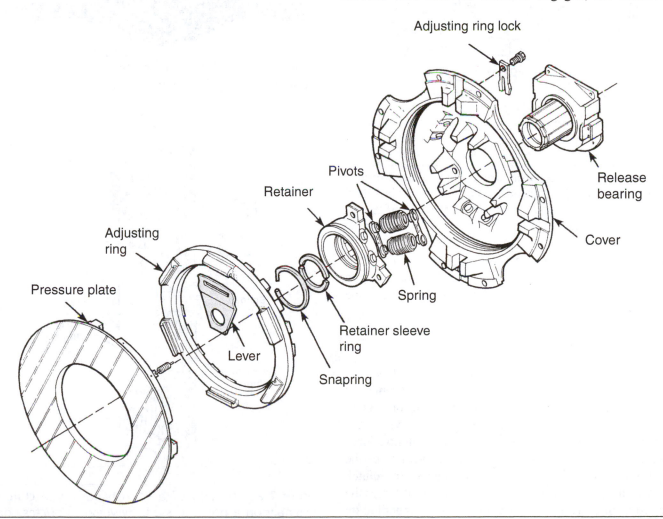

Figure 7-23 Exploded view of an angle-spring clutch assembly. The outside of the adjusting ring is threaded and turns in the clutch cover plate to adjust the clutch. *(Courtesy of Eaton Corp.—Eaton Clutch Div.)*

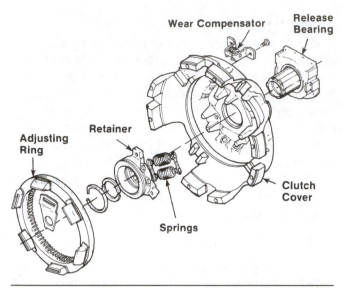

Figure 7-24 Exploded view of the now-obsolete self-adjusting, angle-spring clutch cover assembly. This type of clutch was commonly used into the 1990s. *(Courtesy of Eaton Corp.—Eaton Clutch Div.)*

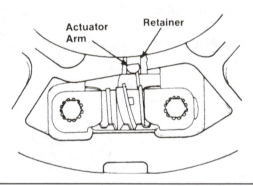

Figure 7-25 The actuator arm of the wear compensator, used in older, self-adjusting clutches, is installed in a slot in the sleeve retainer. *(Courtesy of Eaton Corp.—Eaton Clutch Div.)*

arm rotates the worm gear in the wear compensator. Rotation of the worm gear is transferred to the adjusting ring in the clutch cover, removing slack between the pressure plate and the driven discs.

Release Mechanisms

There are major differences in the way clutches are released or disengaged. All clutches are disengaged though the movement of a **release bearing** or **throw-out bearing.** The release bearing is a unit within the clutch assembly that mounts on the transmission input shaft but does not rotate with it. The movement of the bearing is controlled by a fork attached to the clutch pedal linkage. As the release bearing moves, it forces the pressure plate away from the clutch disc. Depending on the design of the clutch, the release bearing will move in

one of two directions when the clutch disengages. It will either be pushed toward the engine and flywheel, or it will be pulled toward the transmission input shaft.

PUSH-TYPE CLUTCHES

In a **push-type clutch** (see **Figure 7-26** and **Figure 7-27**) the release bearing is not attached to the clutch cover. To disengage the clutch, the release bearing is

Figure 7-26 A Lipe single-plate, push-type clutch. *(Courtesy of Haldex Brake Products, Inc.)*

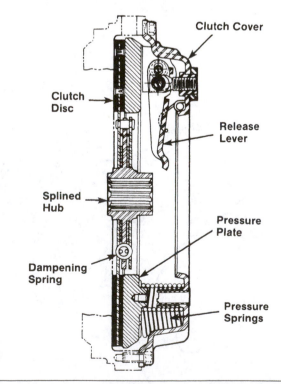

Figure 7-27 Cutaway of a 14-inch push-type clutch assembly on a pot (recessed) flywheel. *(Courtesy of Haldex Brake Products, Inc.)*

pushed toward the engine. When the pedal of a push-type clutch is depressed, there is some free pedal movement between the fork and the release bearing (normally about 1/8 inch). After the initial movement, the clutch release fork contacts the bearing and forces it toward the engine.

As the release bearing moves toward the engine. It acts on release levers bolted to the clutch cover assembly. As the release levers pivot on a pivot point, they force the pressure plate (to which the opposite ends of the levers are attached) to move away from the clutch discs. This compresses the springs and disengages the discs from the flywheel, allowing the disc (or discs) to float freely between the pressure plate and flywheel, breaking the torque between the engine and transmission.

When the clutch pedal is released, spring pressure acting on the pressure plate forces the plate forward once again, clamping the plate, disc, and flywheel together and allowing the release bearing to return to its original position.

Push-type clutches are used predominantly in light- and medium-duty truck applications in which a clutch brake is not required. This type of clutch has no provision for internal adjustment. All adjustments are normally made externally via the linkage system.

PULL-TYPE CLUTCHES

On the **pull-type clutch,** the release bearing is pulled away from the engine toward the transmission. In clutches with angle coil springs or a diaphragm spring, the release bearing is attached to the clutch cover by a sleeve and retainer assembly (see **Figure 7-28** and **Figure 7-29**). When the clutch pedal is depressed, the bearing, sleeve, and retainer are pulled away from the flywheel. This compresses the springs and causes the pivot points on the levers to move away from the pressure plate, relieving pressure acting on the pressure plate. This action

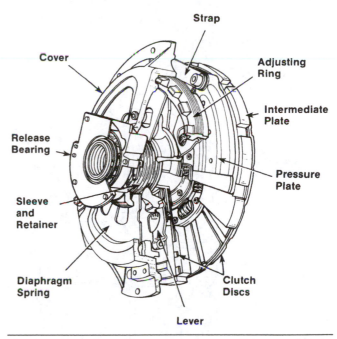

Figure 7-29 Components of a pull-type clutch. *(Courtesy of Arvin Meritor)*

allows the driven disc or discs to float freely between the plate(s) and the flywheel. On pull-type clutches with coil springs positioned perpendicular to the pressure plate (see **Figure 7-30**), the release levers are connected on one end to the sleeve and retainer; on the other end they are connected to pivot points (see **Figure 7-31**). The pressure plate is connected to the levers near the pivot points. So, when the levers are pulled away from the flywheel, the pressure plate

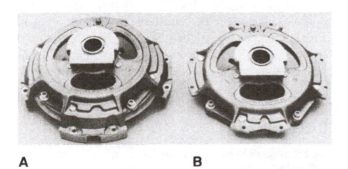

A   **B**

Figure 7-28 Pull-type diaphragm spring clutches; (A) 15 1/2 inch; and (B)14 inch. *(Courtesy of Arvin Meritor)*

Figure 7-30 A single-plate, pull-type clutch. *(Courtesy of Haldex Brake Products, Inc.)*

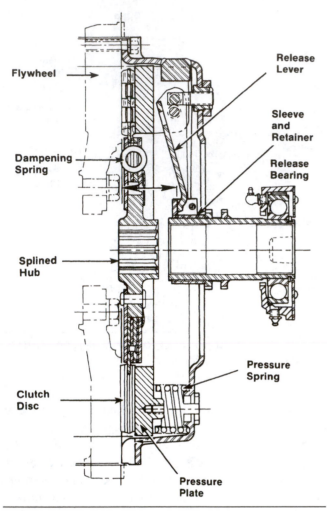

Figure 7-31 Cutaway of a Lipe single-plate, pull-type clutch. *(Courtesy of Haldex Brake Products, Inc.)*

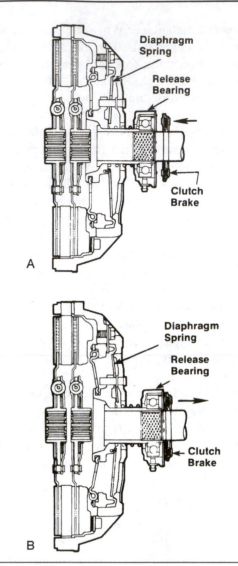

Figure 7-32 Clutch brake: (A) clutch engaged, brake neutral; and (B) clutch disengaged, clutch brake engaged. *(Courtesy of Arvin Meritor)*

is also pulled away from the clutch discs, disengaging the clutch.

When the clutch pedal is released, spring pressure forces the pressure plate forward against the clutch disc, and the release bearing, sleeve, and retainer return to their original position.

Pull-type clutches are used in both medium- and heavy-duty applications and are adjusted internally.

Clutch Brakes

Most pull-type clutches use a component called a **clutch break** not found on push-type clutches. The clutch brake is a disc with friction surfaces on either side. It moves on the transmission input shaft splines between the release bearing and the transmission (see **Figure 7-32A**). Its purpose is to slow or stop the transmission input shaft from rotating to allow gears to be engaged without clashing (grinding). Clutch brakes are used only on vehicles with non-synchronized transmissions.

Only 70 to 80 percent of clutch pedal travel is needed to fully disengage the clutch. The last 1/2 to 1 inch of pedal is used to engage the clutch brake. When the pedal is fully depressed, the fork squeezes the release bearing against the clutch brake, which forces the brake disc against the transmission input shaft bearing retainer (see **Figure 7-32B**). The friction created by the clutch brake facing stops the rotation of the input shaft and countershaft. This allows the transmission gears to mesh without clashing.

Clutch Linkage

Clutches on heavy-duty trucks are usually controlled by a mechanical linkage between the clutch pedal and the release bearing. Some trucks have hydraulic clutch

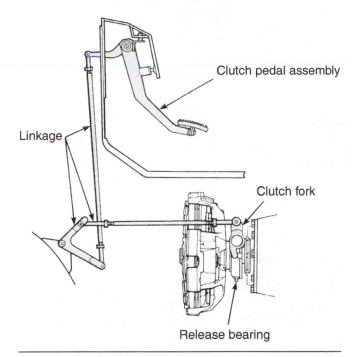

Figure 7-33 The clutch linkage connects the clutch pedal to the clutch release lever and fork. *(Courtesy of Arvin Meritor)*

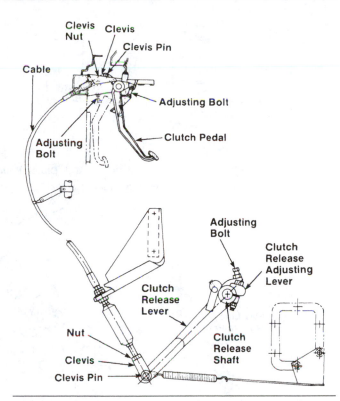

Figure 7-34 Clutch control adjustment arm. *(Courtesy of Eaton Corp.—Eaton Clutch Div.)*

controls. The linkage connects the clutch pedal to the release fork, or yoke (see **Figure 7-33**).

With the clutch pedal fully raised, there should always be 1/8 inch of free play between the fork, or yoke, and the release bearing. This free play should be taken up by the first 1 to 2 inches of clutch pedal travel. Then, as the pedal is depressed farther, the fork acts directly on the release bearing, pulling it back and disengaging the clutch. The last 1/2 to 1 inch of pedal travel will force the release bearing against the clutch brake.

MECHANICAL CLUTCH LINKAGE

Two types of mechanical linkages are used in heavy-duty trucks. The first uses levers to multiply pedal pressure applied by the driver, and the second type links the clutch pedal and release fork by means of a clutch control cable. Examples of both types are shown in **Figure 7-34** and **Figure 7-35.** Components in each type vary, depending on the truck chassis manufacturer.

HYDRAULIC CLUTCH LINKAGE

The typical components of a hydraulic clutch control system can be seen in **Figure 7-36**. The clutch is disengaged by hydraulic fluid pressure, sometimes assisted by an air servo cylinder. The clutch in **Figure 7-36** consists of a master cylinder, hydraulic fluid reservoir, and air-assisted servo cylinder. The components are all connected by rigid and flexible hydraulic lines.

When the driver depresses the clutch pedal, an actuating plunger forces the piston in the master cylinder to move forward. This movement closes off the reservoir and forces hydraulic fluid through the circuit to a reaction plunger and pilot valve in the servo cylinder.

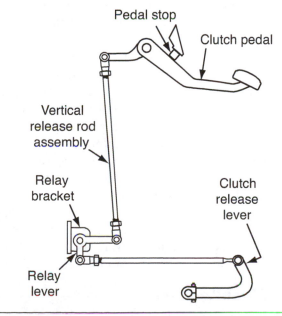

Figure 7-35 Clutch mechanical linkage used on a conventional truck chassis.

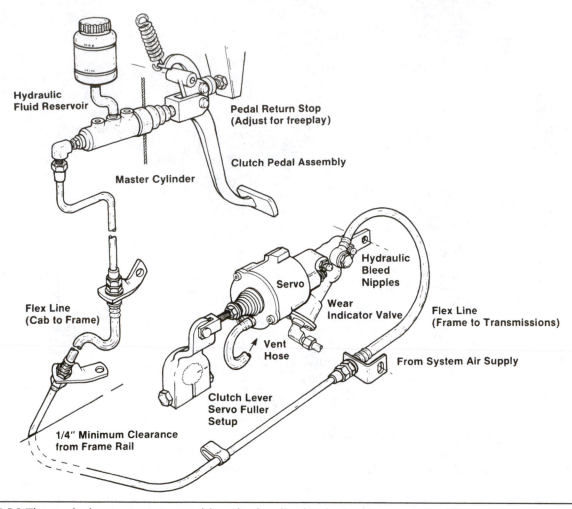

Figure 7-36 The typical components used in a hydraulic clutch circuit.

Hydraulic pressure forces the reaction plunger to move forward to close off an exhaust port and to seat the pilot valve. When the plunger is moved farther, it unseats the pilot valve, which allows air to enter the servo cylinder, exerting pressure on the rear side of the air piston. The movement of the air piston assists in clutch pedal application. As clutch pedal pressure increases, the air piston is moved farther forward, and air pressure overcomes the hydraulic pressure in the reaction plunger. This causes the pilot valve to reseat, preventing any more air from reaching the air piston. The pilot valve and reaction plunger remain in this position until there is a change in the pressure.

When the hydraulic pressure decreases, the return spring returns the reaction plunger and the pilot valve seats itself, which in turn uncovers the exhaust port and allows the air to exhaust from the servo cylinder.

Clutch Maintenance

Clutches should be checked periodically for proper adjustment and lubrication. Actual maintenance varies with the design of the clutch. Some clutches are self-

adjusting, and once the clutch has been installed and the initial adjustment made, no further adjustment to bearing free play should be necessary over the life of the clutch. Other clutches are manually adjusted and must be adjusted periodically as the friction material of the disc is worn away. All clutches require regular inspection.

LUBRICATION

Some manufacturers of clutches use sealed release bearings that do not require lubrication over their service life. Other clutches have release bearings fitted with grease fittings (see **Figure 7-37**), and these must be lubed during a PM service. Frequently these bearings are equipped with a lubrication extension tube assembly that will allow the technician to lubricate the release bearing without removing the expansion cover from the transmission bell housing. These grease fitting extensions help to reduce maintenance and downtime.

A clutch release bearing should be lubricated according to schedule. Typically, this means whenever the clutch is inspected, once a month, every 6,000 to

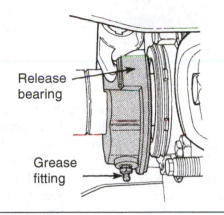

Figure 7-37 Some release bearings have a grease fitting and must be periodically lubricated. *(Courtesy of Arvin Meritor)*

10,000 miles, or whenever the chassis is lubricated, whichever comes first. Off-road or other severe service applications require more frequent service intervals. Good-quality extreme-pressure (EP) grease with a temperature performance range of −10°F to +325°F should be used to lube release bearings

A small amount of grease should be applied between the release bearing pads and the clutch release fork at normal service intervals. Eaton have recently made available an extended-life (EL) release fork equipped with rollers. Because this design significantly reduces friction, it should deliver on the promise of extended service life.

Many clutches have grease fittings in the clutch housing bosses (see **Figure 7-38**) for the clutch cross shaft assembly. These fittings should be greased whenever the release bearing is lubricated.

Whenever the release bearing and other lubrication points on the clutch are serviced, all pivot points on the clutch linkage should also be lubricated (see **Figure 7-39**).

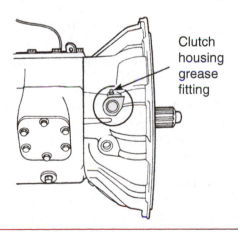

Figure 7-38 Some transmission housings have grease fittings where the clutch release cross shaft passes through the housing bosses. *(Courtesy of Arvin Meritor)*

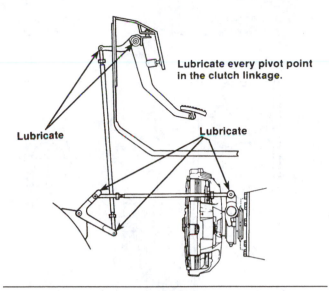

Figure 7-39 Clutch linkage lubrication points. *(Courtesy of Arvin Meritor)*

CAUTION *A replacement release bearing is not pre-packed with grease. They should be lubricated when the clutch is installed in the vehicle, or premature failure of the release bearing will occur.*

CLUTCH ADJUSTMENT

The clutch free pedal, or the initial free travel of the clutch pedal, should be 1-1/2 to 2 inches for either push-type or pull-type clutches. Free pedal is determined by placing your hand or foot on the clutch pedal and gently pushing it down until some resistance to movement is felt. Any movement after this point will cause the release bearing to begin disengaging the clutch.

In a push-type clutch, free pedal is set to 1-1/2 to 2 inches to obtain the desired 1/8-inch free travel clearance between the clutch release bearing and clutch release levers (see **Figure 7-40**), or diaphragm spring, whichever is used.

In a pull-type clutch, the 1/8-inch free travel clearance occurs between the release yoke fingers and the clutch release bearing pads (see **Figure 7-41**). This 1/8-inch free travel at the release bearing should produce 1-1/2 inches of free pedal.

Free pedal dimensions are greater than free travel at the release bearing specifications because, as movement transfers through the linkage, it becomes amplified. Too much free pedal prevents complete disengagement of the clutch. Too little free pedal causes clutch slippage and heat damage and shortens clutch life.

As the friction disc facings wear through normal operation, free pedal will gradually decrease. If

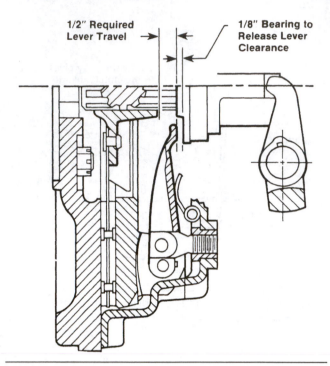

Figure 7-40 In a push-type clutch, the desired free travel clearance is 1/8 inch. Total release bearing travel on a push-type clutch must be approximately 5/8 inch. *(Courtesy of Haldex Brake Products, Inc.)*

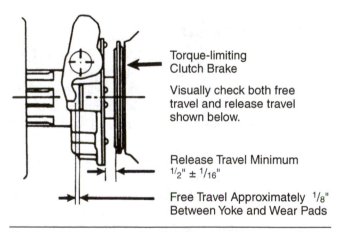

Torque-limiting Clutch Brake

Visually check both free travel and release travel shown below.

Release Travel Minimum $1/2'' \pm 1/16''$

Free Travel Approximately $1/8''$ Between Yoke and Wear Pads

Figure 7-41 On a pull-type clutch, there should be 1/8-inch clearance between the release fork and the boss on the release bearing. *(Courtesy of Eaton Corp.—Eaton Clutch Div.)*

inspection indicates clutch free pedal travel is less than 1/2 inch, adjustment of the clutch is required.

CAUTION *Never wait until no free pedal exists before making this adjustment. Remember that the method of setting free pedal and free travel is different between push-type clutches and pull-types clutches. Use the correct method*

1/2" Required Lever Travel

1/8" Bearing to Release Lever Clearance

for adjusting each type of clutch and refer to the service manual.

PUSH-TYPE CLUTCH ADJUSTMENT

In a push-type clutch, adjusting the external clutch linkage to obtain 1-1/2 to 2 inches of free pedal should result in the specified 1/8-inch clearance between the release bearing and the clutch release forks. Before making a linkage adjustment, check the clutch linkage for wear and damage. If excessive free play is present in the clutch pedal linkage due to worn components, repair as necessary. Excessive wear of the release linkage can give a false impression of release bearing clearance.

Once the free pedal is adjusted between 1-1/2 and 2 inches, it is recommended that the free travel clearance be double-checked. To obtain this adjustment:

1. Set the parking brakes and chock the wheels.
2. Remove the clutch inspection cover from the transmission bell housing.
3. Measure the clearance between the release bearing and the clutch release fork. Clearance must be within specifications for the release bearing to release properly.
4. If this clearance is not present, adjust the linkage until the specified 1/8-inch clearance is obtained. Remember, the linkage must be in good condition to obtain accurate results.

PULL-TYPE CLUTCH ADJUSTMENT

Pull-type clutches may require a two-step adjustment to obtain the specified free travel and free pedal specifications. The first step is a release bearing free travel adjustment that may not be required. The second step is a pedal or linkage adjustment. The free travel adjustment should be performed first. Free travel adjustment is usually an internal adjustment; however, some clutch models are equipped with an external quick adjust mechanism.

Pull-Type Clutch Preadjustment Considerations

Before making adjustments to a pull-type clutch, review the following conditions to ensure optimum clutch performance:

1. Clutch brake squeeze (increase resistance) begins at the point the clutch brake is initially engaged. Optimum clutch brake squeeze begins 1 inch from the end of the pedal stroke or above the floorboard (see **Figure 7-42**). Adjustment is made by shortening or lengthening the external linkage rod.

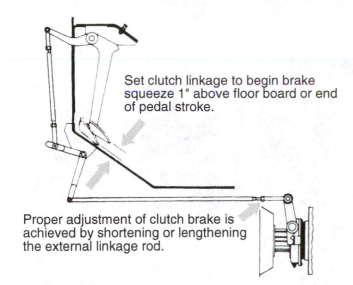

Set clutch linkage to begin brake squeeze 1" above floor board or end of pedal stroke.

Proper adjustment of clutch brake is achieved by shortening or lengthening the external linkage rod.

Figure 7-42 The last inch of clutch pedal travel should squeeze the clutch brake. *(Courtesy of Eaton Corp.— Eaton Clutch Div.)*

2. Optimum free pedal is 1-1/2 to 2 inches (see **Figure 7-43**). This adjustment is made internally in the clutch, never with the linkage.
3. Release travel is the total distance the release bearing moves during a full clutch pedal stroke. A typical release travel distance of 1/2 to 9/16 inch is required to ensure that the release bearing releases sufficiently to allow the friction discs to turn freely, with no clutch drag. Optimum free travel is 1/8 inch (See **Figure 14.32**).
4. Internal adjustment of the adjusting ring should be made before attempting linkage adjustments.

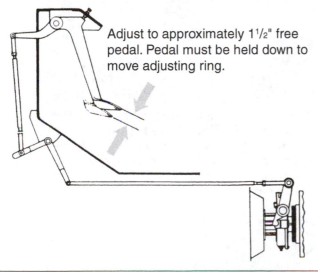

Adjust to approximately 1¹/₂" free pedal. Pedal must be held down to move adjusting ring.

Figure 7-43 The first 1-1/2 inches of pedal travel should take up the clearance between the fork and the release bearing. *(Courtesy of Eaton Corp.— Eaton Clutch Div.)*

5. Internal clutch adjustments should be made with the clutch pedal down (clutch released position).
6. Turning the adjusting ring clockwise moves the release bearing toward the transmission. Turning the adjusting ring counterclockwise moves the release bearing toward the engine.
7. Linkage adjustment on a pull-type clutch should only be made:
   - At initial dealer preparation to set total pedal stroke and yoke throw
   - To compensate for linkage or clutch brake wear
   - When worn or damaged linkage components are replaced

INTERNAL ADJUSTMENT MECHANISMS- ANGLED SPRING CLUTCH

There are three basic types of adjustment mechanisms currently in use on angle coil spring clutches used in heavy-duty truck applications. Two are manually adjusted, and the third category includes several types of self-adjusting mechanisms. **Photo Sequence 5** shows the procedure for adjusting various types of clutches.

LOCKSTRAP MECHANISM

The purpose of the lockstrap mechanism is to lock the clutch adjusting ring. When it is removed, adjustment of free travel can be performed. To adjust a clutch:

1. Remove the inspection plate from the bottom of the clutch housing.
2. Rotate the clutch assembly (bolted to the flywheel) until the lockstrap and its bolt are centered in the opening of the inspection plate as shown in **Figure 7-44.**

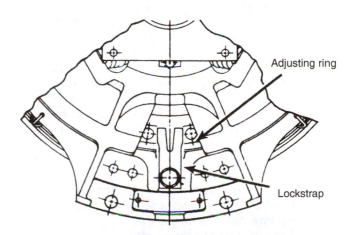

Adjusting ring

Lockstrap

Figure 7-44 For adjustment, the lockstrap and bolt must be centered at the clutch inspection cover opening. *(Courtesy of Eaton Corp.—Eaton Clutch Div.)*

PHOTO SEQUENCE 5 Clutch Adjustment

P5-1 Assess the need for a clutch adjustment by first checking clutch brake squeeze. Clutch brake squeeze should begin 1 inch from the floor at the end of the pedal stroke. Clutch brake squeeze should be verified before attempting an internal adjustment of a clutch. In most cases, clutch brake squeeze will not require adjustment, but it should always be checked.

P5-2 To adjust the clutch brake, either lengthen or shorten the external linkage by loosening the locknut and turning the adjusting rod either clockwise or counterclockwise.

P5-3 Next, check the clutch pedal free play. This is the pedal travel that results before the clutch disengagement occurs. Free travel is always adjusted internally: Attempting to correct this externally will result in incorrect clutch brake squeeze. **Clutch pedal free play should be between 1½ and 2 inches.**

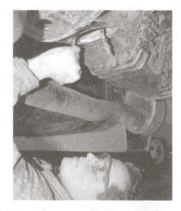

P5-4 Remove the clutch inspection plate to check the condition of the clutch brake, release travel, and cross-shaft yoke free play. Use a trouble light to inspect the clutch brake. Release travel (the total travel of the release bearing through a full stroke of the clutch pedal) should be between ½ inch and $9/16$ inch. Internal free play is the distance between the cross shaft yoke and the release bearing when the clutch is fully engaged. It should measure ⅛ inch.

P5-5 To make an adjustment, fit the adjusting tool to the clutch and have the clutch pedal fully applied by a second person. This releases the clutch, permitting the adjusting ring to be rotated by the adjusting tool. Three notches of CW travel will move the release bearing approximately $1/16$ inch.

P5-6 Install the clutch inspection plate. Drive the truck a short distance to verify the clutch performance after the adjustment.

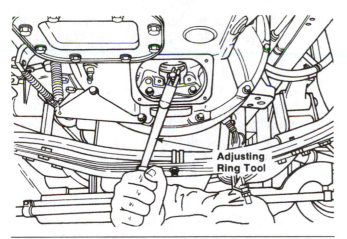

Figure 7-45 Special adjusting ring tools are available for adjusting the clutch. *(Courtesy of Arvin Meritor)*

3. Remove the bolt and lockwasher that fasten the lockstrap to the clutch cover. Remove the lockstrap.
4. Push the clutch pedal to the bottom of pedal travel. Use another person or a block of wood to hold the pedal at full travel. Hold the pedal in this position when the adjusting ring is moved. The pedal should be depressed when turning the adjusting ring, and the pedal must be up when making a measurement.
5. Rotate the adjusting ring to obtain the specified clearance at the release bearing. Use a screwdriver or an adjusting tool as a lever against the notches on the ring to turn the adjusting ring (see **Figure 7-45**). When the adjusting ring is moved one notch, the release bearing will move 0.023 inch. Moving the ring three notches will move the release bearing roughly 1/16 inch. Turning the adjusting ring clockwise moves the release bearing toward the transmission, increasing pedal-free travel. Turning the adjusting ring counterclockwise moves the release bearing toward the engine, decreasing pedal-free travel.
6. Install the lockstrap and torque bolt to specification.
7. Release the clutch pedal.
8. Check the clearance between the yoke and the wear pads. The clearance should be 1/8 inch (see **Figure 7-41**). If out of specification, adjust the clutch linkage according to the vehicle manufacturer's procedures.
9. Reinstall the inspection hole cover and tighten the bolts.

KWIK-ADJUSTMENT MECHANISM.

This manual adjust mechanism (see **Figure 7-46**) allows for the adjustment of free travel without the use of special tools or removing bolts. The

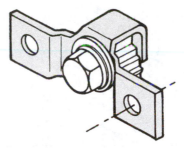

Figure 7-46 Manual adjustment mechanism for quick adjustment of free travel. *(Courtesy of Eaton Corp.—Eaton Clutch Div.)*

adjustment is made using a socket wrench to turn the adjusting bolt:

1. Using a 3/4-inch socket (12 point) or box-end wrench, depress the adjusting nut and rotate to make the adjustment (see **Figure 7-47**). The Kwik-Adjust will reengage at each quarter turn.
2. Make sure the adjusting nut is locked into position with the flats aligned to the bracket.

WEAR COMPENSATOR

Wear compensators were used on clutches in the 1990s, and although they are not very common today, you still might run across one. A wear compensator automatically adjusts for wear of the clutch friction surface each time the clutch is actuated (see **Figure 7-48**). When friction facing wear exceeds a predetermined amount, the adjusting ring is advanced and free pedal travel returned to original specifications. To make a wear compensator adjustment:

1. Manually turn the engine flywheel until the adjuster assembly is in line with the clutch inspection cover opening.

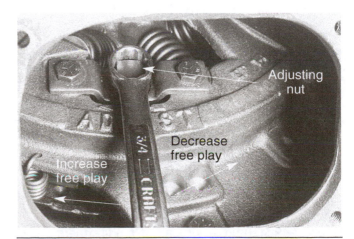

Figure 7-47 Performing the external manual adjustment. *(Courtesy of Eaton Corp.—Eaton Clutch Div.)*

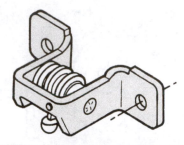

Figure 7-48 Clutch wear compensator used on older, self-adjusting clutches. *(Courtesy of Eaton Corp.— Eaton Clutch Div.)*

2. Remove the right side bolt and loosen the left bolt one turn (see **Figure 7-49A**).
3. Rotate the wear compensator upward to disengage the worm gear from the adjusting ring (see **Figure 7-49B**).
4. Advance the adjusting ring as necessary (see **Figure 7-49C**).

CAUTION *Do not pry on the inside gear teeth of the adjusting ring. Doing this can damage the teeth and prevent the clutch from self-adjusting.*

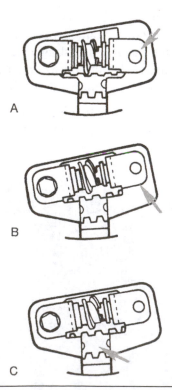

Figure 7-49 Steps in making a clutch adjustment on an older, self-adjusting clutch equipped with a wear compensator. *(Courtesy of Haldex Brake Products, Inc.)*

5. Rotate the assembly downward to engage the worm gear with the adjusting ring. The adjusting ring may have to be rotated slightly to re-engage the worm gear.
6. Install the right-side bolt and torque both bolts to specifications.
7. Visually check that the actuator arm is inserted into the release sleeve retainer. If the assembly is properly installed, the spring will move back and forth as the pedal is full stroked.

Shop Talk: The clutch will not compensate for wear if the actuator arm is not inserted into the release sleeve retainer, or if release bearing travel is less than 1/2 inch.

When a self-adjusting clutch is out of adjustment, check for the following:

■ Actuator arm incorrectly inserted into the release bearing sleeve retainer
■ Bent adjuster arm
■ Seized or damaged clutch components, such as the adjusting ring

After identifying and replacing or repairing the defective component or condition, readjust the bearing setting.

INTERNAL ADJUSTMENT: CLUTCHES WITHOUT CLUTCH BRAKES

The following steps outline the typical internal adjustment procedure for manual and self-adjusting angle-spring clutches not equipped with a clutch brake:

1. Remove the clutch inspection plate below the bell housing.
2. With the clutch engaged (pedal up), measure the clearance between the release bearing and clutch housing.
3. If clearance is not within specifications, typically 1-7/8 inches for a single-plate clutch and 3/4 inch for a two-plate clutch (see **Figure 7-50**), continue with steps 4 and 5 below; otherwise skip to step 6 below.
4. Release the clutch by fully depressing the clutch pedal.
5. Using the internal adjustment procedure outlined previously for lockstrap, Kwik-Adjust, and wear compensator adjustment mechanisms, advance the adjusting ring until the clearance specification is obtained. If the clearance between the release bearing and clutch housing is

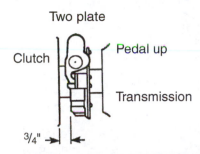

Figure 7-50 Two-plate clutch release bearing clearance on angle-spring clutches without a clutch brake. The ¾-inch specification is used on synchronized transmissions only. On a nonsynchronized transmission, the specification is ½ inch. *Courtesy of Eaton Corp.—Eaton Clutch Div.)*

lower than specification, rotate the adjusting ring counterclockwise to move the release bearing toward the engine. If the clearance is higher than specification, rotate the adjusting ring clockwise to move the release bearing toward the transmission.

6. Apply a small amount of grease between the release bearing pads and the clutch release fork.
7. Check the linkage adjustment.

Internal Adjustment: Clutch Brake Clutches

The following steps outline the typical internal adjustment procedure for manual and self-adjusting angle-spring clutches equipped with a clutch brake:

1. Remove the inspection plate below the clutch housing.
2. With the clutch engaged (pedal up), measure the clearance between the release bearing and the clutch brake. This is the release travel. If clearance (see **Figure 7-51**) is less than 1/2 inch or greater than 9/16 inch (typical), continue with steps 3 and 4 below; otherwise proceed to step 5.

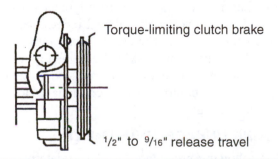

Figure 7-51 Clearance between release bearing and clutch housing on clutches with a clutch brake. *(Courtesy of Eaton Corp.—Eaton Clutch Div.)*

3. Release the clutch by fully depressing the clutch pedal.
4. Using the internal adjustment procedures previously described for lockstrap, Kwik-Adjust, and wear compensator adjustment mechanisms, advance the adjusting ring until a distance of 1/2 to 9/16 inch is attained between the release bearing and the clutch brake with the clutch pedal released.
5. If clearance between the release bearing and the clutch brakes is less than specified, turn the adjusting ring counterclockwise to move the release bearing toward the engine. If the clearance is greater than specification. Rotate the adjusting ring clockwise to move the release bearing toward the transmission.
6. Apply a small amount of grease between the release bearing pads and the clutch release fork.
7. Proceed with linkage adjustment as needed.

CLUTCHES WITH PERPENDICULAR SPRINGS

Pull-type clutches with perpendicular springs use a threaded sleeve and retainer assembly that can be adjusted to compensate for friction disc facing wear. The adjustment procedure is illustrated in **Figure 7-52**:

1. Use a drift punch and hammer or a special spanner wrench to unlock the sleeve lock nut.

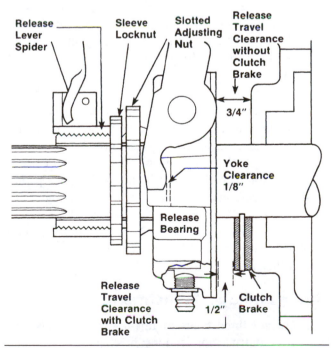

Figure 7-52 Clutch-adjusting mechanism for a clutch with perpendicular springs. *(Courtesy of Haldex Brake Products, Inc.)*

2. Turn the slotted adjusting nut to obtain the release travel clearance of 1/2 inch if the clutch is equipped with a clutch brake and 3/4 inch if it does not have a clutch brake.

3. Securely lock the sleeve lock nut against the release lever retainer, or spider, using the drift punch and hammer or special spanner wrench.

4. Adjust the clutch linkage to obtain a yoke-to-bearing free travel clearance of 1/8 inch if the vehicle is equipped with a nonsynchronized transmission with a clutch brake. On vehicles equipped with synchronized transmissions (without a clutch brake), the yoke clearance should be 1/4 inch and that should result in approximately 3 inches of free pedal.

Clutch Linkage Inspection

Because the clutch will not operate properly if the linkage is worn or damaged, it should be inspected carefully at all PM services. Inspect the linkage according to the following procedure:

1. Depress the clutch pedal and have another person check the release fork for movement. The smallest movement of the clutch pedal should result in movement at the release fork. If the release fork does not move when the clutch pedal moves, locate and correct the free play condition.

2. The linkage must move when the pedal is actuated. Make sure the linkage is not obstructed and every pivot point operates freely. Make sure the linkage is not loose at any point and if it binds, locate and service the cause of the condition.

3. Check that the pedal springs, brackets, bushings, shafts, clevis pins, levers, cables, and rods are not worn or damaged. In a hydraulic clutch, check for leaks and that the reservoir is filled to the specified level. Replace missing or damaged components. Do not attempt to straighten any damaged parts.

4. Lubricate every pivot point in the linkage. Use the lubricant specified by the manufacturer of the vehicle. A NLGI #2 multipurpose lithium grease is typically specified for bushings and other pivot points.

Clutch Linkage Adjustment

There are three types of linkage mechanisms currently used in truck applications: mechanical, hydraulic, and pneumatic. Adjustment methods vary between truck manufacturers, so always follow the procedure

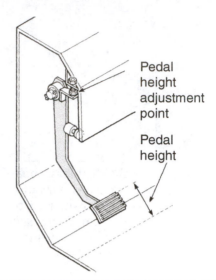

Figure 7-53 Pedal height adjustment point. *(Courtesy of Arvin Meritor)*

listed in the truck service literature. The following is general information on adjusting the clutch linkage.

PEDAL HEIGHT

On some vehicles, the travel height of the clutch pedal is adjusted. The height is set by stop bolts. If the pedal height is not correct, the amount of free pedal will not be correct (see **Figure 7-53**). Consult the manufacturer's service literature for pedal height specifications.

TOTAL PEDAL TRAVEL

The total pedal travel is the complete distance the clutch pedal must move. It can be adjusted with bumpers and stop bolts in the cab or with stop bolts and pads on the linkage. Total travel makes sure that there is enough movement of the pedal to correctly engage and disengage the clutch (see **Figure 7-54**). Consult the manufacturer's service literature for specifications.

FREE PEDAL TRAVEL

The free pedal is the travel of the pedal before the release bearing starts to move. It is typically 1-1/2 to 2 inches. If free travel is more than 2 inches, the clutch might not fully release. The clutch friction discs could be loaded to contact the flywheel continually, causing excessive wear. If free travel is reduced to zero by wear, the clutch could slip and continually load the throw-out bearing.

CHECKING LINKAGE ADJUSTMENT

The following procedure is used to determine if a linkage adjustment is required on manual and self-adjusting angle-spring clutches (see **Figure 7-55**).

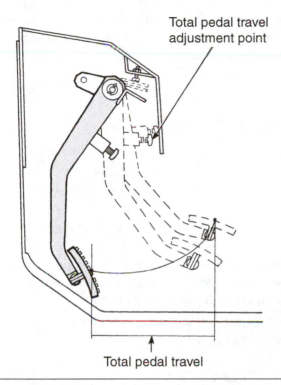

Figure 7-54 Total pedal travel adjustment point. *(Courtesy of Arvin Meritor)*

1. With the clutch disengaged (pedal down), measure the free travel clearance between the release yoke fingers and wear pads on the release bearing. If clearance is either greater or less than 1/8 inch, adjust the external linkage to obtain the 1/8-inch clearance (see **Figure 7-41**).
2. This dimension should correlate to a free pedal of 1-1/2 to 2 inches. If it does not, trim the linkage adjustment to obtain these specifications.
3. Apply a small amount of grease between the release bearing pads and the clutch release fork.
4. Tighten all lock nuts.

ADJUSTMENT

The method for adjusting the release yoke finger to release bearing wear pad clearance can differ, depending on the type of truck and linkage design. A typical procedure is outlined here:

1. Disconnect the lower clutch control rod from the pedal shaft (see **Figure 7-55**) or the bellcrank (see **Figure 7-56**).
2. Place a space block between the pedal stop and the stop bracket or pedal shank. The actual size of the spacer will vary with the linkage design.
3. Force the lower control rod forward until the release bearing contacts the yoke fingers.

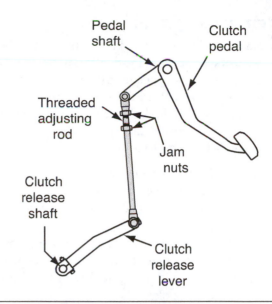

Figure 7-55 This linkage is adjusted by lengthening or shortening the control rod.

4. Loosen the jam nuts on the threaded adjusting rod on the threaded end of the control rod until the holes in the pedal shaft or bellcrank align with the holes in the rod end or clevis.
5. Reconnect the linkage members and remove the spacer block from the pedal stop.
6. Operate the clutch pedal and recheck the yoke-to-bearing clearance. If the clearance is still insufficient it might be necessary to adjust the pedal height or travel. By turning the pedal stop cause the pedal will return farther and the yoke-to-bearing clearance will be increased.

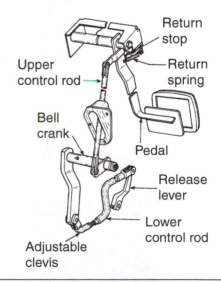

Figure 7-56 Adjustments to this linkage are made by adjusting the clevis on the lower control rod. *(Courtesy of International Truck and Engine Corporation)*

CLUTCH BRAKE SETTING

When checking the setting of the clutch brake, depress the clutch pedal in the cab and note the point at which the clutch brake engages by observation through the inspection cover. With the release travel and free travel settings, clutch brake squeeze should occur approximately 1 inch from the end of the pedal stroke (see **Figure 7-43**). To check this:

1. Insert a 0.010 inch thickness gauge between the release bearing and the clutch brake.
2. Depress the clutch pedal to squeeze the thickness gauge.
3. Let the pedal up slowly until the gauge can be pulled out and note the position of the pedal in the cab. It should be 1/2 to 1 inch from the end of the pedal stroke.
4. To adjust the clutch brake setting, shorten or lengthen the external linkage according to service literature procedure. If the specified adjustment cannot be obtained, check the linkage for excessive wear and pedal height.
5. Reinstall the inspection cover.

AUTOMATIC TRANSMISSION MAINTENANCE

For this section, we will focus on Allison transmissions, as they tend to be the most common ones used in North American trucks and buses. The procedures outlined here are those required for most hydromechanical Allison four- and five-speed transmissions.

Inspection and Maintenance

The transmission should be kept clean to make inspection and servicing work easier. Clean the transmission with a pressure washer, making sure to keep the stream of water away from the breather. Avoid using solvents that could damage the aluminum housing of the transmission. Inspect the transmission for loose bolts, loose or leaking oil lines, oil leakage, and the condition of the control linkage and cables. The transmission oil level should be checked regularly.

Transmission Oil Checks

As with manual transmissions, maintaining the correct fluid level in the transmission is very important. Transmission oil or fluid plays the following roles in an automatic transmission:

- It acts as the drive medium (transfers torque inside the torque converter).
- It acts as hydraulic medium to apply clutches.
- It lubricates the transmission components.
- It cools the transmission components.

If the oil level is tool low or too high, a number of problems can occur.

Low Oil Level

When the transmission oil level is low, oil will not completely cover the oil filter. This pulls air into the pump inlet along with oil, which is then routed to the clutches and converter. The resulting air in the hydraulic system is known as aeration. Air is compressible, so it compromises the operation of hydraulic circuits. The result is converter aeration, irregular shifting, overheating, and poor lubrication. Aeration of transmission oil alters its viscosity and changes its appearance to that of a thin frothy liquid.

High Oil Level

At normal oil levels (FULL mark on the dipstick with oil at operating temperature), the sump oil level should be slightly below the planetary gearsets (see **Figure 7-57**).

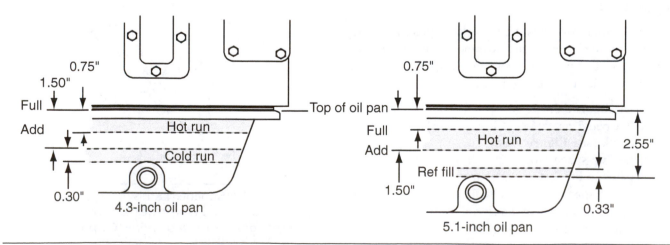

Figure 7-57 Oil levels on a 4.3-inch transmission oil pan and a 5.1-inch oil pan.

When the oil level is maintained above the FULL mark on the dipstick, the oil level in the sump rises so that the planetary gears run in oil, a condition that can cause foaming and aeration. Once again, aerated transmission fluid results in converter aeration, irregular shifting, overheating, and poor lubrication. If accidental overfilling occurs during servicing, the excess oil should be drained.

> **CAUTION** *It should be noted that a defective oil filler tube seal ring will allow the oil pump to draw air into the oil from the sump, which will result in aeration of the oil.*

Interpreting Oil Level Readings

Transmission input speed and oil temperature significantly affect the oil level. An increase in input speed will lower the oil level, whereas an increase in oil temperature will raise it. For these reasons, always check the oil level with the engine at its specified idle speed with the transmission in neutral. Both cold and hot level check should be taken.

A cold level check is required to ensure there is sufficient oil in the transmission until normal operating temperature is reached. The hot check is made when the transmission oil reaches normal operating temperature (160°F to 200°F) and is the more reliable of the two checks.

Foreign material should never be allowed to enter the filler tube when you are checking or adding oil to the transmission. Clean around the filler tube opening before removing it. When testing transmission oil level, park the vehicle on a level surface and apply the parking brakes.

Shop Talk: You should check the transmission oil level at least twice to ensure that an accurate reading is made. If the dipstick readings are inconsistent (some high, some low), check for proper venting of the transmission breather or oil filler tube. A clogged breather can force oil up into the filler tube and cause inaccurate readings. If the filler tube is non-vented, the vacuum produced will cause the dipstick to draw oil up into the tube as it is pulled from the tube. Again, the result will be an inaccurate reading.

Cold Fluid Level Check

Run the engine for around 1 minute to purge air from the system. Shift the transmission into drive, then reverse, and then back into neutral. This charges the clutch cavities and hydraulic circuits with oil. The oil temperature should be between 60°F and 120°F for an accurate cold check. Wipe the dipstick clean and take the reading. If the oil level registers in the REF FILL (COLD RUN) band of the dipstick, the oil level is sufficient to run the transmission until a hot check can be made. Familiarize yourself with the appearance of the Allison dipstick as illustrated in **Figure 7-58.**

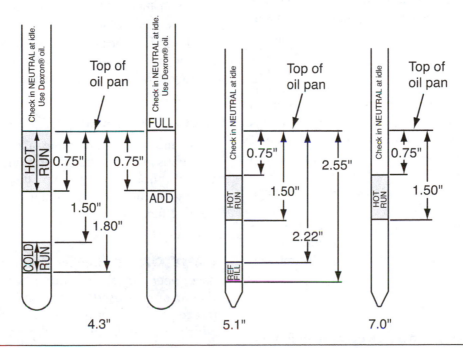

Figure 7-58 (A) Typical current and early 4.3-inch oil pan dipstick markings; (B) typical current 5.1-inch and 7.0-inch oil pan dipstick markings.

Shop Talk: The REF FILL (COLD RUN) level is an approximate level and can vary with specific transmissions. To ensure proper operating levels, a hot oil level check must be performed.

If the oil level registers at or below the lower line of the REF FILL (COLD RUN) band, add oil to bring the level within the REF FILL (COLD RUN) band. Do not fill above the upper line of this band. If the oil level is above the upper line of the REF FILL (COLD RUN) band, drain some oil to correct the level. Next, you can operate the vehicle until the hot check temperate is reached. Then continue with the hot fluid level check.

Hot Fluid Level Check

The oil temperature should be between 160°F and 200°F to make this test. With the engine at idle and the transmission in neutral, wipe the dipstick clean and check the oil level. If the oil level registers in the HOT RUN band (between ADD and FULL), the oil level is correct. If the oil level registers on or below the bottom line of the HOT RUN band or the ADD line, add oil to bring the level to the middle of the band. Note that one quart of oil will raise the level from the bottom of the band to the top of the band in most transmissions (from ADD Line to the FULL line).

Hydraulic Fluid Recommendations

Check the OEM hydraulic fluid specifications. For example, several automatic transmission manufacturers recommend Dexron, Dexron ll, Dexron III, and type C-4 (ATD approved SAE 10W or SAE 30) oils for their automatic transmissions. Type C-4 fluids are the only fluids usually approved for use in off-highway applications. Type C-4 SAE 30 is specified for all applications in which the ambient temperature is consistently above 86°F. Some but not all Dexron II fluids also qualify as type C-4 fluids. If type C-4 fluids are to be used, check that the materials used in auxiliary equipment such as tubes, hoses, external filters, and seal are C-4 compatible.

Allison currently recommends the use of **Trans-Synd** synthetic oil in all their transmissions. Trans-Synd, formulated jointly by Allison and Castrol, can extend oil drain intervals by three times.

Cold Startup

The transmission should not be operated in forward or reverse gears if the transmission oil falls below a certain temperature. Minimum operating temperatures for recommended fluids are as follows:

TransSynd	−10°F
Dexron (I, II, III)	−10°F
Type C-4 SAE 10W	10°F
Type C-4 SAE 30	32°F

When the ambient temperature is below the minimum fluid temperatures listed and the transmission is cold, preheat is required either by using auxiliary heating equipment or by running the engine in neutral at idle a minimum of 20 minutes to reach the minimum operating temperature. Failure to observe the minimum temperature limit can result in transmission malfunction or reduced transmission life.

Oil and Filter Changes

The change interval specified by Allison must also take into account the chassis application. Typical filter and oil change intervals are 25,000 miles or 12 months for on-highway trucks and 1,000 hours or 12 months for off-highway applications. When the Allison-recommended TransSynd synthetic oil is used, oil change intervals can be extended by around 300 percent, depending on application.

Some operating conditions may require shorter oil and filter change intervals. Oil should be changed when there is visible evidence of contamination or high-temperature breakdown of the oil. High-temperature breakdown is indicated by discoloration and burned odor and can be corroborated by oil analysis. Severe operating conditions might require more frequent service intervals.

OIL CHANGE PROCEDURES

The most important thing about changing automatic transmission fluid is cleanliness. Prevent contaminants from getting into the transmission by handling the oil in clean containers and using clean filler and funnels.

CAUTION *Containers or transfer devices that have been used for engine coolant solutions should not be used for transmission fluid. Antifreeze contains ethylene or propylene glycol, which, if transferred to the transmission, can cause clutch plate failure.*

Place the dipstick on a clean surface when filling the transmission and keep replacement filters and seals in their packaging until ready for installation. Before you drain the transmission fluid, it should be at an operating temperature of between 160°F and 200°F. Shift the gear selector to neutral before draining the oil from the sump pan.

Allison transmissions are equipped with either a standard or heavy-duty oil pan (see **Figure 7-59**). When changing oil and filters on a standard pan, the entire oil pan is removed from the base flange of the transmission to access the filter. The heavy-duty oil pan does not have to be removed, because the filter is accessed through an opening in the side of the oil pan.

Standard Oil Pan

The oil and filter change on an Allison transmission with a standard oil pan should be performed as follows:

1. Remove the oil drain plug and gasket from the right side of the oil pan. Allow the oil to drain.
2. Remove the oil pan, gasket, and oil filler tube from the transmission. Discard the gasket. Thoroughly clean the oil pan.

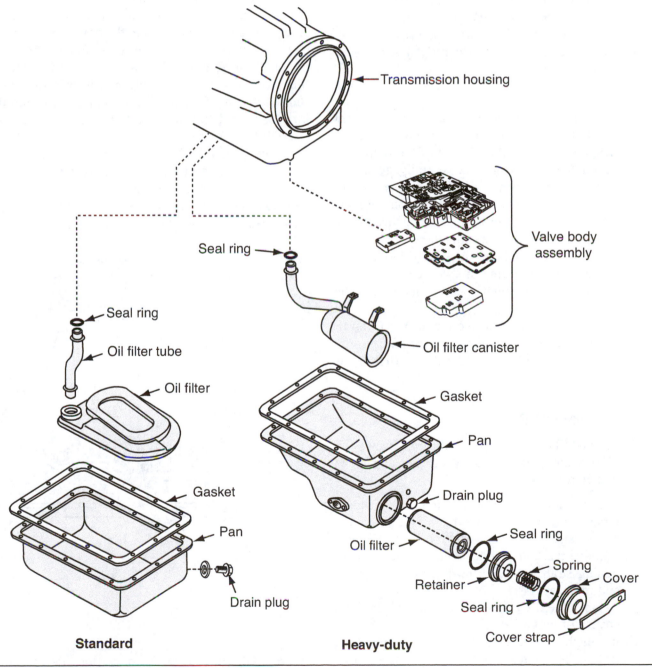

Standard   **Heavy-duty**

Figure 7-59 Typical transmission housing with a standard and a heavy-duty oil filter and oil pan configuration.

3. Remove the oil filter retaining screw. Remove the oil filter and discard it. The oil filter intake pipe is not part of the filter and should be kept for reinstallation.

4. Install the filter tube into the new filter assembly. Install a new seal ring onto the filter tube. Lubricate the seal ring with transmission oil. Install the new oil filter, inserting the filter tube into the hole in the bottom of the transmission. Secure the filter with the screw torqued to 10–15 feet pound.

CAUTION *Do not use gasket-type sealing compounds anywhere inside the transmission. If grease is used for internal assembly of the transmission components, use only greases approved by Allison.*

5. Place the oil pan gasket on the oil pan. Sealant should be applied only to the area of the oil pan flange outside the raised bead of the flange.

6. Install the oil pan gasket, carefully guiding it into position. Ensure that no dirt or other material enters the pan during installation. Turn the oil pan retaining screws by hand and then torque them sequentially to specification.

7. Install the filler tube at the side of the pan and tighten the tube fitting to specification.

8. Install the drain plug and gasket and tighten the plug.

9. Fill the transmission with oil to the proper level. Follow the OEM specifications for oil capacity. Recheck the oil level as described earlier in this chapter.

Heavy-Duty Oil Pan

The oil and filter change on an Allison transmission with a heavy-duty, high-capacity oil pan should be performed as follows:

1. After getting the transmission to operating temperature, remove the oil drain plug from the rear or side of the pan. Allow the oil to drain.

2. Remove the bolt and nut securing the filter retaining strap. Remove the filter cover, spring, and retainer.

3. Remove the filter. Discard the filter retainer and cover O-ring gaskets.

4. Install the seal ring on the filter retainer, then install the filter and retainer into the external access canister.

5. Install the cover seal ring onto the lip of the oil pan.

6. Install the spring onto the cover and place it over the filter retainer.

7. Install the strap, bolt, and nut to the pan and torque to specification.

8. Install the drain plug and tighten.

9. Fill the transmission with oil to the proper level. Follow the OEM specification for oil capacity. Recheck the oil level.

Governor Filter Change

Allison recommends that the governor filter screen be inspected or replaced at every oil change. A pipe plug can be used to retain the governor oil screen in older model Allison transmissions, as shown in **Figure 7-60.** Remove the pipe plug and inspect the filter. If it is undamaged, clean it in mineral spirits and reinstall it. If it is damaged, replace it. Install the filter open end first into the transmission cover and reinstall the pipe plug.

In later generation hydromechanical transmissions, a hexagon plug and O-ring seal are used to retain the governor oil filter in the transmission rear cover, as shown in **Figure 7-61.** On others, the hexagon plug and filter location is on the output retarder. In both cases, the

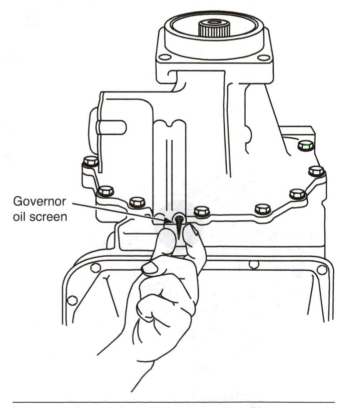

Governor oil screen

Figure 7-60 Location of the governor filter on early model transmissions.

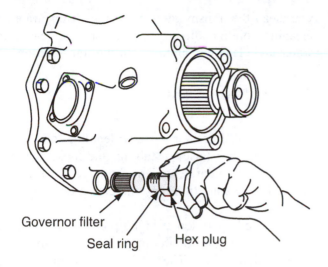

Governor filter

Seal ring   Hex plug

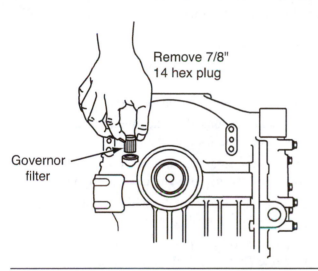

Remove 7/8"
14 hex plug

Governor
filter

Figure 7-61 Location of the governor filter on later model transmissions.

O-ring and filter should be changed at every oil change. Remember to torque the plug to specifications.

Oil Contamination

At each oil change, examine the oil drained from the transmission for evidence of dirt or water. Trace condensation is normal; this will emulsify in the oil during operation. However, if there is visible evidence of water in the oil, check the transmission cooler for internal leakage between the coolant and oil circuits. Oil in the coolant circuit of the transmission cooler is another indication of leakage.

Metal Particles

When draining the oil from the transmission, inspect for any metallic flakes in the oil and on the magnetic drain plug (except for those minute particles normally trapped in the oil filter); these may indicate transmission

damage. When larger metallic particles are found in the sump, the transmission should be disassembled and inspected to locate the source of the breakdown. Beyond locating the cause, metal contamination requires a complete disassembly of the transmission and cleaning of all internal and external circuits, cooler, and all other areas where the particles could lodge.

CAUTION *Metal contamination normally requires the replacement of all the transmission bearings.*

Coolant Leakage

If engine coolant leaks into the transmission oil circuit, action should be taken to prevent malfunction

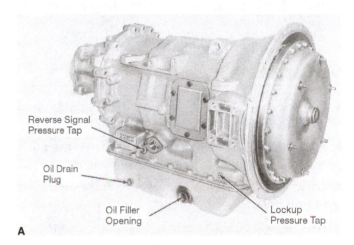

Reverse Signal
Pressure Tap

Oil Drain
Plug

Oil Filler
Opening

Lockup
Pressure Tap

A

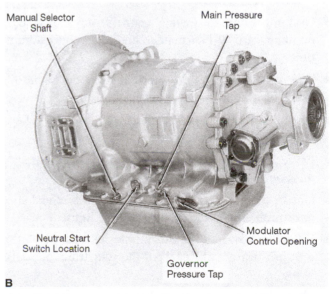

Manual Selector
Shaft

Main Pressure
Tap

Neutral Start
Switch Location

Modulator
Control Opening

Governor
Pressure Tap

B

Figure 7-62 Breather location on top of a transmission housing; also note the position of the manual selector shaft, neutral start switch, modulator control opening, pressure taps, plugs, and other key components.

and complete failure. The transmission should be disassembled, inspected, and cleaned. All traces of the coolant and varnish deposits that result from coolant contamination should be removed.

Test kits can be used to detect traces of glycol in the transmission oil. You should note, however that certain additives in some transmission oil can produce a positive reading. If this is the case, a lab oil analysis should be performed.

Breathers

The **breather** is located at the top of the transmission housing (see **Figure 7-62**). It prevents pressure buildup within the transmission and should be kept clean and unplugged. Exposure to dust and dirt will determine the frequency with which the breather requires cleaning. Use care when cleaning it whenever you clean the transmission. Spraying steam, water, or cleaning solution directly on the breather can force water and cleaning solution into the transmission.

External Lines and Oil Cooler Inspection

Inspect all lines for loose or leaking connections, worn or damage hoses or tubing, and loose fasteners. Examine the radiator coolant for traces of transmission oil. The condition may indicate a defective heat exchanger.

Extended operation at high operating temperatures can cause clogging of the oil cooler and can lead to transmission failure. The oil cooler system should be thoroughly cleaned after any rebuild work is performed on the transition.

Summary

- Proper maintenance of the drive axle depends upon using the correct lubricant.

- Synthetic oils provide so many advantages that most component manufacturers endorse their use.

- Drive axle fluid should be level with the bottom of the fill hole.

- All efforts should be made to avoid mixing different gear lubes in the axle housing.

- Draining lubricants when warm ensures that contaminants are still suspended and also reduces drain time.

- The wheel bearings are lubricated by the same oil that is in the differential carrier housing.

- The driveshaft assembly is made of a U-joint, yokes, slip splines, and driveshafts.

- Driveshafts transmit engine torque from the transmission to the rear drive axles.

- One of the most common causes of U-joint and slip joint problems is the lack of proper lubrication.

- When lubricating U-joints, pump grease slowly into the zerk fitting until each of the four trunnion seals pops.

- Slip splines can be lubricated with the same grease that is used on the U-joints.

- Hanger bearings are usually lubricated for life by the manufacturer and are not serviceable.

- Only the lubricants that the manufacturer recommends should be used in the transmission.

- Transmission manufacturers almost exclusively recommend the use of synthetic gear lubes.

- Standard transmission lubricant should be exactly even with the filler plug opening.

- The function of a clutch is to transfer torque from the engine flywheel to the transmission.

- There are two styles of clutch adjustment mechanism: manually adjusted and self-adjusting clutches.

- All clutches are disengaged through the movement of a release bearing or throw-out bearing.

- The purpose of a clutch brake is to slow or stop the transmission input shaft from rotating, to allow gears to be engaged without clashing.

- Maintaining proper fluid levels in an automatic transmission is critical to the life and performance of the transmission.

- When draining the oil from the transmission, inspect for any metallic flakes in the oil and on the magnetic drain plug.

- The breather is located at the top of the transmission housing. It prevents pressure buildup within the transmission.

Review Questions

1. Technician A says that most OEMs recommend extended rear axle oil drain intervals when synthetic lubricants are used. Technician B says that the differential carrier should be flushed with kerosene at every oil change. Who is correct?

   A. Technician A

   B. Technician B

   C. Both technician A and B are correct.

   D. Neither technician A nor B is correct.

2. Which of the following would indicate that the level is correct when checking lube oil level in a differential carrier housing with the fill plug removed?

   A. Oil spilled from fill hole

   B. Oil level exactly even with bottom of fill hole

   C. Oil level within one finger joint of fill hole

   D. Oil level within 1 inch of fill hole

3. Which gear lube viscosity rating is usually recommended for use in a differential carrier for a truck operating in a temperature of −30°F?

   A. 75W

   B. 80W

   C. 80W-90

   D. 80W-140

4. Which type of lubricant is recommended when hypoid gearing is used in a differential carrier?

   A. API GL-4

   B. Synthetic lube only

   C. Any EP-rated lubricant

   D. Any API-rated lubricant

5. After a differential carrier overhaul, which of the following methods would best ensure that the wheel bearings receive proper initial lubrication?

   A. Fill the differential carrier to the interaxle differential fill hole.

   B. Fill the differential carrier through the axle breather.

   C. Jack the axle up at each side for one minute, then level and check fill hole level.

   D. Remove each axle shaft rubber fill plug and fill to the indicated oil level.

6. When planning to drain and replace the lubricant in a differential carrier, which of the following is preferred?

   A. Drain when the lubricant is at operating temperature.

   B. Flush with kerosene.

   C. Drain after allowing to cool overnight.

   D. Flush with solvent.

7. When a U-joint is lubricated, fresh grease appears at all four trunnion seals. Which of the following is probably true?

   A. The bearings are worn and should be replaced.

   B. The seals are worn and should be replaced.

   C. The trunnions are worn and the U-joint should be replaced.

   D. The U-joint has been properly lubricated.

8. The OEM recommendation for U-joint lubrication interval scheduling on a truck used for an on-highway operation should be:

   A. every 10,000 to 15,000 miles.   C. 3,000 to 5,000 miles.

   B. once per year.   D. every 50,000 miles.

9. Technician A says that when lubricating a U-joint and grease does not exit from one of the four trunnion seals, the U-joint must be immediately replaced. Technician B says that backing off the bearing cap on a U-joint sometimes helps a trunnion to take grease. Who is correct?

   A. Technician A   C. Both technician A and B are correct.

   B. Technician B   D. Neither technician A nor B is correct.

10. What is the function of a driveshaft slip spline joint?

    A. It accommodates variations in   C. It allows the driveshaft to change length while
       driving angles.   transmitting torque.

    B. It multiplies drive torque.   D. All of the above

11. Which of the following OEM-approved lubricants would produce the longest service life in a standard transmission?

    A. Synthetic E-500 gear lube   C. Multipurpose or EP gear lube

    B. SAE 50 grade heavy-duty   D. SAE 50 grade transmission fluid
       engine oil

12. After breaking in a new transmission filled with E-500 lube on a truck used in a linehaul application, when should the first oil change take place?

    A. Between 3,000 and 5,000   C. After 100,000 miles of linehaul service
       miles of service
                                           D. After 500,000 miles of linehaul service
    B. After 25,000 miles of
       vocational service

13. Which of the following would indicate that transmission oil in a truck's standard transmission is at the correct level?

    A. Oil is visible through the   C. Oil is level with the bottom of the filler hole.
       filler hole.
                                           D. Oil is at the high level on the dipstick.
    B. Oil is reachable with the first
       finger joint through the filler
       hole.

14. Which of the following conditions can cause aerated oil in automatic transmissions?

    A. Low oil level   C. A clogged breather

    B. High oil level   D. Both a and b

15. On most Allison transmissions, when the oil level reads at the FULL mark on the dipstick with the transmission at operating temperature, which of the following indicates the actual level of oil in the transmission?

    A. If it is slightly below the   C. If it is slightly above the ring gears
       planetary gearsets
                                           D. If it is at the axis of the sun gears
    B. If it is slightly above the top
       of the oil pan

16. Which of the following could result in an inaccurate dipstick reading?

   A. Failure to wipe the dipstick clean before taking the reading

   B. A clogged breather

   C. An unvented oil filler tube

   D. All of the above

17. When performing an automatic transmission cold check level, at what temperature should the oil be?

   A. Between 0°F and 60°F

   B. Between 60°F and 120°F

   C. Between 120°F and 160°F

   D. Between 160°F and 200°F

18. What effect will an increase in oil temperature have on the transmission oil level?

   A. Lower the oil level

   B. Raise the oil level

   C. No effect

   D. Aerate the oil

19. The _____ is located at the top of the transmission housing, and its function is to prevent pressure buildup within the transmission housing.

   A. auxiliary filter

   B. governor

   C. breather

   D. release valve

20. How often should a clutch be inspected and lubricated?

   A. Every month

   B. Every 6,000 to 10,000 miles

   C. Any time the chassis is lubricated

   D. All of the above

21. When adjusting the clutch linkage, technician A says that pedal free travel should be about 1-1/2 to 2 inches. Technician B says that free travel should be less than 1/2 inch. Who is correct?

   A. Technician A

   B. Technician B

   C. Both technician A and B are correct.

   D. Neither technician A nor B is correct.

8 Tire Hub Wheel and Rim Inspection

Objectives

Upon completion and review of this chapter, the student should be able to:

- Identify the different wheel configurations used in the trucking industry.
- Perform wheel inspections on the different wheel configurations used in heavy-duty trucks.
- Explain the difference between standard and wide-base wheel systems and stud- and hub-piloted mountings.
- Explain the importance of proper matching and assembly of tire and rim hardware.
- Describe brake drum mounting configurations.
- Explain the proper mounting procedures for the wheel configurations used on heavy-duty trucks.
- Perform wheel runout checks and adjustments.
- Explain the proper techniques for front and rear wheel bearing adjustment.
- Outline the procedures for installing preset bearing wheels.
- List the different types of tires used in the trucking industry based upon construction.
- Properly match tires in dual and tandem mounting.
- Explain inspection procedures for tires.
- Identify tire wear conditions and causes.

Key Terms

axial runout

bias ply

ConMet PreSet hub

footprint

hub-piloted

radial tires

scrubbing

scuff

stud-piloted

Technical
    and Maintenance
    Council (TMC)

tire square

Unitized Hub System

unsprung weight

zipper

INTRODUCTION

Wheel assemblies can be broken down into wheels, rims, tires, and hubs. Each of these components must be properly maintained, serviced, and inspected by the servicing technician. Because of the critical nature of wheel assemblies, there is no room for error, making it imperative that technicians be properly trained in all aspects of wheel inspection and service assembly. A new technician should never be afraid to ask questions of more experienced and qualified technicians. Any misdiagnosis or improperly repaired wheel component can mean potential disaster for the vehicle and all those who share the road with it.

SYSTEM OVERVIEW

This chapter begins by explaining the difference between the different styles of wheel assemblies that are available in the trucking industry and also explains mounting and proper torquing. This chapter also explains the reasons for retorquing wheel assemblies.

This chapter then explains the TMC recommendations for adjusting wheel bearings on trucks and trailers.

Tires are an expensive consumable item that must be serviced properly to achieve maximum service life. This chapter will explain how technicians identify the tire's construction type and size so as not to mismatch tires on the vehicle. It then discusses how to inspect the tire's serviceability and how to diagnose tire wear conditions and ends with the out-of-service criteria for tires.

WHEELS AND RIMS

The term *wheel* with regard to trucks can be used to describe the hub assembly to which the rims and brake drum (or rotor) are attached.

Today's trucks use one of two types of wheel configurations that are used along with several methods of attaching tire assemblies to the wheels. The two general types of wheels used are:

- Cast spoke wheels
- Disc wheels

With spoke wheels, the rim and wheel are separate components. On a disc wheel, the rim is a distinct section, a wheel assembly. Disc wheels can also be divided into the two ways in which they are fastened to the hub assembly:

- Stud-piloted
- Hub-piloted

CAST SPOKE WHEELS/ DEMOUNTABLE RIM SYSTEMS

The cast spoke wheel has been around for many years and was used almost universally, but this has changed as operators have come to understand the advantages of disc wheels. This being said, the cast spoke wheel has not outlived its usefulness and is by no means obsolete. These wheels have the disadvantage of being heavy, but they are tough and can withstand more punishment than disc wheels, making them a common choice for dump, construction, and refuse use, and for many leased trucks and trailers.

A spoke wheel consists of a one-piece casting that includes an integral hub and spokes (see **Figure 8-1**). Generally they are manufactured from cast steel, and the hub and wheel mounting surfaces are machined after casting. The tires of this assembly are mounted on a separate rim or rims. In dual applications (two tire/ rim assemblies mounted on a wheel) a spacer band is positioned between the inner and outer rims in the manner shown in **Figure 8-2A**. This space band is clamped between the two tires and rims to provide exact spacing between them (see **Figure 8-2B**).

The spoke wheel may be found in three-, five-, and six-spoke configurations. The six-spoke designs are often used on heavily loaded front axle applications. Five

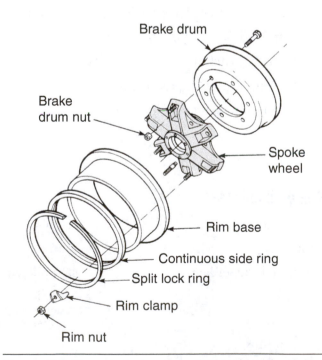

Figure 8-1 Components of a cast spoke wheel and multi-piece rim. *(Courtesy of Daimler Trucks North America)*

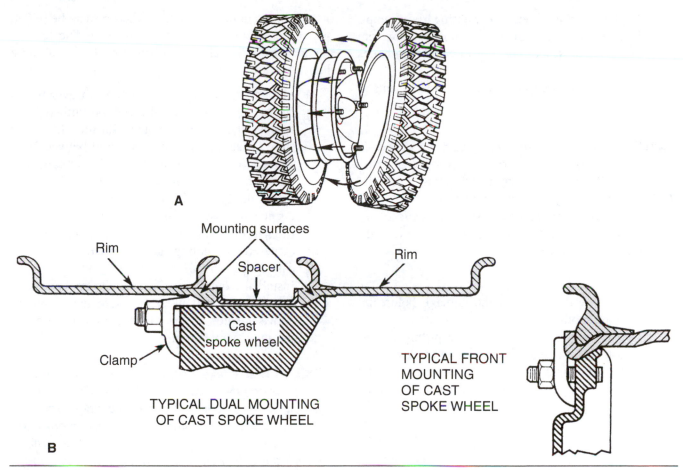

Figure 8-2 (A) Position of the spoke wheel dual mounting spacer band, and (B) cross-section view of mounted dual wheels. *(Courtesy of Navistar International Corp.)*

or six spokes are used on drive axle duals, but six-spoke designs are often preferred because of the added wheel clamping force on the rim, which reduces the chance of rim slippage. Three-spoke wheels have wide spokes, using two wheel clamps per spoke. Trailers are their most popular application.

From the diagram in **Figure 8-2**, you can see that there are many components that make up the spoke wheel assembly. They can use a multi-piece rim or single-piece rim that clamps to the spokes with wheel clamps.

Maintenance on these wheels is very critical, because if the clamps are not installed correctly, the tire/rim will produce a wobble condition known as **axial runout.** Axial wheel runout is more of a problem in cast spoke wheels than with disc wheels, but it can be minimized by following proper installation procedures and torquing sequences, and by checking runout after assembly. Although cast spoke wheels experience greater alignment and balance challenges because of their extra weight and rim mounting, they usually encounter fewer outright failure problems

and, notably, fewer ''wheel-off'' incidents than disc wheel assemblies.

Wheel Inspections: Spoke Wheel/ Demountable Rim Systems

When servicing a vehicle, inspect the wheel system for the following:

1. Check that valve caps are installed. Replace if missing.
2. Check all metal surfaces thoroughly, including space bands and tire side or rims. Watch for:

   - Excessive rust or corrosion buildup
   - Cracks in metal
   - Bent flanges or components
   - Loose, missing, or damaged nuts, clamps
   - Bent, broken, or stripped studs
   - Incorrectly matched rim parts
   - Check valve locators for slip damage and improper location

3. Correct any problem that is discovered.

4. Replace broken studs and missing rim clamps. Uneven rim clamp advance or bottomed-out rim clamps indicate a problem. Loosen all rim clamps and retorque.
5. Replace any assembly that has damaged rims or components.

CAUTION *Excessively corroded or cracked rims are dangerous during the removal of the assembly. Deflate the tire (both tires if dual) before attempting to remove the rim from the vehicle. Verify deflation by installing a piece of mechanic's wire into the tire valve to assure debris has not prevented deflation.*

6. Determine the cause of the damage before installing another rim.
7. Replace any rim with excessive pitting or corrosion that has reduced the metal thickness.
8. Inflate the tires to only the recommended air pressure, being sure not to exceed the rim inflation rating in accordance with OSHA standards. It is a good practice to mark defective parts for destruction to ensure that they will not be used by mistake. Remember that a leak in a tubeless tire assembly can be caused by a cracked rim. Do not put a tube in a tubeless assembly to correct this problem. Cracked rims should be destroyed (use a torch) to avoid accidental reuse. Never attempt to weld or otherwise repair cracked, bent, or out-of-shape components: it is illegal as well as dangerous.

Shop Talk: Determine the cause of any damage before replacing a component to avoid the damage reoccurring.

Spoke Wheel/Demountable Rim Installation

When mounting the tire rim assembly onto the wheel, ensure that you have the correct studs, nuts, and clamps. Spoke wheels use rim studs. Rims studs are threaded on both ends with an anaerobic locking compound that essentially glues them into the spoke of the wheel. If a stud had been damaged, it must be removed with a stud remover and replaced.

Rim lug nuts should be kept tight and checked on a regular basis. Checking alignment of the rim/wheel installation is important because the rims can be pulled out of alignment if improperly tightened. The following are general installation instructions for installing a dual set of tire/rims to cast spoke wheels:

1. Start by installing the inner tire/rim over wheel spokes and push it all the way on until it stops against the tapered mount. Ensure the valve stem is facing out and is centered between two spokes. It should clear the disc brake caliper if installed on a disc brake assembly.
2. Install the spacer ring over the wheel spokes and tap it into position against the inner tire with a tire hammer. Check the spacer, making sure it is centered by rotating the spacer ring around the cast spoke wheel.
3. Install the outer tire/rim over the spoke wheel, making sure that the valve stem faces inboard and is located in the same relative position as the inner valve stem.

CAUTION *Take precautions when mounting wheel assemblies. Watch your fingers to prevent getting them pinched, and it is also good practice to wear gloves. Use proper lifting techniques to protect your back. Many professional wheel installers lift the wheel assembly from behind their body to prevent back injuries.*

4. Install all the wheel clamps and nuts. Turn the nuts on their studs until each nut is flush with the end of each stud. Tap the outside tire/rim with a tire hammer to keep it centered while you are doing this.
5. Using the diagrams in **Figure 8-3** as a guide, turn the top nut 1 until it is snug.
6. Rotate the wheel and rum until nut 2 is at the top position and snug the nut.
7. Rotate the wheel and rim until nut 3 is at the top position and snug the nut.
8. Rotate the wheel and rim until nuts 4, 5, and 6, respectively, are at the top and snug these nuts. Because the entire weight of the tire and rim assembly is on the top spoke, this crisscross sequence will help ensure an even application of force at all points in the rim, keeping the rim in alignment.
9. Repeat the sequence of tightening the nuts incrementally to the OEM torque.
10. After operating the vehicle for approximately 50 miles (80 km), check the lug nuts for tightness in the same tightening sequence. Check the wheel nut torque frequently and at all service intervals.

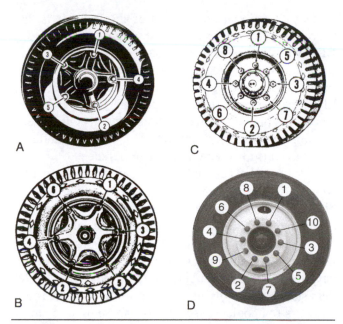

Figure 8-3 Wheel nut tightening sequence: (A) six lug, (B) five lug, (C) twelve lug, and (D) ten lug. (*Courtesy of Mack Trucks, Inc.*)

*onto the cast spoke hub (see **Figure 8-4**). There are wheel designs with different numbers of rim clamps and of various shapes. Each spoke wheel requires rim clamps designed for the specific spoke wheel and designated rim. Dual rims are mounted using a space band, which holds the two rims apart and provides proper dual spacing for the tires, so it is imperative that the correct space band, rim, and clamp combinations be used for the application.*

RECOMMENDED MOUNTING TORQUE FOR DEMOUNTABLE RIMS

Stud Size	Torque Level ft.-lb. Dry
5/8"–11"	160–200
3/4"–10"	200–260

CAUTION *Correct components must be used. Spoke/demountable wheels use rim clamps to secure rims (which have no center bolting disc) to a hub or cast spoke wheel, which may have either three, five, or six spokes. The rim clamps, fastened by hex nuts, wedge the rim*

Spoke Wheel/Demountable Rim Runout

Whenever a spoke wheel is reinstalled, it must be checked for runout after it is torqued to specifications. To check runout, position a wooden block on the floor or tire hammer stood on end and position about 1/2 inch (13 mm) from the tire as shown in **Figure 8-5.**

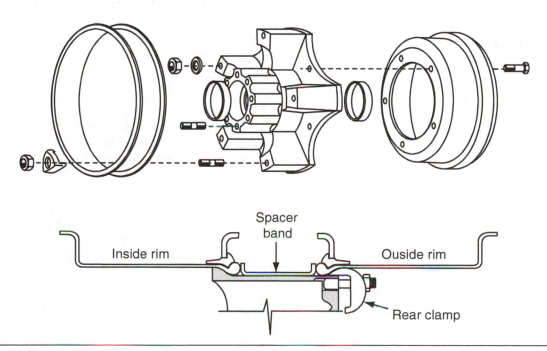

Figure 8-4 Spoke wheel end.

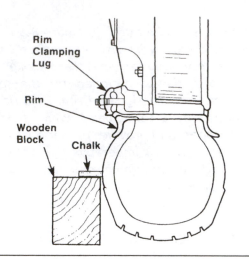

Figure 8-5 Checking tire runout using a wood block and chalk. *(Courtesy of Mack Trucks, Inc.)*

If the wheel runout exceeds 1/8 inch (3 mm), it should be corrected.

Position a piece of chalk on the wood block as shown and rotate the wheel through one revolution so that the chalk marks the tire high spots. The high and low (unmarked) areas indicate which lug nuts have to be loosened and which have to be tightened to correct the condition. Slightly loosen the lug nuts 180 degrees opposite to the chalk marks and tighten those on the chalk-marked side of the tire. Do not overtorque the nuts. Recheck runout and repeat until runout is within 1/8 inch (3 mm). If runout cannot be corrected in this way, inspect for part damage or dirt between the mating parts.

DISC WHEELS

Disc wheels differ from spoke wheels in that the tire rim forms the wheel and mounts to the hub assembly as shown in **Figure 8-6.** Stud holes in the center of the disc wheel allow them to be mounted to the hub studs with nuts. These hub studs pass through both rim and disc wheels.

One of the real advantages of disc wheels is that they run true and produce few alignment problems. Disc wheels use one-piece steel or aluminum construction, which makes them lighter, reducing vibration (and tire running temperatures) and extending tire life. Because of their ability to extend tire life, disc wheels are popular with cost-conscious fleet managers. By improving driver comfort and vehicle handling, they have also become a driver preference.

Because wheel assemblies are part of the **unsprung weight** (the weight that is not supported by the vehicle's suspension system), there is always a handling advantage to reducing their weight. The other advantage to reducing this weight is that payload can be increased, so aluminum disc wheels have steadily gained in popularity. Aluminum is not only lighter than steel but also dissipates heat faster, with the result that tires run cooler. Disc wheels can be used in single and dual configurations. There are also two different mounting systems for disc wheels: **stud-piloted** and **hub-piloted.**

Stud-Piloted Wheels

A stud-piloted mounting system is illustrated in **Figure 8-7A,** a single configuration. The wheel simply mounts onto studs on the hub and is secured using single cap nuts. **Figure 8-7B** shows a dual stud-piloted configuration for disc wheels.

In a dual mounting system, an inner cap nut, often called a cone, secures the inner wheel. The tapered shape of the nut secures the inner rim in addition to helping to center the wheel. The outer wheel is installed over the inner cone and a cap nut is installed onto the inner cone; again the tapered shape of the nut centers the outer disc wheel. In this type of mounting system, all the weight of the vehicle is placed between the studs and the wheel rims. The torquing procedures for stud-piloted wheels will be discussed later in this chapter. Stud-piloted disc wheels have been used less often following the introduction of hub-piloted wheels, which use a simpler mounting procedure.

Figure 8-6 Components of a typical disc wheel. *(Courtesy of the Budd Company)*

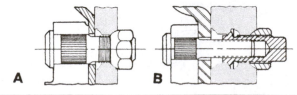

Figure 8-7 Aluminum disc wheel stud-piloted mounting configurations: (A) single wheel, and (B) dual wheel. *(Courtesy of Alcoa Wheel Products International)*

Hub-Piloted Wheels

The hub-piloted system is a much simpler mounting system for disc wheels compared to stud-piloted wheels. Both the inner and outer wheels are secured to the hub with only one nut per stud. The wheel hub has pilot pads around the circumference of the hub. The weight of the vehicle is transferred through these pilot pads to the wheels and not through the studs. The one-piece flange nut applies a clamping force to secure the wheel to the hub.

In stud-piloted systems, a loose inner nut can easily go undetected, eventually pounding out the nut ball seat. With hub-piloted systems, both the inner nut and its ball seats are eliminated.

> **CAUTION** *Correct components must be used. It is important to note that some hub-piloted and stud-piloted wheels may have the same bolt circle pattern. Therefore, they could mistakenly be interchanged.*

Each mounting system requires its correct mating parts. It is imperative that the proper components be used for each type of mounting and that the wheels are fitted to the proper hubs.

If hub-piloted wheel components (hubs, wheels, fasteners) are mixed with stud-piloted wheel components, loss of torque, broken studs, cracked wheels, and possible wheel loss can occur, since these parts are not designed to work together.

Mixing hub-piloted and stud-piloted wheels will not allow the inner cap nut to fit into the inner wheel and will result in the inner cap nut interfering with the outer wheel (see **Figure 8-8**).

Ball seat, stud-piloted wheels should not be used with flange nuts because they have larger bolt holes and

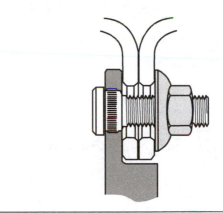

Figure 8-9 Using incorrect hardware to mount a wheel.

do not have sufficient area near the bolt hole to support the flange nut. Slippage may occur. Also, the center hole is too large to center the wheel (see **Figure 8-9**).

It is also important to note that hardware for stud and aluminum wheels cannot be randomly mixed. If stud and aluminum wheel hardware are mixed, loss of clamp load, broken studs, cracked wheels, or possible wheel loss can occur, since these parts are not designed to work together (see **Figure 8-10**).

If combinations of stud and aluminum wheels are used in a dual application, consult your wheel manufacturer for hardware recommendations.

Wide-Base Wheels

Wide-base wheels can also be referred to as high flotation, super singles, wide body, duplex, or jumbo

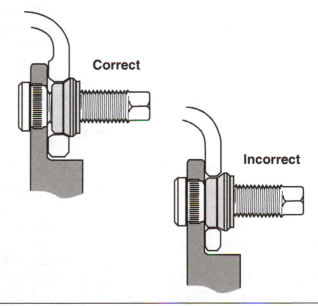

Figure 8-10 Using both correct and incorrect mounting hardware.

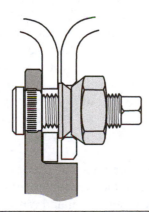

Figure 8-8 Using incorrect hardware to mount a wheel.

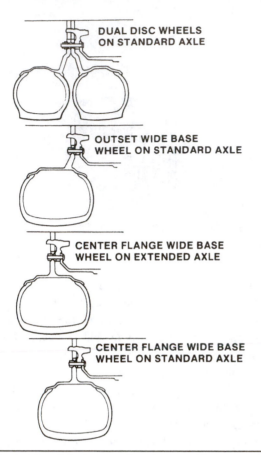

Figure 8-11 Wide-base wheel mounting compared to dual configurations. *(Courtesy of Navistar International Corp.)*

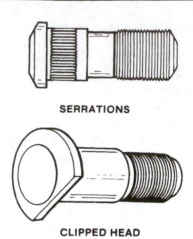

Figure 8-12 Headed wheel stud shanks for disc wheels. *(Courtesy of Daimler Trucks North America)*

on the passenger (curb or right) side of the vehicle. "L" studs are mounted on the road (driver's or left) side of the vehicle. This helps ensure that wheel rotation does not loosen the fasteners. Other systems use right-handed threads only. Whatever type of fastener is used, ensure that all hardware is in good condition.

> **Note:** The right and left sides of the vehicle are always identified as if you were sitting in the driver's seat facing forward.

wheels. One wide-base wheel and tire replaces traditional dual wheels and tires. You can see some of the available configurations in **Figure 8-11.** If wide-base wheels were to be used for a tractor trailer combination, instead of a total of 18 tires, only 10 would be required: a pair of traditional steer tires would be used at the steering axle, and eight wide-base wheels used everywhere else on the rig, replacing the duals. The advantage of the wide-base single is that they are significantly lighter compared to steel dual wheels and tires, allowing an increase in payloads and fuel efficiency.

Disc Wheel Installation

Disc wheels are mounted to the wheel hub with threaded studs and nuts or headed studs. A headed wheel stud either has serrations on the stud body or a single flat groove machined into the head of the stud to prevent it from spinning in the hub. **Figure 8-12** shows the two different stud styles. In some disc wheel systems, the end of the stud that faces away from the hub is stamped with an "L" or "R" indication that the stud uses left- or right- hand threads. "R" studs are mounted

STUD-PILOTED DISC WHEEL INSTALLATION

The following are some general installation instructions for installing stud-piloted dual disc wheels that use conical (cones) nuts:

1. Slide the inner wheel (dual configuration) into position over the studs and push back as far as possible, taking care not to damage the threads of the studs and ensuring valve stem has clearance at the caliper (disc brakes).
2. Install the outer wheel nut for a single or the conical nut in a dual. Run the nuts on studs until they come into contact with the wheel. Rotate the wheel a half turn to allow parts to seat naturally.
3. Draw up the stud nuts alternately following the sequence (crisscross pattern), as illustrated in **Figure 8-13.** Do not fully tighten the nuts at this time.
4. Continue tightening the nuts to torque specifications using the same alternating method.
5. Install the outer wheel and repeat the preceding method. Be sure that both inner and outer tire valve stems are accessible.

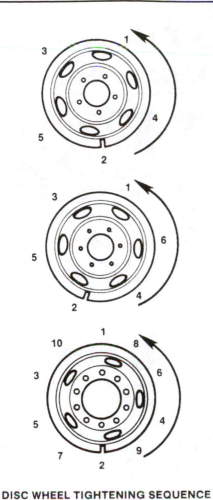

DISC WHEEL TIGHTENING SEQUENCE
(Arrows illustrate 1/2 turn
to seat parts.)

Figure 8-13 Torquing sequence for disc wheels. (*Courtesy of Navistar International Corp.*)

HUB-PILOTED DISC WHEEL INSTALLATION

Figure 8-14 illustrates a cross-section of a hub-piloted installation (see **Photo Sequence 6**). Follow these procedures when installing hub-piloted disc wheels:

1. Check the wheel nuts. Ensure that multi-piece nuts turn smoothly on their flanges. Discard all nuts with damaged threads.
2. Apply two drops of oil to a point between the nuts and flange and two drops to the last two or three threads at the end of each stud. Also, lightly lubricate the pilots on the hub to ease wheel installation and removal. Do not get lubricant on the mounting face of the drum or wheel.
3. Install the inner wheel of a dual configuration or the single steer wheel over the studs and push back as far as possible. Be sure not to damage the threads of the studs during installation.
4. Position the outer rear tire and wheel in place over the studs and push back as far as possible, again using care not to damage the studs.
5. Run the nuts onto the studs until the nuts contact the wheel(s). Rotate the wheel assembly a half turn to permit parts to seat.
6. Draw up the nuts alternately following the crisscross sequence illustrated in **Figure 8-14**. Do not fully tighten the nuts at this time. This allows uniform seating of the nuts and ensures even face-to-face contact of wheel and hub (see **Figure 8-15**).
7. Continue tightening the nuts to torque specifications using the same alternating sequence.

6. After the vehicle is operated for approximately 50 miles (80 km), the wheel should be retorqued. This is because the wheel will settle once the vehicle is operated; some nuts may become loose, so all nuts should be retorqued.

RETORQUING STUD-PILOTED DUALS

To retorque a dual configuration, both the inner and outer nuts must be checked. The problem is that you do not want to allow the wheels to relax during the process, or you will have accomplished nothing. To do this:

- Loosen the outer nut on every other stud.
- Check the torque on the inner cones.
- Retorque the outer nuts.
- Loosen the remaining outer nuts.
- Check the torque on those inner nuts.
- Retorque the outer nuts.

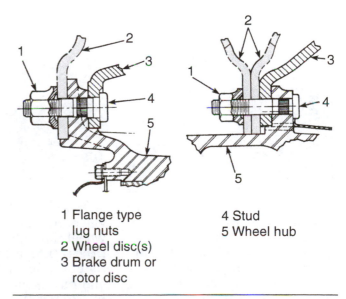

1 Flange type lug nuts	4 Stud
2 Wheel disc(s)	5 Wheel hub
3 Brake drum or rotor disc	

Figure 8-14 Cross-section of a hub-piloted disc wheel. (*Courtesy of Navistar International Corp.*)

Install a Set of Hub-Piloted Duals to a Wheel Assembly

PHOTO SEQUENCE 6

P6-1 Before beginning the disassembly of the wheel, perform a visual inspection. Check that the wheel nuts are properly engaged and look for damaged studs.

P6-2 Remove rust and road dirt from the wheel studs with a wire brush.

P6-3 Use an air gun to remove the wheel nuts.

P6-4 Carefully remove the wheels, ensuring that they are not dragged over the studs damaging the threads.

P6-5 Before reassembling the wheel, clean the hub, wheel, and hub/brake drum mounting faces of rust, dirt, and loose paint. Visually inspect all the studs. Do not paint or apply any other substance on the mounting faces.

P6-6 If an outboard mounted drum is used, make sure that the brakes are released and lift the drum into position, ensuring that it is not sitting on the pilot ledge. Lubricate the fasteners. (Hubpiloted studs must be lightly lubricated to ensure that the correct amount of clamping force is achieved.) Mount the wheels to the hub.

P6-7 Tighten the wheel nuts in sequence in three stages using a torque wrench. First-stage torque value should be 50 ft.-lb. Second-stage torque should be between 50 and 80 percent of the final torque value. Final torque values are listed at right.

P6-8 As a final step, check the wheel runout by rotating the wheel through a full revolution. Do not attempt to correct a runout problem by loosening and retorquing the nuts: disassemble the wheel assembly and locate the problem.

Typical Torque Values for Steel Disc Wheels

Socket Size	Thread Size	Torque Specification
1½" or 38 mm	M22 × 1.5	450–500 lb.-ft.
30 mm	M20 × 1.5	280–330 lb.-ft.

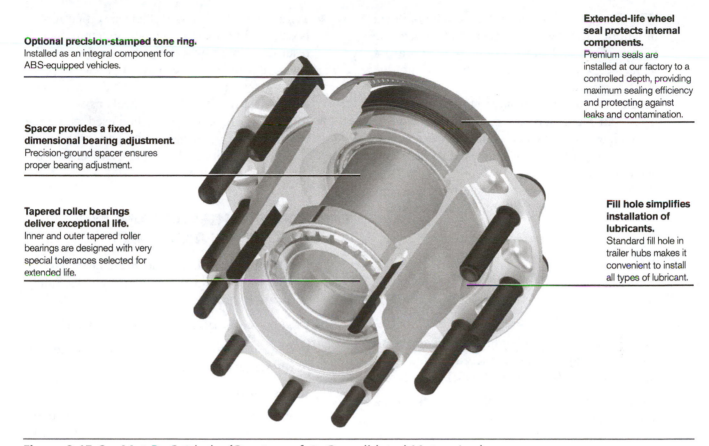

Optional precision-stamped tone ring.
Installed as an integral component for ABS-equipped vehicles.

Spacer provides a fixed, dimensional bearing adjustment.
Precision-ground spacer ensures proper bearing adjustment.

Tapered roller bearings deliver exceptional life.
Inner and outer tapered roller bearings are designed with very special tolerances selected for extended life.

Extended-life wheel seal protects internal components.
Premium seals are installed at our factory to a controlled depth, providing maximum sealing efficiency and protecting against leaks and contamination.

Fill hole simplifies installation of lubricants.
Standard fill hole in trailer hubs makes it convenient to install all types of lubricant.

Figure 8-15 ConMet PreSet hub. *(Courtesy of ® Consolidated Metco, Inc.)*

RETORQUING HUB-PILOTED WHEELS

After the vehicle is operated for approximately 50 miles (80 km), the wheel should be retorqued. The retorquing process for hub-piloted disc wheels is a simpler task in dual configurations as opposed to stud-piloted disc wheels.. Simply take a torque wrench and, using the same crisscross pattern as previously mentioned, torque all the wheel studs.

Shop Talk: Whenever the vehicle is in to the shop for service, the torque of the wheel studs should be rechecked.

Wheel Inspections: Disc Wheels

When servicing a vehicle, inspect the wheel system for the following:

1. Check that valve caps are installed. Replace if missing.
2. Missing or worn nuts: If missing or worn, re-place. Retighten all other nuts to the proper torque. The inner nuts on both sides of a broken inner cap nut should also be replaced.
3. Broken studs: If a stud is broken, replace it and the stud on each side of the broken one. If two or more studs are broken, replace them all. Use a press to install studs and be sure the hub flange is supported. Aluminum hubs require different stud installation procedures. Consult the OEM for procedures. Stud heads can be bent from hammer blows, which will prevent the stud from seating properly and can result in stud failure.
4. Rust streaks extending from the bolt holes: This indicates either worn, poor-quality, or loose wheel nuts. Make sure the correct nuts for the wheel system are being used, inspect wheel bolt holes for wear or damage, and then tighten fasteners to the correct torque. Then remove rust streaks.
5. Cracks in washers of hub-piloted wheel nuts: Replace damaged nuts and tighten to the proper torque.
6. Variations in the number of stud threads that protrude beyond the nuts indicate loose, backed-off wheel nuts. Check all components for damage.

7. Wheels not seated on pilots of hub-piloted mounting: Remove the wheels and inspect the wheels, hub, and brake drum. If they are not damaged, reinstall on the vehicle using the correct procedures.

8. Cracks or damage on any wheel component: Check all metal surfaces thoroughly, including both sides of the wheels and between duals. Replace any wheel that is cracked or has damaged components.

9. Replace any rim with excessive pitting or corrosion that has reduced the metal thickness.

Shop Talk: Determine the cause of any damage before replacing a component to avoid the damage reoccurring.

10. Inflate the tires to only the recommended air pressure, being sure not to exceed the rim inflation rating in accordance with OSHA standards.

It is a good practice to mark defective parts for destruction to ensure that they will not be used by mistake. Remember that a leak in a tubeless tire assembly can be caused by a cracked rim. Do not put a tube in a tubeless assembly to correct this problem. Cracked rims should be destroyed (use a torch) to avoid accidental reuse. Never attempt to weld or otherwise repair cracked, bent, or out-of-shape components; it is illegal as well as dangerous.

RECOMMENDED MOUNTING TORQUE FOR DISC WHEELS

Mounting Type	Nut Thread	Torque Level ft.-lb. Oiled*
Hub-piloted with flange nut	11/16"–16"	300–400
	M20 X 1.5	280–330
	M22 X 1.5	450–500
		ft.-lb. Dry
Stud-piloted, double cap nut standard type (7/8" radius)	3/4"–16"	450–500
	1-1/8"–16"	450–500
Stud-piloted, double cap nut heavy duty type (1-3/16" radius)	15/16"–12"	750–900
	1-1/8"–16"	750–900
	1-5/16"–12"	750–900

*See Hub-Piloted Disc Wheel Installation step 2

Note: If using specialty fasteners, consult the manufacturer for recommended torque levels.

■ Tightening wheel nuts to their specified torque is extremely important. Undertightening, which results in loose wheels, can damage wheels, studs, and hubs and can result in wheel loss. Overtightening can damage studs, nuts, and wheels and result in loose wheels as well.

■ Regardless of the torque method used, all torque wrenches, air wrenches, and any other tools should be calibrated periodically to ensure the proper torque is applied.

OUT-OF-SERVICE CRITERIA FOR WHEELS, RIMS, AND HUBS

(Taken from the North American Standard Out-of-Service Criteria by the Commercial Vehicle Safety Alliance, April 1, 2006)

1. **Lock or Slide Ring (multi-piece wheel)**
   - ■ Bent, broken, cracked, improperly seated, sprung, or mismatched ring(s)

2. **Rim Cracks**
   - ■ Any circumferential crack except on intentional manufactured crack at a valve stem hole

3. **Disc Wheel Cracks**
   - ■ Any single crack 3" (76 mm) or more in length
   - ■ A crack extending between any holes, including hand holes, stud holes, and center hole
   - ■ Two or more cracks any place on the wheel

4. **Stud Holes (Disc Wheels)**
   - ■ Fifty percent or more elongated stud holes (fasteners tight)

5. **Spoke Wheel Cracks**
   - ■ Two or more cracks more than 1" (25 mm) long across a spoke or hub section
   - ■ Two or more web areas with cracks

6. **Tubeless Demountable Adapter Cracks**
   - ■ Cracks at three or more spokes

7. **Fasteners**
   - ■ Loose, missing, broken, cracked, or striped (both spoke and disc wheels) ineffective as follows: for ten fastener positions—three anywhere or two adjacent; for eight fastener positions or fewer (including spoke wheels and hub bolts)—two anywhere.

8. **Welds**
   - Any cracks in welds attaching disc wheel to rim
   - Any crack in welds attaching tubeless demountable rim to adapter
   - Any welded repair on aluminum wheel(s) on a steering axle
   - Any weld repair other than disc to rim attachment on steel disc wheels(s) mounted on the steering axle

9. **Hubs**
   - When any axle bearing (hub) cap is missing or broken, allowing an open view into hub assembly
   - Smoking from wheel hub assembly due to bearing failure

....

Note: Not to be associated with smoke from dragging brake.

....

WHEEL BEARING ADJUSTMENT

There has been an unacceptable number of heavy truck wheel-off incidents in the United States and Canada, some of which have been the result of improperly adjusted wheel bearings. For this reason, all the manufacturers of wheel-end hardware have approved a single method of wheel bearing adjustment. This method was agreed to through a meeting of the **Technical and Maintenance Council (TMC)** committee of the American Trucking Associations, and the trucking industry has embraced this single standard throughout the continent. Wheel bearing adjustment is a simple but highly critical procedure. It is recommended that technicians learn the procedure, follow it precisely, and use no other method of adjusting wheel bearings. The TMC-recommended procedure is reprinted here, word for word.

CAUTION *The TMC adjustment procedure explained here does not apply to hubs with preset bearings and seal assemblies.*

TMC Wheel Bearing Adjustment

This procedure was developed by the TMC's Wheel End Task Force, and it is important to remember that it represents the combined input of manufacturers of wheel and components. **Photo Sequence 7** runs through the procedure outlined below:

- Step 1. Bearing Lubrication. Lubricate the wheel bearing with clean lubricant of the same type used in the axle sump or hub assembly.
- Step 2. Initial Adjusting Nut Torque. Tighten the adjusting nut to a torque of 200 ft.-lb; while rotating the wheel.
- Step 3. Initial Back-Off. Back the adjusting nut off one full turn.
- Step 4. Final Adjusting Nut Torque. Tighten the adjusting nut to a final torque of 50 ft.-lb, while rotating the wheel.
- Step 5. Final Back-Off, use **Table 8-1.**
- Step 6. Jam Nut Torque, use **Table 8-2.**
- Step 7. Acceptable End Play. The dial indicator should be attached to the hub or brake drum with its magnetic base. Adjust the dial indicator so that its plunger is against the end of the spindle with its line of action approximately parallel to the axle of the spindle. Grasp the wheel or hub assembly at the three o'clock and nine o'clock positions. Push and pull the wheel-end assembly in and out while oscillating the wheel approximately 45 degrees. Stop oscillating the hub so that the dial indicator tip is in the same position as it was before oscillation began. Read the bearing end play as the total indicator movement. Acceptable end play is 0.001–0.005 inch.

This is a simple hands-on procedure that you must become familiar with before working on the shop floor. Observing the TMC wheel-end procedure takes only a little longer than the numerous shortcut methods that result in hit-or-miss adjustments that make our roads unsafe.

TABLE 8-1: FINAL BACK-OFF		
Axle Type	Threads Per Inch	Final Back Off
Steer (single nut)	12	1/6 turn*
	18	1/4 turn*
Steer (double nut)	14	1/2 turn
	18	1/2 turn
Drive	12	1/4 turn
	16	1/4 turn
Trailer	12	1/4 turn
	16	1/4 turn

Wheel-End Procedure: TMC Method of Bearing Adjustment

PHOTO SEQUENCE 7

P7-1 Torque the adjusting nut to 200 ft.-lb. to seat the bearing. Ensure that the wheel is rotated during torquing.

P7-2 Torque wrench specification should read 200 ft.-lb.

P7-3 Now back off the adjusting nut one full turn; this will leave the wheel assembly loose.

P7-4 Now torque the wheel adjusting nut to 50 ft.-lb. while rotating the wheel.

P7-5 To establish endplay, back off the wheel adjusting nut: The amount of rotation required to back off the nut will depend on the tpi (threads per inch). For a typical 12-tpi axle spindle, the adjusting nut should be backed off $1/4$ of a turn.

P7-6 Use a jam nut and locking plate on double nut systems. Torque the locking nut to specification. Generally, jam nuts 2 $5/8$ inch(hex diameter) or less are torqued to 200–300 ft.-lb. and those larger than 2 $5/8$ inch (hex diameter) are torqued to 250–400 ft.-lb. Check OEM specifications.

P7-7 Now verify that endplay exists. Install a dial indicator on the axle spindle and apply a rocking force to the wheel assembly. The endplay reading on the dial indicator must be between 0.001 inch and 0.005 inch. If not within specification, the complete procedure must be repeated.

> Remember, the consequences of not observing correct wheel-end procedures are wheel-off incidents that can kill.

TABLE 8-2: JAM NUT TORQUE		
Axle Type	Nut Size	Torque Specification
Steer (double nut)	Less than 2 5/8" 2 5/8" and over	200–300 ft.-lb. 300–400 ft.-lb.
Drive	Dowel type washer Tang type washer	300–400 ft.-lb. 200–275 ft.-lb.
Trailer	Less than 2 5/8" 2 5/8" and over	200–300 ft.-lb. 30–400 ft.-lb.

CAUTION *A truck wheel-off incident can have fatal consequences. When such an incident occurs at highway speeds, any other vehicle on the road is not the match for the weight of a bouncing wheel assembly. Wheels have been falling off trucks since trucks first traveled our highways. However, our roads are more congested today, and the consequences tend to be more severe. The media have rightly placed some focus on such incidents and forced the trucking industry to take action. The first thing that can be said about wheel-off incidents is that they are almost always related to a service practice error. In few cases is the cause sourced to equipment failure. For this reason, anybody working in the trucking industry must become aware of what is required to safely work on wheel ends.*

Figure 8-16 Dana Spicer UHS. *(Courtesy of Dana Spicer® UHS Unitized Hub)*

Preset Hub Assemblies

In response to wheel-off incidents, most OEMs have made preset (bearing) and unitized hub systems available. A preset hub is a preadjusted but serviceable assembly, where a unitized hub is preadjusted and nonserviceable. A couple of examples are the **ConMet PreSet hub** assembly and the Dana Spicer **Unitized Hub System** (UHS). These types of hub systems are projected to become adopted almost universally over the coming years, if for no other reason than they limit the liability fleets and service facilities have assumed over wheel-off incidents.

The ConMet PreSet hub uses a precision-dimensioned spacer between the tapered roller bearings in the axle hub that allows the assembly to be torqued to 300 ft.-lb. on installation, eliminating bearing adjustment by the technician. The ConMet PreSet hub is shown in **Figure 8-15.** They are available for truck steer axle, drive axle, and trailer axle hubs.

The Dana Spicer Unitized Hub System more recently introduced represents an improvement on the low maintenance system (LMS) for hubs that was similar to the ConMet PreSet hub. It unitizes the key hub components and lubricates them with synthetic grease for life. This eliminates the need for bearing adjustment, seal replacements, and even periodic lubrication top-ups. The USH hub is shown in **Figure 8-16.**

WHEEL END SUMMARY

Make sure you know exactly what you are working with when servicing wheel ends. If you are not sure, ask questions. Learn how to identify wheel ends that require bearing adjustments and the newer preset hubs. After having made a wheel bearing adjustment, make sure that you lock and set the adjusting nut with the appropriate locking/jam mechanism. **Figure 8-17** shows some examples of wheel-end bearing setting mechanisms.

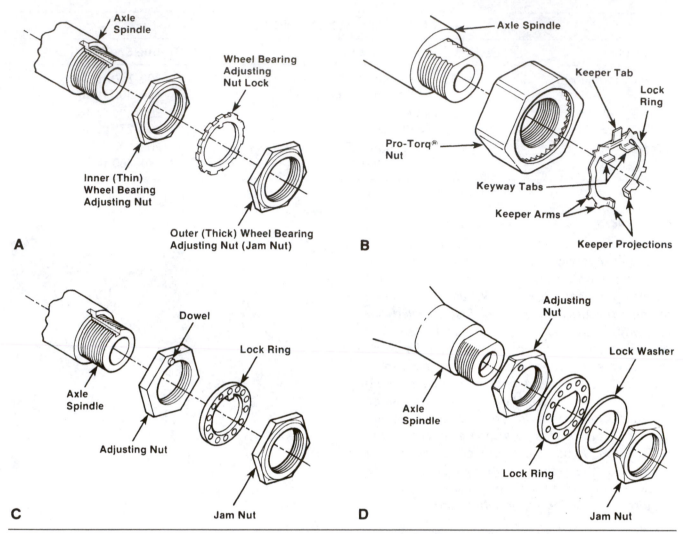

Figure 8-17 Bearing setting hardware: (A) jam nut and D-shaped lock ring, (B) Pro-Torq nut, (C) Eaton axle with adjusting nut and jam nut, and (D) Rockwell axle with adjusting nut and jam nut. *(Courtesy of Daimler Trucks North America)*

TIRES

The trucking industry generally uses one of two basic types of tire construction: **bias ply** and **radial tires.** It is important that the technician be able to distinguish the difference between the two because radial and bias ply tires should never be mixed on the same axle. Radial and bias ply tires differ in their tread profile, surface contact, and handling characteristics. Tire manufacturers recommend using one type of tire construction on all axles on a vehicle.

Dual wheels should never have mismatched tires. The tires on an axle should be of the same construction, tread pattern, and nominal size. Mismatched tires on opposite sides of the same axle can cause drive axle failure by continually working the differential. This can happen even if the bias ply and radial tires are the same size; the tire side walls don't flex at the same

rate, causing the radial diameter to become smaller under load and producing a different dynamic **footprint.** Footprint is the tread-to-road contact area of a tire at a given moment of operation. Radial tires tend to have a constantly changing footprint in operation, and they tend to bounce down the road. The action can cause cyclic overloading of bias ply tires if mixed with radial tires. If the vehicle has two or more drive axles, the tires on the drive axles should be either all bias ply or all radial.

Figure 8-18 shows a comparison of the static contact footprint of bias ply and radial tires.

Tire Size

The size of any type of tire can be obtained by the information printed on the sidewall of the tire. It is

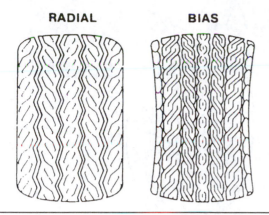

RADIAL   BIAS

Figure 8-18 Comparison of footprint between a radial and bias ply-type truck tire. *(Courtesy of Bridgestone Firestone Tire Sales Co.)*

important that the technician be able to compare the sidewall information to ensure that all the tires are the same size. See **Figure 8-19** for a breakdown on sidewall information.

Matching of Dual Wheels

Mismatched tire sizes for dual wheel assemblies has the same effects on tire wear as does running with one wheel overloaded or underinflated. An underinflated tire on a dual assembly shifts its share of the load to its mate, which then becomes overloaded and may fail prematurely. A difference of 15 psi inflation may result in the lesser inflated tire supporting 500 pounds less than the tire with the proper inflation. A similar action occurs when one tire's diameter is smaller that its mate's. A difference of 1/4 inch in diameter may result in the larger tire carrying 600 pounds more than the smaller. The shift in load becomes greater as the difference in diameter increases.

Improperly matched duals are subject to rapid tread wear because the larger tire carries more load and will wear fast and unevenly. Even though mismated duals have differing diameters, they must rotate at the same speed. This will cause the smaller tire to wear because the outside diameter of the tire travels a shorter distance,

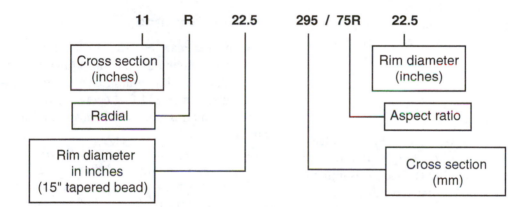

More recently the trend has been towards low profile tires. These are usually tubeless tires designed for either 22.5" or 24.5" diameter wheels. The most common low profile tires are listed below showing conventional sizes which they normally replace.

Low profile sizes

295/75R22.5 (275/80R22.5)
285/75R24.5 (275/80R24.5)

Conventional sizes

10.00R20,   11R22.5
10.00R22,   11R24.5

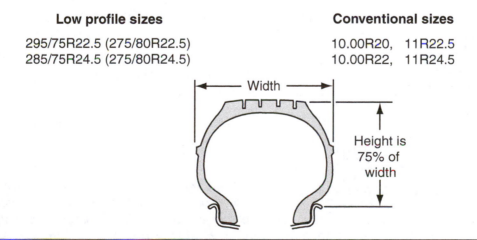

Figure 8-19 Breakdown of tire sidewall information.

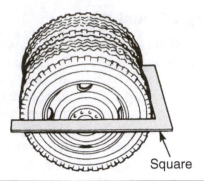

Figure 8-20 Using a square to check dual tire matching. *(Courtesy of Daimler Trucks North America)*

which will cause it to **scuff** (skid) over the road. The overall result is abnormal and unequal tread wear for both tires.

Checking Dual Mating

Check dual tire assemblies for proper mating with a **tire square.** This is the standard method of checking dual diameter matching on the vehicle (see **Figure 8-20**). The square must be placed parallel to the floor to avoid the tire "bulge." Measure the distance (if any) between the tire tread and the square arm with a ruler. It should not exceed 1/4 inch.

Straightedge

A straightedge can be placed across the four tires on a dual axle to compare tire diameters. Measurements should be taken from the straightedge to the tire tread where gaps show (see **Figure 8-21**). The measurement should be doubled to obtain the diameter difference. A taut string can be used in place of the straightedge.

Tape Measure

A flexible tape measure can be used to check the circumference of an unmounted tire (see **Figure 8-22**). Make sure the tape runs around the tread centerline of

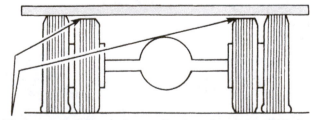

Measure distance between tire tread and straightedge

Figure 8-21 A straightedge positioned across the tires will detect difference in tire size. *(Courtesy of Daimler Trucks North America)*

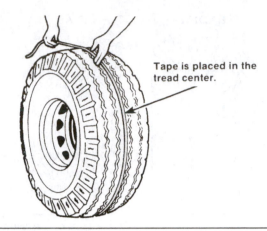

Tape is placed in the tread center.

Figure 8-22 Measuring tire circumference with a flexible tape measure. *(Courtesy of Daimler Trucks North America)*

the tire. A difference of 3/4 inch in circumference is normally acceptable in mated tires.

The smaller of the two tires should always be mounted on the inside wheel position of the dual tire assembly.

Tire Maintenance

The proper maintenance of tires is one of the easiest things to maintain but unfortunately one of the most neglected. Maintaining correct tire pressure not only ensures optimized fuel economy but also reduces tire wear, which is one of the most expensive areas to maintain.

Regular inspection of tires is the first step in increasing fuel economy and tire longevity. Regular inspection will help to identify underinflation, overinflation, and misalignment before they can cause damage. Minor damage that can be detected and repaired during an inspection can sometimes save a tire that would otherwise blow out.

TIRE PRESSURE CHECK

Tire pressure should be checked weekly, every time the vehicle enters the shop, or on every PM service. Many PM forms require you to record the pressure on every tire. If a tire on any axle is more than 20 percent below its rated pressure (80 psi on a 100 psi tire), it should be considered flat and be inspected, as in the following section of this chapter.

CAUTION *Never air up a flat tire during a PM inspection while the tire is on the vehicle. If it has been run flat, it could have sidewall damage that could cause the tire to explode when inflated.*

Adjust air pressure to the recommended level, keeping your body away from the tire's sidewall during inflation. Better still, use a remote inflation gauge with a foot valve.

CHECK VALVE STEMS AND CAPS

On the PM service, always check for damaged or inaccessible valves.

Valve stems should be aligned at 180 degrees (or as close as possible) on dual wheels.

INSPECT TREAD DEPTH

Tire tread depth should be measured and recorded on every PM service.

Measure tread depth in three places on each tire at equal intervals. Take each measurement in the major tread groove nearest the center of the tire. Do not measure on a wear bar. Record the smallest tread depth measurement that you find. If tread depth measures less than 2/32 inch (1.6 mm) on steer tires or 1/32 inch (0.8 mm) on any other tire, replace it.

Inspection Procedures for Tires Suspected of Having Been Run Underinflated or Overloaded

INSPECT DEFLATED SUSPECT TIRES MOUNTED ON THE RIM

Look for
- Cuts, snags, or chips exposing body cords or steel
- Distortions or a wavy appearance by using an indirect light to cause shadows left by any sidewall irregularities

Feel for
- Soft spots in the sidewall flex area
- Distortions or wavy appearance
- Protruding filaments indicating broken cords

Listen for
- Any popping sound when feeling for soft spots or when rolling the tire

If you identify any of the above conditions, the tire should be scrapped and taken out of service. If no other condition is present and a tire contains cuts, snags, or chips exposing body cords or steel, it must be referred to a full-service repair facility to determine if it is repairable and not a source of a potential **zipper**. A zipper rupture is a circumferential rupture in the mid sidewall of a steel cord radial tire. It can be caused by weakened steel cables in the tire's sidewall, caused by running underinflated or flat (defined as a tire that carries less than 80 percent of proper inflation). The resulting rupture and

air blast can explode with the force of as much as three-quarters of a pound of dynamite, leaving a 10- to 36-inch gash in the sidewall that looks like a zipper.

If none of these conditions are present, place the tire/rim/wheel assembly in an approved inflation safety cage.

> **CAUTION** *Remain outside of the tire's trajectory and do not place hands in safety cage while inspecting tire or place head close to safety cage.*

INSPECT SUSPECT TIRE INFLATED TO 20 PSI

Look for
- Distortions or a wavy appearance

Listen for
- Any popping sound

If any of these conditions are present, the tire should be made unusable and scrapped.

If none of these conditions are present, dismount the tire to visually and manually inspect it both inside and outside.

INSPECT SUSPECT TIRES AFTER DISMOUNTING

Look for
- Bead rubber torn to the fabric or steel
- Cuts, snags, or chips exposing body cords or steel
- Distortions or a wavy appearance using an indirect light source that produces shadows left by any sidewall irregularities
- Creasing, wrinkling, cracking, or possible discoloration of the inner liner
- Any other signs of weakness in the upper sidewall

If any of the above conditions are found, the tire should be scrapped. If no other condition is present and a tire contains cuts, snags, or chips exposing body cords or steel, it must be referred to a full-service repair facility, to determine if it is repairable and not a source of a potential zipper.

If none of these conditions are present, the tire may be returned to service.

Inspection Procedures for All Tires Returning to Service

These procedures include used, retread, or repaired tires, regardless of being suspect or not suspect of being run underinflated or overloaded.

INSPECT DISMOUNTED TIRES (INCLUDING USED, RETREADED, OR REPAIRED)

Look for

- Bead rubber torn to the fabric or steel
- Cuts, snags, or chips exposing body cords or steel
- Distortions or wavy appearance using an indirect light source that produces shadows left by any sidewall irregularities
- Creasing, wrinkling, cracking, or possible discoloration of the inner liner
- Any other signs of weakness in the upper sidewall

Feel for

- Soft spots in the sidewall flex area
- Distortions or a wavy appearance
- Protruding filaments indicating broken cords

Listen for

- Any popping sound when feeling for soft spots, or when rolling the tire

If any of the above-mentioned conditions are present, the tire should be scrapped. If no other condition is present and a tire contains tears cuts, snags, or chips exposing body cords or steel, it must be referred to a full-service repair facility, to determine if it is repairable and not a source of a potential zipper.

If none of these conditions are present, place the tire/rim/wheel assembly in an approved inflation safety cage.

CAUTION *Remain outside of the tire's trajectory, and do not place hands in safety cage while inspecting tire or place head close to safety cage. After properly seating the beads, with the valve core removed, adjust the tire to 20 psi using a clip-on chuck with a pressure regulator and extension air hose.*

INSPECTION OF MOUNTED TIRES INFLATED TO 20 PSI

Look for

- Distortions or a wavy appearance

Listen for

- Any popping sound

If any of the above-mentioned conditions are present the tire should be scrapped.

If none of these conditions are present, with valve core still removed, inflate the tire to 20 psi over the recommended operating pressure. During this step, if any of the above conditions appear, **immediately stop inflation.**

INSPECT MOUNTED TIRES INFLATED 20 PSI OVER OPERATING PRESSURE

Look for

- Distortions or a wavy appearance

Listen for

- Any popping sound

Any tire suspected of having been underinflated and/or overinflated must remain in the safety cage at 20 psi over operating pressure for 20 minutes.

If any of these conditions are present, the tire should be scrapped.

If none of these conditions are present, before removing the tire/rim/wheel assembly from the safety cage, reduce the inflation pressure to the recommended operating pressure. **emain outside of the tire's trajectory.**

Tire Wear Conditions and Causes

Irregular wear reduces the useful life of a tire and can lead to tire failure. Irregular tread wear is a symptom of a problem with a vehicle, a tire, or a tire maintenance practice.

Tire tread wears away because of friction with the pavement. Tread should wear down uniformly all the way around the circumference of the tire and all the way across the tread face. When this does not occur, the tire has irregular wear.

When you see a tire with irregular wear, you should find the underlying cause. If you only replace the tire, you have not solved the problem that caused the irregular wear in the first place. To solve the underlying problem, you must see the irregular wear when the tire is still on the vehicle.

Some irregular wear patterns look the same all the way around the tread of the tire. Other wear patterns are not consistent all the way around, but occur in various spots. The underlying causes of the two categories are different.

During routine maintenance, you can check for tire wear caused by misalignment by running your hand across the surface of the treads. This can be compared to running your hand over the surface of a cheese grater. If the treads have a sharp edge as you move your hand in one direction and not in the other, it is an indication that the tire is **scrubbing** (being

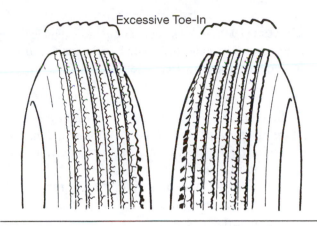

Figure 8-23 Excessive toe-in wear pattern. *(Courtesy of Mack Trucks, Inc.)*

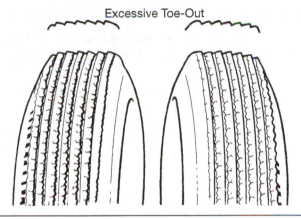

Figure 8-24 Excessive toe-out wear pattern. *(Courtesy of Mack Trucks, Inc.)*

dragged sideways) in the direction of the sharp edge (see **Figure 8-23** and **Figure 8-24**).

For more information along with pictures on tire wear conditions and causes, go to www.kaltire.com and select truck tires and wear conditions.

Tire Out-of-Service Criteria

(Taken from the North American Standard Out-of-Service Criteria by the Commercial Vehicle Safety Alliance, April 1, 2006)

ANY TIRE ON A FRONT STEERING AXLE(S) OF POWER UNIT

1. With less than 2/32 inch (1.6 mm) tread when measured in any two adjacent major tread groves at any location on the tire
2. When any part of the breaker strip or casing ply is showing in the tread
3. When sidewall is cut, worn, or damaged to the extent that the ply cored is exposed
4. Labeled "Not for Highway Use" or carrying other markings that would exclude use on steering axles

5. Visually observable bump, bulge, or knot apparently related to tread or sidewall separation

Exception: A bulge due to a section repair is allowed, not to exceed 3/8 inch (1 cm) in height. This bulge may sometimes be identified by a blue triangular label in the immediate vicinity.

6. Tire has noticeable (can be heard or felt) leak, or has 50 percent or less of the maximum inflation pressure marked on the tire sidewall.

Note: Measure tire pressure only if there is evidence the tire is underinflated.

7. So mounted or inflated that it comes in contact with any part of the vehicle
8. Front Steer Axle(s): Weight carried exceeds tire load limit. This includes overloaded tire resulting from low air pressure.
9. Passenger Carrying Vehicle: Regrooved, recapped, or retreaded tires on front steering axles

ALL TIRES OTHER THAN THOSE FOUND ON THE FRONT STEERING AXLE(S) OF A POWER UNIT

1. Tire has noticeable (can be heard or felt) leak, or has 50 percent or less of the maximum inflation pressure marked on the tire sidewall.

Note: Measure tire air pressure only if there is evidence the tire is underinflated.

2. Bias Ply Tire: When more than one ply is exposed in the tread area or sidewall or when the exposed area of the top ply exceeds 2 square inches (13 sq. cm)
3. Radial Ply Tire: When two or more plies are exposed in the tread area or damaged cords are evident in the sidewall or when the exposed area exceeds 2 square inches (13 sq. cm) in the sidewall
4. Any tire with visually observable bump or knot apparently related to tread or sidewall separation

Exception: A bulge due to a section repair is allowed not to exceed 3/8" (1 cm) in height. The bulge may sometimes be identified by a blue triangular label in the immediate vicinity.

5. So mounted or inflated that it comes in contact with any part of the vehicle. (This includes any tire contacting its mate in a dual set.)
6. Weight carried exceeds tire load limit. This includes overloaded tire resulting from low air pressure.

Exception: Does not apply to vehicles being operated under the special exclusion found in Federal Motor Carrier Safety Regulation.

7. So worn that less than 1/32 inch (0.8 mm) tread remains when measured in any two adjacent major tread grooves at three separate locations on the tire.
8. Seventy-five percent or more of the tread width loose or missing in excess of 12 inches (30 cm) in circumference

Summary

- Wheel assemblies can be broken into wheels, rims, tires, and hubs.
- Trucks use one of two types of wheel configurations: cast spoke wheels or disc wheels.
- Disc wheels can also be divided into stud piloted and hub piloted.
- Cast spoke wheels are tough and can withstand more punishment than disc wheels.
- When mounting the tire rim assembly onto the wheel, ensure that you have the correct studs, nuts, and clamps.
- Whenever a spoke wheel is reinstalled, it must be checked for runout.
- Disc wheels differ from spoke wheels in that the tire rim forms the wheel and mounts to the hub assembly.
- After operating the vehicle approximately 50 miles (80 km), the wheel should be retorqued.
- With stud-piloted wheels, the weight of the vehicle is transferred through to the wheel by the mounting studs.
- With hub-piloted wheels, the weight of the vehicle is transferred through to the wheel by the pilot pads. The nuts and studs just keep the wheel clamped to the hub.

- Wide-base wheels can also be referred to as high flotation, super singles, wide body, duplex, or jumbo wheel.
- Disc wheels are mounted to the wheel hub with threaded studs and nuts or headed studs.
- Wheel bearing adjustment is a simple but highly critical procedure. Always follow the TMC-recommended procedures.
- A preset hub is a preadjusted but serviceable assembly, where a unitized hub is preadjusted and nonserviceable.
- The trucking industry generally uses one of two basic types of tire construction: bias ply and radial tires.
- The size and type of tire can be obtained by the information printed on the sidewall of the tire.
- Mismatching the tire sizes in dual wheel assemblies has the same effects on tire wear as does running with one wheel overloaded or underinflated.
- Tire pressure should be checked weekly, every time the vehicle enters the shop, or on every PM service.
- If a tire on any axle is more than 20 percent below its rated pressure, it should be considered flat.
- Tire tread depth should be measured and recorded on every PM service.

Review Questions

1. The presence of rust on a wheel could indicate:

   A. cracks.

   B. poor mating of inner and outer hub-piloted wheels.

   C. Loose lug nuts.

   D. All of the above

2. Which of the following statements is true?

   A. Dual tires should be matched by tread design and, ideally, tire casings should be matched by manufacturer.

   B. The smaller tire in a dual position should be on the outside of the vehicle.

   C. Steer tires can be of different design as long as the manufacturer is the same.

   D. None of the above

3. According to the TMC a steer with a measured tread depth of _____ or less must be replaced.

   A. 1/4 inch (6.4 mm)

   B. 2/32 inch (1.6 mm)

   C. 4/32 inch (3.2 mm)

   D. 6/32 inch (4.8 mm)

4. Which of the following wheel types uses tires mounted to rims that mount to the wheel using clamps?

   A. Disc wheels

   B. Cast spoke wheels

   C. Hub-piloted

   D. Stud-piloted

5. If a set of dual tires were properly matched, which of the following would be true?

   A. Both tires would have the same nominal size.

   B. Both tires would have the same tread design.

   C. Both tires would have equal wear.

   D. All of the above

6. What is the maximum allowable runout permitted on cast spoke wheels?

   A. 1/16 inch

   B. 1/8 inch

   C. 3/16 inch

   D. 1/4 inch

7. Which type of wheel assembly presents the fewest maintenance and alignment problems?

   A. Stud piloted disc wheels

   B. Cast spoke wheels

   C. Hub-piloted wheels

   D. Any wheels using radial clamps

8. Which of the following methods are used to lock the bearing setting on the axle or spindle?

   A. Jam nuts

   B. Castellated nuts with cotter pins

   C. Split forging nuts

   D. All of the above

9. Which is the preferred method of repairing a wheel seat leak that has glazed the brake shoe linings?

   A. Replace the wheel seal only.

   B. Replace the wheel seal and bearings.

   C. Replace the wheel seal and damaged brake shoes.

   D. Replace the wheel seal and brake shoes on both sides of the axle.

10. Technician A says that an advantage of disc wheels over cast spoke is that they produce fewer wheel alignment problems. Technician B says that, in stud-piloted disc wheel systems, a loose inner nut can easily go undetected, eventually pounding out the ball seat. Who is correct?

    A. Technician A only

    B. Technician B only

    C. Both technician A and B are correct.

    D. Neither technician A nor B is correct.

11. Technician A always inflates severely underinflated tires in a safety cage. Technician B uses a clip-on air chuck with a remote in-line valve and gauge. Who is correct?

    A. Technician A   C. Both technician A and B are correct.

    B. Technician b   D. Neither technician A nor B is correct.

12. Technician A says that wheel runout on a cast spoke wheel can be eliminated only by removing the wheel rims and breaking down the tires. Technician B adjusts cast spoke wheel runout by chalk-marking high points, then loosening the lug nuts on the chalk-marked side and tightening those 180 degrees opposite. Who is correct?

    A. Technician A   C. Both technician A and B are correct.

    B. Technician B   D. Neither technician A nor B is correct.

9 Braking System Inspection

Objectives

Upon completion and review of this chapter, the student should be able to:

- Explain the importance of changing brake fluid at manufacturer's specified time.
- Describe how to change brake fluid in a hydraulic braking system.
- Explain how to inspect and service a typical master cylinder.
- Describe how to inspect and service drum brakes.
- Explain how to inspect and check rotor runout.
- Describe the inspection process for brake lines and hoses.
- Explain how to inspect brake linings and pads and take accurate measurements of pad and lining thickness to determine serviceability.
- List and explain three different methods for bleeding a hydraulic brake system.
- Explain how to test the operation of the parking brakes.
- Describe how to test a truck's service brakes.
- List some of the out-of-service criteria for hydraulic braking systems.
- Explain some of the safety precautions when working with air brake systems.
- Describe the procedures to service a truck's air supply system.
- Explain service and inspection procedures for a typical air dryer.
- Explain how the air dryer operates and how to perform a leakage test.
- List and explain in sequential order the procedures for testing the components of a typical air brake system.
- Describe how to perform a check on the manual parking brakes or emergency parking brakes.
- Explain the inspection and testing procedure for checking the foundation brakes.
- List some of the out-of-service criteria for air brake systems.

Key Terms

anti-lock braking system (ABS)	brake fade	governor cutout pressure
applied stroke	freestroke	hygroscopic
bleeding	governor cut-in pressure	inboard

loaded   primary reservoir   supply reservoir (wet tank)

outboard   secondary reservoir

preload

INTRODUCTION

Along with the vehicle's steering system, the braking system is the most important maintenance issue to be addressed by the technician to keep the truck operating safely on the road. Work on braking systems should only be performed by a qualified technician or under the supervision of a qualified technician. All checks and repairs to the vehicle's braking system should be performed at specific intervals in strict accordance with the vehicle manufacturer's procedures.

As a new technician, never be afraid to ask questions of more experienced and qualified technicians. Any misdiagnosis or improperly repaired brake component can mean potential disaster for the vehicle and all those who share the road with it.

SYSTEM OVERVIEW

This chapter starts by discussing the importance of replacing the hydraulic brake fluid on a regular basis in order to maintain an efficient hydraulic braking system and also explains how to change the fluid. From there it discusses servicing procedures for master cylinders and bleeding hydraulic braking systems, as well as service and inspection of brake drums, rotors pads, and linings and test procedures for both parking and service brakes. The section on hydraulic brakes finishes with listing some of the out-of-service criteria for hydraulic brakes.

The section on air brakes starts with air supply system, including compressor service and maintenance, air reservoir maintenance, governor checks, and air dryer testing and maintenance. It then guides you through a step-by-step procedure for testing the brake components and manual and emergency parking brakes tests. The chapter finishes with checks for the foundation brake system and out-of-service criteria for air brake systems.

HYDRAULIC BRAKE SYSTEMS

When servicing hydraulic brake systems, always follow the manufacturer's procedures. There are differences in maintenance and repair practices between manufacturers. For this reason it is good practice to consult the manufacturer's service literature when performing repairs. This section will explain some general maintenance and repairs commonly performed on hydraulic brake systems.

Heavy-Duty Brake Fluid

When it comes to brake fluid for trucks, it is important to use the manufacturer's specific fluid to ensure proper operation of the braking system. The fluid is designed to retain a proper consistency at all operating temperatures. It is also designed not to damage any of the internal rubber components or seals, and to protect the metal parts of the brake system against corrosion and failure.

Never mix incompatible brake fluids. Use of the wrong brake fluid can damage the cup seals of the wheel cylinder and can result in brake failure.

Never leave the cap off a brake fluid container any longer than necessary to pour out fluid and seal the container again. Brake fluid is **hygroscopic** (absorbs moisture from the air) and will become saturated. Once brake fluid is opened, it has a short shelf life, and it is good practice not to mix brake fluids of the same type if it is not known how old they are. Different types of brake fluid should absolutely never be mixed.

CAUTION *Never use brake fluid from a container that has been used to store any other liquid. Mineral oil, alcohol, antifreeze, cleaning solvents, and water, in trace quantities, can contaminate brake fluid. Contaminated brake fluid will cause pistons cups and valves in the master cylinder to swell and deteriorate. One way of checking brake fluid is to place a small quantity of brake fluid drained from the system in a clear glass jar. Separation of the fluid into visible layers is an indication of contaminated or mixed types of brake fluid. It is generally regarded as good practice to discard used brake fluid that has been bled from the system. Contaminated fluid usually appears darker. Brake fluid drained from the bleeding operation may contain dirt particles or other contamination and should not be reused.*

Changing Brake Fluid

Brake fluid should be changed regularly according to the vehicle manufacturer's specific maintenance schedules. Due to the fact that brake fluid is hygroscopic, over time the boiling point of the fluid will be reduced. Brake fluid should also be changed whenever a major brake repair is performed. The system can be flushed with clean brake fluid, isopropyl alcohol, or rubbing alcohol. A simple flushing technique is to pour the flushing agent into the master cylinder reservoir and open all bleed screws in the system. The brake pedal is then pumped to force the flushing agent through the system. Ensure that containers are located at each bleed screw nipple to catch the fluid. Replace all the rubber components (seals, cups, hoses, and so on) in the system before adding fresh brake fluid to a flushed, clean system.

Bleeding Brakes

Bleeding of the brakes is an essential component of servicing hydraulic brakes. The process involves purging air from the hydraulic circuit. Bleeding of the brakes may be performed manually, in which case no special equipment is required, or by using a pressure bleeding process. Both methods are outlined here.

MANUAL BLEEDING

Two people or one person and a one-person bleeder kit are required to perform this task. Each wheel cylinder must be bled separately. In most cases, the wheel cylinder farthest from the master cylinder is bled first. This means that the right rear wheel cylinder would be first and then the left rear. After the rear brakes are bled, the right front brake and then the left front should follow. Ensure that the master cylinder reservoir is kept topped up with brake fluid during the bleeding procedure. The flowing procedure is used:

1. Loosen the master cylinder line nut to the rear brake circuit one complete turn.
2. Push the brake pedal down slowly by hand to the floor of the cab. This will force air trapped in the master cylinder to escape at the fitting.
3. Hold the pedal down and tighten the fitting. Release the brake pedal.
4. Repeat this procedure until bubble-free fluid exits at the fitting, and then tighten the line fitting.
5. To bleed a wheel cylinder, loosen the bleed fitting on the wheel cylinder or the caliper with a line or box-end wrench. Attach a clear plastic or rubber drain hose to the bleed fitting, ensuring the end of the tube fits snugly around the fitting.
6. Submerge the free end of the tube in a container partially filled with brake fluid, as shown in **Figure 9-1.** Loosen the bleed fitting three-quarters of a turn.

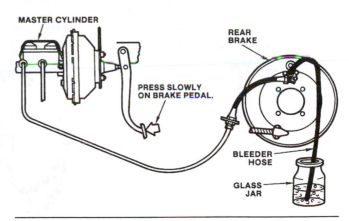

Figure 9-1 Manual brake bleeding technique.

7. Push the brake pedal all the way down and hold. Close the bleed fitting and return the pedal to the fully released position.
8. When the fluid exits the bleed tube completely free of air bubbles, close the bleeder fitting and remove the tube.
9. Repeat this procedure at the wheel cylinder or caliper on all the remaining wheels, keeping in mind to start at the farthest wheel from the master cylinder and work toward it.
10. Keep the master cylinder reservoir topped up throughout the bleeding process.
11. When the bleeding process is complete, top up the master cylinder with brake fluid to the appropriate level as indicated on the reservoir.

PRESSURE BLEEDING BRAKES

Pressure bleeding equipment must be of the diaphragm type to prevent air, moisture, oil, and other contaminants from entering the hydraulic system. The adapter must limit pressure to the reservoir to 35 psi (241 kPa).

1. Clean all dirt from the top of the master cylinder and remove the reservoir caps.
2. Install appropriate master cylinder bleeder adapter. A typical pressure bleeder is shown in **Figure 9-2.** Connect the hoses from the bleeder equipment to the bleeder adapter and open the release valve on the bleeder equipment.
3. Using the correct size box-end or line wrench over the bleeder screw, attach a clear plastic or rubber hose over the screw nipple. Insert the other end of the hose into a glass jar containing enough fluid to cover the end of the hose.
4. Bleed the wheel cylinder or caliper farthest from the master cylinder first, and then work your way back through the remaining cylinders or calipers to the one closest to the master.

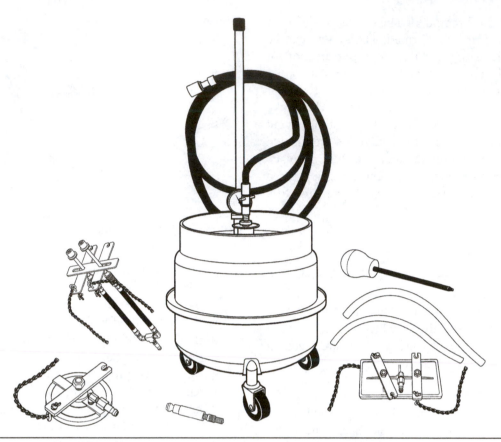

Figure 9-2 A typical pressure bleeder. *(Courtesy of Wagner Brake, div. of Federal-Mogul Corp.)*

5. Open the bleeder screw and observe the fluid flow at the end of the hose.

6. Close the bleeder screw as soon as the bubbles stop exiting the hose and the fluid flows in a solid stream.

7. Remove the hose and wrench from the bleeder screw.

8. Fill the master cylinder to the correct level in the reservoir.

BLEEDING AN AIR-OVER HYDRAULIC SYSTEM

To bleed a typical air-over hydraulic system, follow this procedure:

1. Adjust the air pressure in the air reservoir to 28–43 psi (193–296 kPa). If the air pressure drops below this range during the bleeding operation, start the engine.

2. When the air pressure is 28–43 psi (193–296 kPa), stop the engine.

3. Bleed the air from each air bleeder screw in the following order:

   - Front brakes and air booster hydraulic cylinder plug
   - Right front wheel air bleeder screw
   - Left front wheel air bleeder screw
   - Rear brakes and air booster hydraulic cylinder plug
   - Right rear wheel air bleeder screw
   - Left rear wheel air bleeder screw

4. Fill the brake reservoir with fluid. Refill frequently during bleeding so no air enters the brake lines.

5. Attach a clear plastic tube to the bleeder screw. Insert the other end of the tube into a bottle partially filled with brake fluid.

6. Depress the brake pedal and at the same time turn the bleeder screw open.

7. Tighten the bleeder screw when the pedal reaches the full stroke. Release the pedal.

8. Repeat until no bubbles are observed.

Master Cylinder Inspection

- Check the fluid level of the master cylinder. The brake fluid should be at the halfway mark, or above a minimum level. If not, make note to follow up.
- Brake fluid should look, smell, and feel like new fluid. If it does not, make note to follow up.

Note: Dirty brake fluid can have an adverse affect on **anti-lock braking system (ABS)** operation. If the fluid has a scum-like appearance, do not be too concerned, because this is probably a result of silicone on the master cylinder at the time of assembly that won't mix with the brake fluid.

■ Keep in mind that the cleanest brake fluid will be in the master cylinder and the dirtiest will be at the wheel ends.
■ Inspect the master cylinder for any external leaks both inside the cab and in the engine compartment.

Servicing a Master Cylinder

A master cylinder should be removed from the truck in the following sequence in order to service it:

1. Disconnect the negative battery cable.
2. Disconnect the pressure differential (brake light warning) switch.
3. Disconnect and cap the hydraulic lines from the master cylinder to prevent dirt from getting in them. (On some trucks using remote reservoirs, it may be necessary to disconnect the lines connecting the reservoirs with the master cylinder.)
4. Remove the master cylinder from the brake booster assembly.

Clean the master cylinder and any other parts to be reused in clean alcohol. Petroleum-based products should not be used to clean the components. Inspect all the disassembled components to determine whether they can be reused or not. Master cylinder bodies can be manufactured from either cast iron or aluminum. The master cylinder should be replaced if the pitting and scoring cannot be cleaned up.

Note: Rebuilding of master cylinders should be done only be qualified technicians due to their complexity and importance from a safety point of view.

Servicing Brake Drums

When a brake drum is removed to perform a check on the brake shoe linings or wheel cylinder, make a habit of evaluating the other components in the braking system. Before inspecting drum brakes, release the parking brake. Evidence of grease or oil at the center of the brake assembly indicates that the axle seal needs replacing. If brake fluid drips out of the wheel cylinders when the rubber dust boots are removed, the wheel cylinder needs to be replaced. Check for any cracked or corroded brake lines.

Drum Removal and Inspection

Brake drums may be segregated into two groups, **inboard** and **outboard.** Outboard brake drums are easily replaced without removing the hubs of the vehicle. Inboard brake drums require the removal of the entire hub assembly in order to remove the bolts securing the brake drum to the wheel assembly. To remove an outboard brake drum, all that needs to be done is to remove the nuts holding the wheels on, as they also secure the drums in place. If it is determined that the drum is to be replaced, the new drum must be washed to remove the oily residue from the manufacturing process. Always use a light to make this inspection; many problems are missed due to inadequate lighting conditions. Remove any dust shields as necessary to establish the condition of the brake drum.

To inspect a brake drum, proceed as follows:

1. Visually check the drum for cracks, scoring, pitting, or grooves.
2. Check the edge of the drum for chipping and fractures.
3. Drums that are blue, glazed, or heat-checked have been overheated and should not be reused; severe heat-checking in drums will cause rapid lining wear, and even when machined, the drums will wear prematurely.

Note: Severe overheating is caused by drums that are machined too thin, improper lining to drum contact, incorrect lining friction ratings, or vehicle overloading. A probable driver complaint would be **brake fade.** Brake fade is the reduction in stopping power that can occur after repeated application of the brakes, especially in high-load or high-speed conditions.

4. If the brake linings appear to be unevenly worn, check the drum for a barrel-shaped or taper condition. (see **Figure 9-3**). Consult the OEM specifications for the maximum drum taper specifications. Also, check the shoe heel-to-toe dimensions to ensure that the arc is not spread.

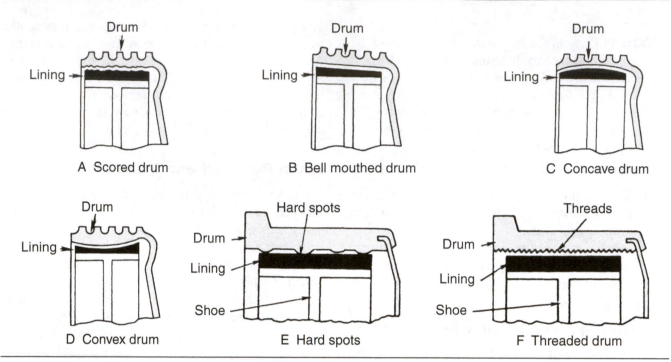

A Scored drum   B Bell mouthed drum   C Concave drum

D Convex drum   E Hard spots   F Threaded drum

Figure 9-3 Brake drum wear conditions.

5. Check for drum out-of-round condition using an inside micrometer or drum gauge. Measure the drum at inboard and outboard locations on the machined surface at four locations. Compare to the OEM specifications. An out-of-round condition produces a pulsating brake pedal condition.
6. Clean the inside of the drum with a water-dampened cloth. New drums may be corrosion-protected with a light layer of grease, which must be removed with solvent from the friction surface of the drum before installation.

CAUTION *Never paint the outside of the drum. Paint insulates the drum, preventing the dissipation of heat. This heat must be dissipated to the atmosphere around the drum to prevent brake fade.*

Installation of the brake drums is accomplished by reversing the order of removal. If reusing brake drums, it is good practice to install them in their original locations. Torque the fasteners to the specifications recommended in the service manual.

Disc Brake Rotor Inspection

To inspect the rotors, it might be necessary to remove the hub, rotor, or dust shield assembly. This will depend on the configurations used by the OEM. It also may be necessary to separate the rotor from the hub. Always use a light to make this inspection; many problems are missed due to inadequate lighting conditions. The following checks are required:

1. Visually inspect the rotor surface for scored, cracks, and heat-checking. This check is critical and will determine whether those steps that follow are required. Reject any rotor with evidence of heat-checking.
2. Next, check the rotor or hub and rotor assembly for lateral runout and thickness variation. To check lateral runout:

   ■ Mount a dial indicator on the steering arm or anchor plate with the indicator plunger contacting the rotor 1 inch (25 mm) from the edge of the rotor.
   ■ Adjust the wheel bearings so they are just loose enough for the wheel to turn; that is, set them with a little **preload.**
   ■ Check the lateral runout of both sides of the rotor, zero the dial indicator, and measure the total indicated runout (TIR).
   ■ The lateral runout TIR is always a close-fitting dimension, seldom exceeding 0.015 inch (0.38 mm).

3. If the lateral runout exceeds the specification, the rotor will have to be either machined or replaced.
4. Thickness variations also should be measured at 12 equally distant points with a micrometer at about 1 inch (25 mm) from the edge of the rotor.

5. If the thickness measurements vary by more than the specified maximum, usually around 0.002 inch (0.0508 mm), the rotor should be machined or replaced. This measurement can be made using two dial indicators or a micrometer. If using dial indicators, position them opposite each other and zero the dials. Rotate the rotor and observe the indicator readings for the proper tolerance. Some light scoring or wear is acceptable. If cracks are evident, the hub and rotor assembly should be replaced.

6. With the wheel bearings adjusted to zero end play, check the radial runout. Rotor radial runout should not exceed the OEM TIR specification, usually around 0.030 inch (0.762 mm). Set a dial indicator on the outer edge of the rotor and turn through 360 degrees.

7. Next, inspect the wheel bearings and adjust them using the OEM-specified procedure.

8. Wear ridges on rotors can cause temporary improper lining contact if the ridges are not removed, so it makes sense to skim rotors at each brake job.

Shop Talk: Excessive rotor runout or wobble increases pedal travel because of opening up the caliper piston and can cause pedal pulsation and chatter.

Most rotors fail because of excessive heat. Dishing and warpage are caused by extreme heat. These conditions almost always require the replacement of the rotor. Excessive rotor radial runout can cause noise from caliper housing-to-rotor contact.

Brake rotors have a minimum allowable thickness dimension cast on an unmachined surface of the rotor. This is the replacement thickness. It is illegal to machine a rotor that will not meet the minimum thickness specifications after refinishing. In other words if a rotor will not "true up" during skimming before the minimum allowable thickness is obtained, it must be replaced.

Brake Lines and Hoses

When performing the service inspection, visually inspect the condition of the brake lines and hoses. Replace any brake hose that is seeping or swells when the brake pedal is depressed or if the hose is abraded through the outer cover-to-fabric layer. When brake hoses have to be disconnected, ensure that the lock clips are removed and that only properly fitting line wrenches are used to separate the hose nuts, as shown

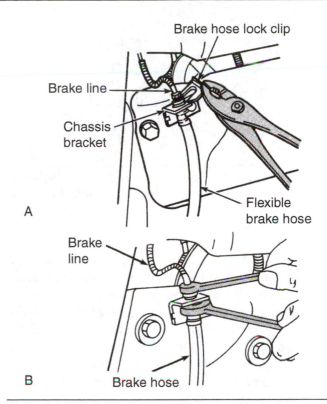

Figure 9-4 Brake hose replacement. (A) Remove the lock clip from the chassis bracket; (B) disconnect the female end first.

in **Figure 9-4.** Steel brake lines are usually double flare or ISO flare, as shown in **Figure 9-5.** Note the different seats and the fact that they cannot be interchanged. **Figure 9-6** shows how the double flare is made using a double-flare anvil and cone.

Brake Lining/Pad Inspection

Inspect the brake linings or pads. Make note of lining cracks or voids as well as sections of the lining that may be missing. Make sure the lining segment is

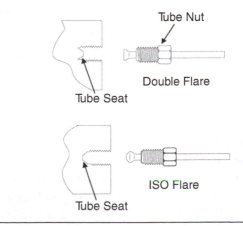

Figure 9-5 Two common types of line flares and their seats.

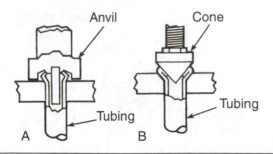

Figure 9-6 Making a double flare brake line. (A) Anvil folds tubing; (B) cone performs second fold and doubles seat thickness.

securely fastened. Check for evidence of oil seepage into or out of the brake lining/drum interface area, indicating an oil leak at the wheel seal.

Using a brake lining wear gauge, measure the thickness of the lining or pad and record. These tools are calibrated in 32nds of an inch. The minimum brake pad/ lining thickness for vehicles with hydraulic brakes is 1/16 inch (1.6 mm), but most companies will replace them before they wear to this point.

Shop Talk: Grease on the lining edge, back of shoe, or drum edge and oil stains with no evidence of fresh oil leakage are not conditions for out-of-service.

Measuring Brake Lining/Pad Thickness

Most brake lining measurements can be obtained without removing wheels/hubs, but with some vehicles it may be necessary to remove the wheels to obtain the brake lining measurements. The thickness of the brake lining should be measured at its thinnest point.

DISC BRAKES

- After removing the wheel and without removing pads from the caliper, insert the tips of the wear gauge between the edge of the disc brake rotor and the steel plate on the pad.
- Read and record the measurement.

DRUM BRAKES

- Check the thickness of the lining at the center of the shoe. Rest one of the tips of the wear gauge tool at the point where the lining and shoe are joined. Expand the tips until the other end meets the edge of the lining.
- Read and record the measurement.

Parking Brake Test

The parking brake is another part of the hydraulic brake system that must be checked on a regular basis, including any service. To test the parking brake on a medium size truck, first apply it and check its holding power by trying to move the vehicle with no more than about a 1/4 throttle. Ensure that the warning light on the dash is illuminated, indicating that the parking brake is applied. Release the parking brake and make sure the vehicle can be moved freely (parking brake not dragging). The warning light on the dash must be off. On vehicles with hydraulic park control, turn the ignition to the run position and open the driver's door with the park brake released. The "Apply Brake" warning lamp must illuminate.

To test the parking brake on a heavy duty truck, make sure the parking brake is set, and place the transmission in second gear. Try to move the vehicle with no more that 1/4 throttle. Ensure that when the park brake is released, the vehicle moves freely.

Service Brake Operation

The service brakes should be tested when the vehicle is being driven before being brought into the shop. Drive the vehicle at approximately 5 mph (8 kph) and apply the service brakes. Listen for any unusual noises (grinding or squealing). As the brakes apply, the steering wheel should track straight ahead and not pull to the left or right. Note any problems on your inspection sheet.

Testing Service Brakes (Vacuum Assist)

- For this test, shut off the vehicle and make several brake applications to deplete all vacuum in the system.
- Apply the service brakes with firm pressure and start the engine. The brake pedal should fall farther toward the floor as the engine builds vacuum.
- Release your foot from the brake pedal and then reapply. If the brake pedal travels more than half of its total possible stroke, make note to inspect further.
- Hold the pedal down with the engine running and check for a firm feel, with no pedal travel.

Testing Service Brakes (Hydraulic Assist)

- Shut off the vehicle engine and make several brake applications.

- Apply the service brakes with firm pressure and start the engine. The brake pedal must fall farther as the engine starts. An initial "kickback" will be felt from the pedal.
- Release your foot from the brake pedal and then reapply the brake. If the brake pedal travels more than half of its total possible stroke, make a note to inspect further.
- Hold the pedal down with the engine running and check for a firm feel, with no pedal travel.

Testing Service Brakes (Electro-Hydraulic Assist)

- With the engine shut off, make a brake application. The pedal should drop and you should also hear an electric pump. Remove your foot from the brake pedal, and all pumps, lights, and buzzers must stop operating.
- Position the ignition key in the run position and apply the foot brake. The pedal should drop, you should hear an electric pump, and the dash warning light will be on.
- With the brake pedal held down, start the vehicle engine. The brake pedal should drop slightly. If not, shut off engine and re-test with more foot pedal pressure. After the engine has started, all dash lights and buzzers must be off.
- Release your foot from the brake pedal and then reapply the brake. If the brake pedal travels more than half of its total possible stroke, make a note to inspect further.
- Hold the pedal down with the engine running and check for a firm feel, with no pedal travel.

Defective Brakes

The number of defective brakes is equal to or greater than 20 percent of the service brakes on the vehicle or combination. A defective brake includes any of the following out-of-service criteria. See defective brake chart later in this chapter.

OUT-OF-SERVICE CRITERIA FOR HYDRAULIC BRAKES

(Taken from the North American Standard Out-of-Service Criteria by the Commercial Vehicle Safety Alliance, April, 1, 2006)

(Including: Power Assist over Hydraulic and Engine Driven Hydraulic Booster)

1. No pedal reserve with engine running
2. Master cylinder less the 1/4 full

3. Power assist unit fails to operate
4. Seeping or swelling brake hose(s) under application of pressure
5. Missing or inoperable breakaway braking device
6. Hydraulic hose(s) abraded (chafed) through outer cover-to-fabric layer
7. Fluid lines or connections restricted, crimped, cracked, or broken
8. Hydraulic system: Brake failure light/low fluid warning light on and/or inoperative
9. Absence of effective braking action upon application of the service brakes (such as brake linings failing to move or contact braking surface upon application)
10. Missing or broken mechanical components including shoes, linings, pads, springs, anchor pin, or spider
11. Brake linings or pads (except on power unit steering axles:)
    - Cracked, loose, or missing lining
    - Lining cracks or voids of 1/16 inch (1.6 mm) in width observable on the edge of the lining
    - Portions of a lining segment missing such that a fastening device (rivet or bolt) is exposed when viewing the lining from the edge
    - Cracks that exceed 1-1/2 inch (38 mm) in length
    - Loose lining segments: Approximately 1/16 inch (1.6 mm) or more movement
    - Complete lining segment missing
12. Evidence of oil seepage into or out of the brake lining/drum interface area. This must include wet contamination of the lining edge accompanied by evidence that further contamination will occur—such as oil running from the drum or a bearing seal.
13. Hydraulic and electric brakes: Lining with a thickness 1/16 inch (1.6 mm) or less at the shoe center for disc or drum brakes
14. Missing brake on any axle required to have brakes
15. 20 percent or more of the service brakes are defective.

OUT-OF-SERVICE CRITERIA FOR VACUUM SYSTEM

(Taken from the North American Standard Out-of-Service Criteria by the Commercial Vehicle Safety Alliance, April, 1, 2006)

1. Insufficient vacuum reserve to permit one full brake application after engine is shut off

2. Vacuum hose(s) or line(s) restricted, abraded (chafed) through outer cover-to-cord ply, crimped, cracked, broken, or has collapse of vacuum hose(s) when vacuum is applied

AIR BRAKE SYSTEMS

Air Brake Safety

Technicians working on air brake system must keep in mind that compressed air can be dangerous if not handled with caution. The potential energy of compresses air can be compared to the potential energy in a compressed coil spring. If the force is released improperly, it can cause serious injury.

While working on an air brake system, make sure the wheels are blocked to prevent the vehicle from moving. This is done with the use of wheel chocks or wedges placed on both sides of a wheel, not on a steering axle. Many tasks performed on an air brake system will require the air to be drained from one or more tanks. Always wear eye protection when draining an air tank and point the discharged air away from yourself and other people. Never disconnect a hose under pressure, as it can whip around as air escapes and cause an injury.

To avoid injury, always follow OEM-recommended procedures when working on any air device that when released might be subject to mechanical (spring) or pneumatic force.

Air Supply Circuit Service

The supply circuit of an air brake system includes all the components that compress the air, manage the cycling of the compressor, regulate the system air pressures, remove contaminants from the compressed air, and store it in the **supply reservoir (wet tank).**

Air Compressor Service

Part of a scheduled maintenance program should include service to the air compressor to ensure its proper operation and to extend it service life. The OEM maintenance manual should be referred to for specific maintenance instructions. The following would be typical:

- Every 5,000 miles (8,000 km) or monthly, service, clean, and replace (as necessary) the air cleaner filter elements. A typical air induction system for compressor is shown in **Figure 9-7.**
- Every 25,000 miles (40,000 km) or every 3 months, perform governor cutout, low pressure warning, and pressure buildup test outlined

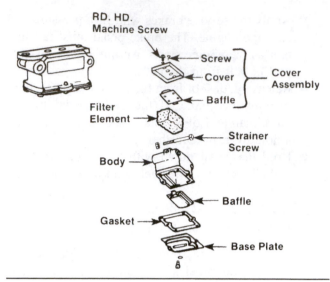

Figure 9-7 Typical air induction system for an air compressor.

later in this chapter to verify that the system meets FMVSS 121 requirements.
- Every 50,000 miles (80,000 km) or every 6 months, inspect the compressor discharge port, inlet, and discharge lines for restrictions, oil, and carboning.
- Check external oil supply (if applicable) and return lines for kinks and flow restrictions.
- Check for noisy compressor operation.
- Check pulley and belt (if equipped) alignment and tension.
- Check compressor mounting bolts and retorque if necessary.

The compressor must have an unrestricted supply of clean filtered air. Part of the preventive maintenance program will be to check the compressor induction system. More frequent maintenance will be required when the vehicle is operated in dusty or dirty environments.

Draining Air Tanks

Draining the contaminants from the air tanks is a task that must be performed at every level of service. Drivers are instructed to drain the air tanks daily. The contaminants in air brake reservoirs consist of water condensed from the air during the compression process as well as a small amount of oil from the compressor. The water and oil normally pass into the air tank in the form of vapor because of the heat generated during compression. There is probably no simpler yet important maintenance task than reservoir draining. If water is allowed to collect in any air tank, the storage capacity of the tank is reduced due to the volume of the water. An air tank with insufficient reserve capacity

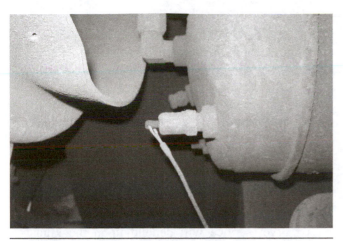

Figure 9-8 Manual drain valve. *(Courtesy of John Dixon)*

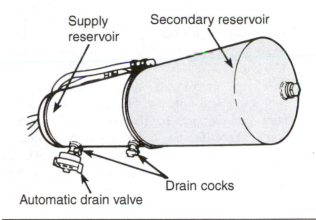

Figure 9-10 Supply, secondary air tank equipped with an automatic drain valve and manual drain cocks.

means that the air pressure drips off too rapidly when air demand on the system is highest.

The result can be inadequate braking in high-demand situations.

For a technician to drain the air tanks, he must first determine what type of drain valve device is used in their application. A common type of drain valve is a cable-release, spring-loaded valve. This enables the technician to simply pull on the cable for a couple of seconds, allowing the contaminants to be purged from the tanks. Once the technician can no longer see contaminants coming out of the valve, they simply release the cable. (see **Figure 9-8**).

Another style of drain valve is an automatic drain or spitter valve and the manual drain cock. (see **Figure 9-9**). This is the least popular, as it requires the technician or driver to climb under the vehicle chassis to open and close it. However, it is the most effective valve for ensuring that the system is completely drained. Automatic drain valves are most popular among drivers because

they do not require daily draining. All automatic drain valves (spitter valves) should be checked periodically for proper operation. **Figure 9-10** shows a typical supply, secondary air tank equipped with an automotive drain valve (supply tank or wet tank) and a manual drain cock (secondary tank).

Note: All air tanks that are FMVSS 121 compliant must have a means of mechanically draining them. Most automatic drain valves are equipped with a mechanical dump valve, but it can be difficult to locate.

Governor Cutout, Low Pressure Warning, and Pressure Buildup Test

To perform this test, do the following:

1. Completely drain all air tanks to 0 psi.
2. Start the truck engine and run at a fast idle. The low pressure warning light and an audible buzzer in the cab should be on. On some trucks equipped with an anti-lock braking system (ABS), the warning light will also come on momentarily when the ignition is turned on.
3. The dash warning light and buzzer should go off at or slightly above 60 psi.
4. Note and observe the time it takes to build pressure from 85 to 100 psi. This should take no more than 40 seconds. This is an FMVSS 121 requirement, and the tests are often used by enforcement officers to verify the performance of the truck air supply circuit because it can be performed quickly. This standard does vary in some jurisdictions.

Figure 9-9 Automatic drain valve. *(Courtesy of John Dixon)*

5. Record the **governor cutout pressure** (the compressor unloads) and check it to specifications. This value typically will be around 120 to 130 psi.
6. Discharge air from the system by pumping the foot valve and note the **governor cut-in pressure** (air compressor starts to build air pressure). The difference between cut-in and cutout pressure should never be more than 25 psi. As well, the difference should not be smaller that 20 psi to prevent frequently cycling of the compressor.

Reservoir Air Supply Leakage Test

The air supply leakage test should be performed regularly on a PM service. The test is performed by running the vehicle engine until the air system pressure is at the governor cutput pressure and then shut the engine off. Proceed as follows:

1. Allow pressure to stabilize for at least 1 minute.
2. Watch the air pressure gauge on the dash for 2 minutes and note any drop in pressure.

   ■ Pressure drop for a tractor or straight truck: 2 psi drop within 2 minutes is the maximum allowable for either circuit service tank (primary and secondary).
   ■ Pressure drop for a tractor/trailer combination: 6 psi drop within 2 minutes is the maximum allowable in all the service tanks.
   ■ Pressure drop for a tractor/trailer train (multi trailers): no more than 8 psi drop within 2 minutes is allowable in the trailer service tanks.

Locating an air leak on the vehicle's air supply system is a task that can easily be performed by an entry-level technician. This can be done by applying a simple soap and water solution to suspected air leaks. In many cases, if the vehicle in a quiet area, a technician may be able to locate the leak by sound alone. The cause of the air leak is likely to come from one or more of the following:

■ Supply lines and fittings (tighten)
■ Supply tank
■ Safety (pop-off) valve in supply reservoir
■ Governor
■ Compressor discharge valves
■ Air dryer and its fittings
■ Alcohol injection components

Shop Talk: Maximum allowable leakage/dropoff rates are often defined by local jurisdictions (state and provincial governments). These regulations may differ from the test values used here. Contact your local transportation enforcement office to identify the specifications used in your area.

Air Dryer Service

Air dryer service will depend upon the kinds of ambient conditions that the vehicle is operated in and experience can be a valuable guide in determining the bets maintenance interval. Most OEMs recommend a service interval around 25,000 miles (40,000 km) linehaul or every 3 months. The service requirements for a typical air dryer are shown in **Figure 9-11:**

1. Check for moisture in the air brake system by opening reservoirs, drain cocks, or valves and checking for water content. If moisture is present, the desiccant may require replacement. However, the following conditions also can cause water accumulation and should be considered before replacing the desiccant:

   ■ An outside air source has been used to charge the system. This air may not have been passed through the air dryer.
   ■ Air usage is higher than normal for a typical highway vehicle. This can be caused by accessory air demands or some usual air requirement that does not allow the compressor to spend sufficient time in its unloaded cycle. Also check for air system leakage.
   ■ The air dryer is newly installed in a system that in the past had no air dryer. The entire air system may be wet through with moisture, and several weeks of operation may be required to dry it out.
   ■ The air dryer is too close to the air compressor, resulting in the air being too hot to condense the moisture.
   ■ In areas that have a 30°F (16°C) range in temperature in the same day, small amounts of water can accumulate in the air brake system because of condensation. Under these conditions, the presence of small amounts of moisture is normal and should not be considered as an indication that the dryer is not performing properly.

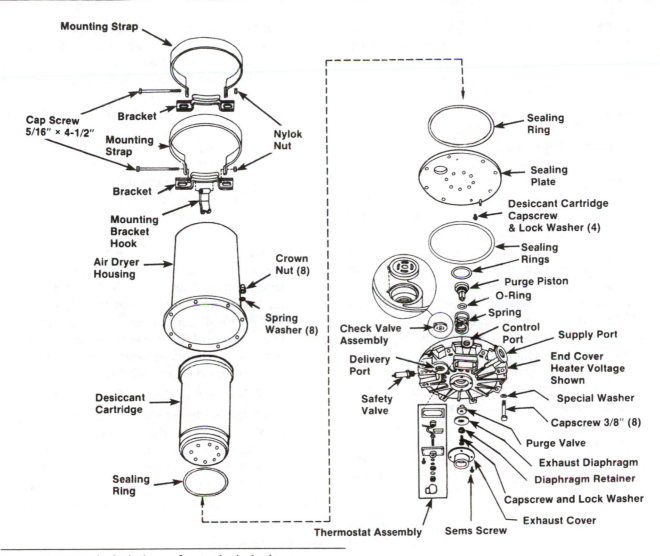

Figure 9-11 Exploded view of a typical air dryer.

Shop Talk: A small amount of oil in the system may be normal and should not, in itself, be considered a reason to replace the dryer desiccant. An oil-stained desiccant can function adequately.

2. Check mounting bolts for tightness. Retorque to specifications.
3. A major service event should be conducted every 300,000 miles or 36 months. This requires the air dryer to be disassembled and rebuilt. The overhaul procedure would usually include replacing the desiccant cartridge.

Note: The desiccant change interval may vary from vehicle to vehicle. Although typical desiccant cartridge life is 3 years, many will perform adequately for a longer period of time. In order to take maximum advantage of desiccant life and ensure that replacement occurs only when necessary, perform an operational and leakage test to determine the need for service.

Air Dryer Operation and Leakage Test

- Test the outlet check valve unit by raising the air pressure in the system to governor cutout and observe a test air pressure gauge installed in the supply tank. Rapid loss of pressure could indicate a failed outlet check valve.
- Confirm the failed outlet check valve by reducing the system air pressure and removing the check valve assembly from the end cover, subjecting air

pressure to the unit, and applying a soap solution to the check valve side. Leakage should not exceed a 1 inch (25 mm) bubble in 1 second.

■ Test for excessive leakage around the dryer purge valve. With the compressor in effective cycle (loaded mode), apply a soap solution to the purge valve housing assembly exhaust port and check that leakage does not exceed a 1 inch (25 mm) bubble in 1 second. If it does, service the purge valve housing assembly.

■ With all reservoir drain cocks closed, build up the system pressure to governor cutout pressure and check that the purge valve dumps an audible volume of air. Use the foot valve to pump down the service brakes to reduce system air pressure to governor cut-in. Note that the system once again builds to cutout pressure and is followed by an audible purge at the dryer.

■ Test the operation of the safety valve by pulling the exposed stem when the compressor is **loaded** (compressing air). Air should exhaust when the stem is held, and the valve should reseat when the stem is released.

■ Check all lines and fittings leading to and from the air dryer for leakage and integrity.

■ Test the operation of the end cover heater and thermostat (see **Figure 9-12**) assembly during cold weather operation as follows:

1. **Electrical Supply to the Dryer.** With the ignition key on, use a digital multimeter (DMM) or test light to check for voltage to the heater and thermostat assembly. Unplug the electrical connector at the air dryer and place the DMM

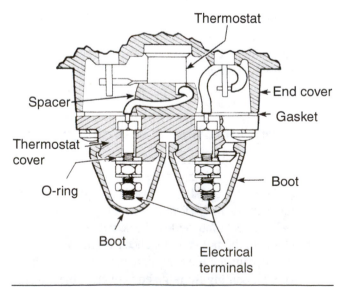

Figure 9-12 A typical air dryer thermostat assembly.

test leads on each of the pins of the male connector. If there is no voltage, look for a blown fuse, broken wires, or corrosion in the vehicle wiring harness. Check to see whether a good ground path exists.

2. **Thermostat and Heater Operation.** Turn off the ignition switch and cool the end cover assembly to a temperature below 40°F. (4.5°C). Using an ohmmeter, check the resistance between the electrical pins in the female connector. Compare the readings with the values specified by the OEM. If the resistance is higher than the maximum specified, replace the purge valve housing assembly, which includes the heater and thermostat. Now heat the end cover assembly to a temperature greater than 90°F (32°C) and check the resistance. The resistance should exceed 1,000 ohms. If the resistance values obtained are within specified limits, the thermostat and heater assembly are operating properly. If the resistance values obtained are outside specifications, replace the purge valve housing assembly.

Some current air dryers are integral with the supply tank and also may have the governor built into the assembly. Each type of air dryer requires a different rebuild procedure. Although this is not especially difficult, you should always use the manufacturer's service literature to guide you through the procedure.

Air System Test

Perform the following air system test in the same order as listed here. Start by chocking the wheels and releasing the parking brakes:

1. **Check for Air Pressure Leaks**
   ■ With the trailer glad hands secured to the dummy fittings, run the engine until the air system is fully charged and shut the unit off.

Tech Tip: The glad hands must seal for this test; if the glad hand brackets will not seal the glad hands, dead-head glad hands or some other means must be used.

■ Release the brakes on both the tractor and trailer brakes at the dash knobs (red and yellow) and apply the foot brakes fully.

■ When the system stabilizes (indicated by the gauges on the dash), continue full foot brake application to check for any air loss.

- The maximum air loss is 3 psi per minute.
- Continue with the glad hands still sealed.

TEST ONE-WAY CHECK VALVE

- Check the air pressure to the dash gauges.
- Locate the supply reservoir (it is the tank supplied directly from the air drier) and drain all the air from the tank.
- Recheck the pressure on the air pressure gauge on the dash.
- A drop in pressure of either the primary or secondary gauge indicates a problem with a one-way check valve at the **primary** or **secondary reservoir.**

TEST DOUBLE CHECK VALVE AND LOW AIR PRESSURE WARNING (PRIMARY SIDE)

- With the air pressure at a minimum of 100 psi indicated on the dash gauges and the engine off, open the reservoir drain on the primary reservoir and note the pressure on the dash gauge when the low air warning light and buzzer activate. The warning light and buzzer must operate before the primary air gauge drops to 55 psi (380 kPa).
- Check the secondary air gauge; it should not change while the primary reservoir is draining. If an air loss on the secondary side is found, the double check valve is defective and should be repaired. If the parking brakes apply when the primary reservoir is drained, it should be further inspected.
- Close the drain valve on the primary air reservoir.

TEST THE SPRING BRAKE INVERSION VALVE IF EQUIPPED

- With the primary air reservoir drained from the previous step, apply the service brakes by activating the foot valve. The pushrods of the spring brake chambers should extend, and at the same time you should hear air exhausting from the chambers. Release the foot valve and ensure the pushrods retract.

TEST TRACTOR PROTECTION VALVE

- To test the tractor protection valve, you must have a minimum of 100 psi in the secondary reservoir.
- Remove the emergency supply (red) and service (blue) glad hands from the dummy bracket or seal, allowing the air to escape.

- Watch the dash gauge air pressure when the dash control valve (red) button pops out. This button must pop out before the dash air gauges show 45 psi.

TEST THE AIR DRIER PURGE CYCLE OPERATION

- Run the truck engine to build up air pressure to governor cutout pressure. When the cutout pressure is reached, listen for the drier purge valve to exhaust moisture and contaminants from the air drier.
- The drier purge valve should cycle only at the governor cutout pressure.

TEST DOUBLE CHECK VALVE AND LOW AIR PRESSURE WARNING (SECONDARY SIDE)

- Run the truck engine to build up air pressure and shut off the engine. Turn the key to the on position.
- Open the reservoir drain on the secondary reservoir and note the pressure on the dash gauge when the low air warning light and buzzer activate. The warning light and buzzer must operate before the primary air gauge drops to 55 psi (380 kPa).
- Check the primary air gauge; it should not change while the secondary reservoir is draining. If an air loss on the primary side is found, the double check valve is defective and should be repaired. If the parking brakes apply when the secondary reservoir is drained, it should be further inspected.
- Close the drain valve on the secondary air reservoir.

Manual Parking/Emergency System Test

To perform this test, make sure the air pressure in the system pressure is at normal working pressure, between cut-in and cutout, and proceed as follows.

STRAIGHT TRUCKS AND TRACTORS

Have someone in the vehicle cab apply and release the park/emergency dash valve while watching the movement of the slack adjusters. The brakes should apply and release promptly as the valve is operated.

See **Figure 9-13** for identification of the park/emergency valves.

TRACTOR TRAILER COMBINATIONS

Have someone manually operate the trailer supply/ emergency valve in the tractor cab (it is red and

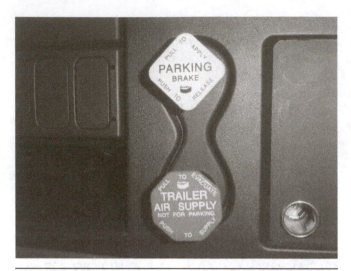

Figure 9-13 Park/emergency valves. *(Courtesy of John Dixon)*

Figure 9-14 Example of a very poor pushrod clevis. *(Courtesy of John Dixon)*

octagonal) while watching the movement of the slack adjusters. The trailer brakes should apply and release promptly as the valve is operated.

Have someone manually operate the system park control button (usually yellow and diamond or square shaped) and check that all parking brakes (tractor and trailer) apply and release promptly.

If performance from either of the two tests is sluggish, check for:

- Dented or kinked air lines
- Improperly installed hose fitting
- A defective trailer combination relay emergency valve
- Defective ABS modulator(s)

If the trailer parking/emergency brakes do not operate when the trailer supply valve is operated, check:

- Tractor protection control
- Trailer spring brake valve/combination relay/ emergency valve

For more information on how these valves are tested refer to *Heavy Duty Truck Systems,* 4th Edition, by Sean Bennett and Ian Andrew Norman.

Air Brake Foundation Inspection and Service

Inspect both the front and rear brake assemblies in the same manner:

- Start by visually inspecting the brake chamber to identify any damage to clamps and brackets.
- Make sure the brake chamber is securely mounted by simply pushing and pulling on the assembly. As well, check for any cracks, bent

or broken parts, and security of chamber mount bolts.

- Inspect the tightness of the jam nut securing the yoke to the pushrod.
- Check to verify that the two clevis pins of the ASA are free to rotate in both the clevis and the ASA. **Figure 9-14** shows a very poor clevis.
- Lubricate the ASA at its grease fitting and at the inner and outer S-cam bushings.
- Check for evidence of oil seepage into or out of the brake lining/drum interface area indicating an oil leak at the wheel seal.

INSPECT AND RECORD (FRONT/REAR) BRAKE LINING/PAD THICKNESS

- Using a brake lining wear gauge, measure the thickness of the lining or pad and record. These tools are calibrated in 32nds of an inch. The minimum brake pad/lining thickness for vehicles is 1/4 inch (6 mm) of remaining lining at the center of the drum. If necessary, pull the wheels to obtain the brake lining measurements.

INSPECT AND RECORD (FRONT/REAR) DRUM/ROTOR CONDITION

- Start by removing any dust shields the may be necessary to adequately inspect the drum or rotor condition.
- Always use a light to make this inspection; many problems are missed due to inadequate lighting conditions. Look for broken or cracked drums or rotors, excessive scoring, and wear.
- Reinstall any dust shields that were previously removed.
- Note drum or rotor condition and write up for replacement if necessary.

ADJUSTMENT OF MANUAL SLACK ADJUSTER

In order to make brake adjustments to vehicles with manual slack adjusters, the vehicle must first be prepared. Start by doing the following:

- Block the wheels to prevent the vehicle from moving.
- All wheels should be off the ground on proper jack stands.
- Brake drums should be cool.
- The brakes must be fully released.

To make the actual adjustment, follow these procedures:

- Clean the adjusting hex and locking collar. A little penetrating fluid or light oil on the locking collar makes it much easier for the locking collar to spring back out to lock the hex adjuster.
- With a box-end wrench, depress the locking collar releasing the lock.
- Spin the wheel and rotate the hex clockwise until the shoes come into contact with the drum.
- Turn the hex back in the opposite direction just until the wheel rotates freely.
- Remove the wrench, ensuring the locking collar returns to the lock position.

....

Tech Tip: If the locking collar will not return to the lock position, sometimes you can lightly tap the slack adjuster with a hammer while pulling the collar with a pair of locking pliers. If all else fails, replace the slack adjuster, because a locking collar that will not lock can allow the brakes to back off.

....

Adjusting Automatic Slack Adjusters

Just like the name implies, this type of slack adjuster automatically compensates for brake lining wear by adjusting drum to lining clearance. Automatic slack adjusters (ASAs) only need to be adjusted when they are originally installed or whenever the brake job is performed. Otherwise, they should never need adjustment unless something within the slack adjuster has failed.

Checking Free Stroke Travel (Freestroke)

When the foundation's brakes are in good condition, checking the length of **freestroke** at each brake to determine that brakes are correctly adjusted.

MEASURING "FREE STROKE" WITHOUT TEMPLATE

Drum Brake: Y Minus X Must = 5/8" – 3/4"
In Service Disc Brake: Y Minus X Must = 3/4" – 1"
Initial Disc Brake: Y Minus X Must = 7/8" – 1-1/8"

Figure 9-15 Measuring freestroke without a template. *(Courtesy of Arvin Meritor)*

Photo Sequence 8 shows the procedure used to check freestroke on S-cam brakes. Measure the freestroke length by using a bar or lever or by pulling on the adjuster by hand, as shown in **Figure 9-15**. Checking the freestroke confirms the clearance between the brake shoes and the drum. Freestroke between 3/8 inch and 5/8 inch indicates normal running clearance.

To check freestroke, proceed as follows:

- Block the wheels of the vehicle to prevent it from moving.
- Release the parking brakes.
- Extend each pushrod using a pry bar and measure the distance traveled.
- Each pushrod should move between 3/8 inch and 5/8 inch.
- Apply a full service brake application to ensure that the brakes fully actuate and release.

In most cases, when the freestroke is correct, the **applied stroke** falls within the adjustment limit. However, do not assume that this is true in all cases. Short freestroke indicates correct brake adjustment only when foundation brakes are in good mechanical condition. Use of this method does not guarantee that applied stroke will fall within the adjustment limit when brakes are in poor condition.

....

Note: Brake freestroke is typically specified to be between 3/8 inch (9 mm) and 5/8 inch (15 mm) regardless of chamber size. Check the OEM specifications.

....

Check Freestroke on S-CAMC Foundation Brakes

P8-1 Block wheels. If adjusting the brakes on a tractor/trailer combination, block one of the drive axle duals on the tractor and one trailer axle.

P8-2 Release the parking brakes on the rig by depressing both the System Park and Trailer Supply dash valves.

P8-3 Ensure that the system air pressure is above the governor cutout pressure (around 100 psi is ideal) after the brakes have been released.

P8-4 If the rig has visual stroke indicators, get somebody in the cab to make a full service brake application and hold. Inspect each pushrod stroke indicator to confirm the correct pushrod stroke.

P8-5 Correct travel when service brakes are applied.

P8-6 Excessive pushrod travel; brakes require adjustment.

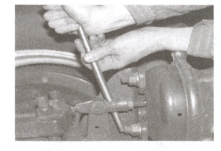

P8-7 If the pushrods on the rig do not have visual stroke indicators, release the parking brakes and then check each brake for pushrod travel using a heel bar.

P8-8 If a pushrod travels between $3/8$ and $5/8$ inch, the brake has the correct amount of free travel and does not require adjustment.

P8-9 If the pushrod travels more than $5/8$ inch, the brakes require adjustment.

Check Pushrod Travel (Applied Stroke)

Checking the pushrod travel using the applied stroke method should be done only after the brake adjustment has been performed. Adhere to the following to check brake stroke:

- Make sure the air pressure in the system is at 100 psi. (This will achieve 80 to 90 psi of applied pressure.)
- With a piece of chalk or soapstone or something similar, mark the pushrods flush with the brake chambers.
- Apply and hold a full foot brake application (this will require the help of another person) until all measurements are taken.
- Measure the pushrod travel at each chamber to the mark previously made on the pushrod.
- From the chart, compare each of the measurements with the specific dimensions for the type of chamber used on the vehicle.
- Regardless of the chamber type, the difference in measurements across an axle should not exceed 1/8 inch (3 mm).

VISUAL STROKE INDICATORS

Visual stroke indicators allow the technician or vehicle operator to confirm correct brake adjustment at a glance without having to get under the vehicle or have another person's assistance. Properly fitted visual stroke indicators also help to prevent unnecessary brake readjustment.

To determine freestroke using visual stroke indicators:

- Block wheels.
- Release parking brakes.
- Make sure reservoir pressure is between 90 and 100 psi.
- Make a full service brake application and hold.
- With the service brakes applied, use the visual indicator to confirm that the pushrod travel is correct.

Air Brake Chamber Identification

Air brake chambers can be segregated into standard size and long stroke. It is important that the technician be able to distinguish the difference between the two, as it will have an affect on the allowable pushrod travel.

STANDARD BRAKE CHAMBERS

In order to identify the chamber size, look for a stamped "Type" number around an air line port hole on a service break chamber, and on the clamp or around the air line port holes of a spring brake chamber. The air line ports of a standard size chamber will be round.

LONG STROKE CHAMBERS

Long stroke chambers may be identified by the following:

- An identification tag on the service chamber clamp band
- Square air line ports on the chamber

OUT-OF-SERVICE CRITERIA

(Taken from the North American Standard Out-of-Service Criteria by the Commercial Vehicle Safety Alliance, April 1, 2006)

The following items are defects that deem a brake to be defective as per the 20 percent criterion:

1. Absence of effective braking action upon application of the service brakes (such as brake linings failing to move or contact braking surface upon application)
2. Missing or broken mechanical components including shoes, linings, pads, springs, anchor pins, spiders, cam rollers, pushrods, and air chamber mounting bolts
3. Loose brake components including air chambers, spiders, and cam shaft support brackets
4. Audible air leak at brake chamber (Example: ruptured diaphragm, loose chamber clamp, etc.)
5. Brake adjustment limits: Bring reservoir pressure between 90 and 100 psi (620 to 690 kPA), turn engine off, and then fully apply the brakes.

   (A) One brake at 1/4 inch (6 mm) or more beyond the adjustment limit (Example: type 30 clamp type brake chamber pushrod measured at 2-1/4 inches [57 mm] would be one defective brake.)

   (B) Two brakes less than 1/4 inch (6 mm) beyond the adjustment limit also equal one defective brake. (Example: Type 40 clamp type brake chamber pushrods measure— Two at 2-1/8 inches [54 mm]. This example would equal one defective brake.)

   (C) Any wedge brake where the combined brake lining movement of both top and bottom shoes exceeds 1/8 inch (3 mm)

6. Break linings or pads (except on power unit steering axles):

   (A) Cracked, loose, or missing lining

   - Lining crack or voids of 1/16 inch (1.6 mm) in with observable on the edge of

Figure 9-16 An example of a brake shoe in very poor condition. *(Courtesy of John Dixon)*

the lining. **Figure 9-16** shows a very bad brake shoe.

- Portions of a lining segment missing such that a fastening device (rivet or bolt) is exposed when viewing the lining from the edge
- Cracks that exceed 1-1/2 inch (38 mm) in length
- Loose lining segments (approximately 1/16 inch [1.6 mm] or more movement)
- Complete lining segment missing

(B) Evidence of oil seepage into or out of the brake lining/drum interface area. This must include wet contamination of the lining edge accompanied by evidence that further contamination will occur—such as oil running from the drum or a bearing seal.

(C) Air Brakes: Lining with a thickness less than 1/4 inch (6 mm) or to wear indicator if lining is so marked, measured at the show center for drum brakes or less than 1/8 inch (3 mm) for disc brakes

7. Missing brake on any axle required to have brakes

Front Steering Axle(s) Brakes

In addition to being included in the 20 percent criterion, the following criteria place a vehicle in an out-of-service condition:

1. Any inoperative brake on either wheel of any steering axle of any vehicle equipped with steering axle brakes, including the dolly and front axle of a full trailer. This includes tractors required to have steering axle brakes.

2. Mismatch across any power unit steering axle of:

(A) Air chamber sizes. Note: Mismatched air chamber size includes long stroke brake chamber versus regular stroke brake chamber and excludes differences in design type such as type 20 clamp versus type 20 rotorchamber.

(B) Slack adjuster length

3. Brake linings or pads on the steering axle of any power unit:

(A) Cracked, loose, or missing lining

- Lining cracks or voids of 1/16 inch (1.6 mm) in width observable on the edge of the lining
- Portions of a lining segment missing such that a fastening device (rivet or bolt) is exposed when viewing the lining from the edge
- Cracks that exceed 1-1/2 inch (38 mm) in length
- Loose lining segments (approximately 1/16 inch [1.6 mm] or more movement)
- Complete lining segment missing

(B) Evidence of oil seepage into or out of the brake lining/drum interface area. This must include wet contamination of that lining edge accompanied by evidence that further contamination will occur—such as oil running from the drum or bearing seal.

Note: Grease on the lining edge, back of shoe, or drum edge and oil stains with no evidence of fresh oil leakage are not conditions for out-of-service.

(C) Lining with a thickness less that 3/16 inch (5 mm) for a shoe with a continuous strip of lining or 1/4 inch (6 mm) for a shoe with two pads for drum brakes or to wear indicator if lining is so marked, or less than 1/8 inch (3 mm) for air disc brakes

SPRING BRAKE CHAMBERS

Any non-manufactured holes or cracks in the spring brake housing section of a parking brake.

TRAILER BREAKAWAY AND EMERGENCY BRAKING

Inoperable breakaway braking system on trailer(s).

PARKING BRAKE

No brakes on the vehicle or combination are applied upon actuation of the parking brake control, including driveline hand-controlled parking brakes.

BRAKE DRUMS OR ROTORS (DISCS)

1. Drums with any external crack or cracks that open upon brake application

Note: Do not confuse short hairline heat check cracks with flexural cracks.

2. Any portion of the drum or rotor missing or in danger of falling away

BRAKE HOSE/TUBING

1. Any damage extending through the outer reinforcement ply. (Rubber-impregnated fabric cover is not a reinforcement ply.) Thermoplastic nylon tube may have braid reinforcement or color difference between cover and inner tube; exposure to second color is out-of-service.
2. Bulge/swelling when air pressure is applied
3. Audible leak other than a proper connection
4. Improperly joined such as a splice made by sliding the hose ends over a piece of tubing and clamping the hose to the tube
5. Damaged by heat, broken, or crimped in such a manner as to restrict air flow

LOW PRESSURE WARNING DEVICE

Low pressure warning device missing, inoperative, or does not operate at 55 psi (397 kPA) and below, or 1/2 of the governor cutout pressure, whichever is less.

Note: If either an audible or visual warning device is working as required, vehicle should not be placed out-of-service.

AIR LOSS RATE

If an air leak is discovered and the reservoir pressure is not maintained when:

1. Governor is cut-in
2. Reservoir pressure is between 80 and 90 psi (551–620 kPA)
3. Engine is at idle
4. Service brakes are fully applied

TRACTOR PROTECTION SYSTEM

Inoperable or missing tractor-protection system components, including a tractor-protection valve and/or trailer supply valve.

AIR RESERVOIR

Air reservoir security separated from its original attachment points.

AIR COMPRESSOR

1. Loose compressor mounting bolts
2. Cracked, broken, or loose pulley
3. Cracked or broken mounting brackets, braces, or adapters

Standard Stroke Clamp Type Brake Chambers

Type	Outside Diameter	Brake Adjustment Limit
6	4-1/2 inch (114 mm)	1-1/4 inch (32 mm)
9	5-1/4 inch (133 mm)	1-3/8 inch (35 mm)
12	5-11/16 inch (145 mm)	1-3/8 inch (35 mm)
16	6-3/8 inch (162 mm)	1-3/4 inch (45 mm)
20	6-25/32 inch (172 mm)	1-3/4 inch (45 mm)
24	7-7/32 inch (184 mm)	1-3/4 inch (45 mm)
30	8-3/32 inch (206 mm)	2 inch (51 mm)
36	9 inch (229 mm)	2-1/4 inch (57 mm)

Note: A brake found at the adjustment limit is not a defect for the purpose of the 20 percent rule.

Long Stroke Clamp Type Brake Chambers

Type	Outside Diameter	Brake Adjustment Limit
12	5-11/16 inch (145 mm)	1-3/4 inch (45 mm)
16	6-3/8 inch (162 mm)	2.0 inch (51 mm)
20	6-25/32 inch (172 mm)	2.0 inch (51 mm)
24	7-7/32 inch (185 mm)	2.0 inch (51 mm)
24*	7-7/32 inch (185 mm)	2.5 inch (64 mm)
30	8-3-32 inch (206 mm)	2.5 inch (64 mm)

*For 3-inch maximum stroke type 24 chambers.

Note: A brake found at the adjustment limit is not a defect for the purpose of the 20 percent rule.

Rotochamber Data

Type	Outside Diameter	Brake Adjustment Limit
9	4-9/32 inch (109 mm)	1-1/2 inch (38 mm)
12	4-13/16 inch (122 mm)	1-1/2 inch (38 mm)
16	5-13/32 inch (138 mm)	2 inch (51 mm)
20	5-15/16 inch (151 mm)	2 inch (51 mm)
24	6-13/32 inch (163 mm)	2 inch (51 mm)
30	7-1/16 inch (180 mm)	2-1/4 inch (57 mm)
36	7-5/8 inch (194 mm)	2-3/4 inch (70 mm)
50	8-7/8 inch (226 mm)	3 inch (76 mm)

Note: A brake found at the adjustment limit is not a defect for the purpose of the 20 percent rule.

DD-3 Brake Chamber Data

Type	Outside Diameter	Brake Adjustment Limit
30	8-1/8 inch (206 mm)	2-1/4 inch (57 mm)

Note: This chamber has three air lines and is found on motorcoaches.

Note: A brake found at the adjustment limit is not a defect for the purpose of the 20 percent rule.

DEFECTIVE BRAKES

The number of defective brakes is equal to or greater than 20 percent of the service brakes on the vehicle or combination. A defective brake includes any brake that meets the out-of-service criteria.

Defective Brake Chart

Total Number of Brakes Required to Be on a Vehicle Combination	Total Number of Defective Brakes Necessary to Place the Vehicle or Combination Out-of-Service
4	1
6	2
8	2
10	2
12	3
14	3
16	4
18	4
20	4
22	5
*	

For a vehicle or combination that exceeds 22 brakes, determine the number of defective brakes by using 20 percent of the total number of brakes. All fractions should be rounded down to the next whole number.

Summary

- It is good practice to consult the manufacturer's service literature when performing repairs.
- Use the manufacturer's specific fluid to ensure proper operation of the braking system.
- Never mix incompatible brake fluids.
- Brake fluid is hygroscopic.
- Never use brake fluid from a container that has been used to store any other liquid.
- Brake fluid should be changed regularly according to the vehicle manufacturer's specifications.

- Bleeding the brakes involves purging air from the hydraulic circuit.
- Rebuilding of master cylinders should be done only by qualified technicians due to their complexity and importance from a safety point of view.
- Brake drums may be segregated into two groups, inboard and outboard.
- Brake fade is the reduction in stopping power that can occur after repeated application of the brakes.
- Never paint the outside of a brake drum.

- Rotor radial runout should not exceed the OEM TIR specification, usually around 0.030 inch (0.762 mm).

- Excessive rotor runout or wobble increases pedal travel because of opening up the caliper piston and can cause pedal pulsation and chatter.

- When performing the service inspection, visually inspect the condition of the brake lines and hoses.

- The thickness of the brake lining should be measured at its thinnest point.

- The number of defective brakes is equal to or greater than 20 percent of the service brakes on the vehicle or combination.

- Compressed air can be dangerous if not handled with caution.

- While working on an air brake system, make sure the wheels are blocked to prevent the vehicle from moving.

- Always wear eye protection when draining an air tank and point the discharged air away from yourself and other people.

- The compressor must have an unrestricted supply of clean filtered air.

- Draining the contaminants from the air tanks is a task that must be performed at every level of service.

- All air tanks that are FMVSS 121 compliant must have a means of mechanically draining them.

- The air supply leakage test should be performed regularly on a PM service.

- Locating an air leak on the vehicle's air supply system is a task that can easily be performed. This can be done by applying a simple soap and water solution to suspected air leaks.

- Maximum allowable leakage/drop-off rates are often defined by local jurisdictions (state and provincial governments).

- Air dryer service will depend upon the kinds of ambient conditions that the vehicle is operated in.

- To perform parking/emergency brake checks, make sure the air pressure in the system pressure is at normal working pressure, between cut-in and cutout.

- Checking the pushrod travel using the applied stroke method is only done after the brake adjustment has been performed.

- The automatic slack adjuster compensates for brake lining wear by adjusting drum to lining clearance.

- Air brake chambers can be segregated into standard size and long stroke. It is important that the technician be able to distinguish the difference between the two, as it will have an effect on the allowable pushrod travel.

Review Questions

1. Two technicians are discussing the brake fluid in the master cylinder. Technician A says the brake master cylinder should be inspected on every PM service, checking the fluid level and condition. Technician B says that if the fluid level is above 1/2 full, don't fill it up. Which technician is correct?

   A. Technician A

   B. Technician B

   C. Both technicians A and B are correct.

   D. Neither technician A nor B is correct.

2. On a light-duty truck with hydraulic brakes, at what point should the linings be replaced for maximum service life?

   A. 3/16 of an inch

   B. 1/4 of an inch

   C. 1/8 of an inch

   D. 9/32 of an inch

3. Technicians are discussing hydraulic brake fluid, and technician A says that the fluid in the master cylinder should smell, look, and feel like new fluid. Technician B says that the brake fluid is always the dirtiest at the wheel end. Who is correct?

   A. Technician A

   B. Technician B

   C. Both technicians A and B are correct.

   D. Neither technician A nor B is correct.

4. When inspecting a vacuum-assisted service brake system:

   A. the brake pedal should fall farther as the engine is started.

   B. the brake pedal should not travel more than half its total stroke.

   C. the electric pump should come on with the key on, or when the brake pedal is applied.

   D. Both A and B are correct.

   E. Both B and C are correct.

5. When testing hydraulically assisted service brakes:

   A. there should be a "kickback" felt from the pedal.

   B. the brake pedal should not travel more than half its total stroke.

   C. the brake pedal should nod fall farther as the engine is started and runs.

   D. All of the above statements are correct.

6. A typical TIR specification for disc rotor lateral runout would most likely be which of the following?

   A. 0.0015 inch (0.038 mm)

   B. 0.0050 inch (0.127 mm)

   C. 0.0100 inch (0.254 mm)

   D. 0.0150 inch (0.381 mm)

7. In a hydraulic drum brake system, what force pulls the shoes away from the drum when the brakes are released?

   A. Hydraulic force at the master cylinder

   B. Retraction spring force at the brake shoe assembly

   C. Hydraulic force at the wheel cylinder

   D. Brake pedal spring force

8. A typical maximum rotor thickness variation specification required by an OEM would most likely be:

   A. 0.002 inch (0.05 mm).

   B. 0.005 inch (0.127 mm).

   C. 0.020 inch (0.508 mm).

   D. 0.050 inch (1.27 mm).

9. A technician is draining the primary air reservoir and notices that the needles on both the primary and secondary dash gauges are dropping. The technician also notices that the spring brakes are applying. Which of the following could be the problem?

   A. The one-way check valve in the secondary tank needs to be replaced.

   B. The double check valve is bad.

   C. The inversion valve is diverting air away from the primary service brakes.

   D. The spring brake inversion valve is defective.

10. When checking for air pressure leaks in the breaking system, the technician should:

   A. hold the brake pedal down for 1 minute.

   B. check and verify if a tractor's air supply drops 3 psi or more in 1 minute.

   C. make sure the tractor and trailer park brakes are released.

   D. All of the above are correct.

11. What is the required amount of slack adjuster free travel on a properly adjusted, manual S-cam brake?

    A. Less than 1/4 inch (6.4 mm)   C. 3/8 to 5/8 inch (10 to 16 mm)

    B. 1/4 to 2/4 inch (6 to12 mm)   D. 3/4 to 1 inch (19 to 25 mm)

12. Which of the following foundation brake components are lubricated with chassis grease, either for component life or as part of a lube job?

    A. S-cam profile   C. S-cam bushings

    B. S-cam roller face   D. Brake shoe friction face

13. When adjusting a manual slack adjuster on an S-cam foundation brake, technician A says that the lock collar should be retracted and the adjusting nut always rotated clockwise to decrease free travel. Technician B says that free travel can be reduced only when the adjusting nut is rotated counterclockwise. Who is correct?

    A. Technician A only   C. Both technician A and B are correct.

    B. Technician B only   D. Neither technician A nor B is correct.

14. When adjusting S-cam brakes on a trailer equipped with manual slack adjusters, which of the following must be true?

    A. The tractor parking brakes   C. The trailer parking brakes must be released.
       must be applied.
                                            D. The tractor service brakes must be applied.
    B. The trailer service brakes
       must be applied.

15. When an S-cam foundation brake slack adjuster is in the fully applied position, ideally the slack adjuster-to-pushrod angle should be:

    A. less than 90 degrees.   C. more than 90 degrees.

    B. exactly 90 degrees.   D. It does not matter.

16. How much freestroke travel should a properly adjusted automatic S-cam brake slack adjuster have?

    A. Marginal preload   C. 3/8 to 5/8 inch (10 to 16 mm)

    B. Zero   D. 1 to 1-1/2 inch (19 to 38 mm)

17. What is the adjustment stroke limit for a 24 series brake chamber?

    A. 1.00 inch (25 mm)   C. 2.25 inches (57 mm)

    B. 1.75 inches (44 mm)   D. 3.00 inches (76 mm)

18. What is the adjustment stroke limit for a standard 30 series brake chamber?

    A. 1.00 inch (25 mm)   C. 2.25 inches (57 mm)

    B. 2.00 inch (51 mm)   D. 3.00 inches (76 mm)

19. While checking an air brake system, the wet tank should be the first reservoir to be drained because:

    A. it has the most air.   C. it is easier to reach than the other tanks.

    B. it will make draining the   D. draining one of the other tanks first could cause moisture and
       primary and secondary tanks   contaminants to be siphoned throughout the rest of the system.
       much easier.

20. To check one-way check valves:

    A. drain the primary tank first, then the secondary tank.

    B. seal both glad hands and push in the yellow and red knobs on the dash.

    C. drain the wet tank.

    D. build air in the system until the governor cuts out.

CHAPTER

10 Chassis/Cab Service and Inspection

Objectives

Upon completion and review of this chapter, the student should be able to:

- List the various components that should be checked within the cab of the vehicle during a vehicle service.

- Explain how to make the actual determination as to whether an item within the cab requires maintenance.

- Explain how to test the HVAC system.

- List the safety items within the cab that must be inspected and account for any mandatory safety equipment.

- List and explain how to maintain cab hardware.

- Explain two different methods of maintaining the vehicle by preventing premature corrosion.

- Perform an inspection of the air-conditioning system.

- Explain how to performance test an air-conditioning system.

- List the different methods for finding potential refrigerant leaks in an air-conditioning system.

Key Terms

commercial vehicle operator's registration (CVOR)

Environmental Protection Agency (EPA)

glazing

heating, ventilating, and air conditioning (HVAC)

permeation

power take off (PTO)

sacrificial anode

supplemental restraint system (SRS)

INTRODUCTION

The vehicle cab is an area of maintenance that cannot be overlooked by the technician. The vehicle operator spends so much of his time in the cab that he will usually notice and report any defects he finds. There are many areas of safety that must be observed and checked by the service technician. The **heating, ventilating, and air conditioning (HVAC)** system is another area of maintenance that must be checked to ensure it is working properly for driver comfort.

Important items for driving safety include a clear field of vision for the driver: an unobstructed windshield and windows, and properly positioned mirrors.

SYSTEM OVERVIEW

This chapter starts out with a look at the components within the cab that must be checked during a service. This includes all cab lighting, switches, and gauges as well as any accessories that may be inside the cab, a system check of the vehicle's HVAC system, and the inspection of any onboard diagnostic systems. Next, onboard safety equipment is looked at. Regulations for commercial vehicles dictate that mandatory safety equipment be on board the vehicle and in proper working order. We will then go through the serviceable items on the cab that must be inspected, lubricated, or maintained.

The air conditioning system should be performance tested on a regular basis, and we will also look at refrigerant leak detection, because leaks are one of the most common causes of HVAC system failures or complaints. Finally, keeping the vehicle body work and vehicle wiring harnesses in good corrosion-free condition requires some type of maintenance and regular inspection. Two types of corrosion prevention systems are discussed here.

INSTRUMENTS AND CONTROLS

During the PM service, the cab must be inspected and maintained. Vehicle operators will usually make note of any cab-related problems because they spend so much time in this area of the vehicle. Inspect the following:

1. Check the condition of the ignition and/or door key. Inspect for any cracks that could allow the key to be broken off inside the ignition or lock cylinder. Test the operation of the ignition switch and the keyed door lock operation.
2. Check the proper operation of any indicator lights, warning lights, and alarms. Most of these can be checked by turning the ignition switch on but not starting the vehicle. Make note of anything not operating.
3. Start the vehicle and check the operation of all instrument/gauges and panel lights. It may be necessary to move the vehicle out of direct sunlight in order to check panel lights. Record oil pressure and charging voltage.
4. Check the operation of electronic **power take off (PTO)** and engine idle speed controls (if applicable).
5. With the vehicle running, check the operation of the climate control system. Start by testing the fan operation to ensure that the fan operates at all speed settings. Next, rotate the air selection switch. Air should be discharged out through the vents in the locations as dictated by the selection switch. Lastly, check the temperature control switch. You may install a thermometer in the vent or use your hand to verify that the

discharge air temperature changes with movement of the temperature control switch.
6. Check the operation of any accessories that the vehicle may have.
7. Using a diagnostic tool or onboard diagnostic system, extract and record engine, transmission, and brake monitoring information and codes.

ONBOARD SAFETY EQUIPMENT

1. Test the operation of both the electric horn and the air horn. Make sure that there is sufficient air in the reservoir to operate the air horn.
2. Inspect the condition of all onboard safety equipment, including roadside flares or reflective triangles, spare fuses, fire extinguisher (check date of inspection and that the indicator is good), and all required decals. Check to make sure the vehicle has a first-aid kit and a pair of latex gloves. Make sure that all mandatory safety equipment is on board the vehicle and in proper working order. Failure to comply can mean stiff penalties by enforcement officers.

Note: In many cases, fleets are choosing reflective triangles over the flares because the flares can get damp and must be replaced at their expiration date. As well, many operators carry a spill kit (bag of floor-drying material) or an absorbent sock (dam).

3. Inspect seat belts and sleeper restraints for proper operation and any tears in the belt material. Check **supplemental restraint system (SRS)** warning light operation (if applicable).
4. Inspect wiper blades for any tears in the material and check to see that the wiper arms are strong enough to keep the blade in constant contact with the windshield throughout their sweep.
5. Check the operation of the windshield wipers. Check for high- and low-speed operation as well as intermittent operation. Check the operation of the washer system. Inspect to make sure that the reservoir is full and that the spray pattern is correct.
6. Check the vehicle for all required vehicle permits, registration, decals, insurance, **commercial vehicle operator's registration (CVOR),** and inspection papers.

CAB HARDWARE INSPECTION AND MAINTENANCE

1. Inspect the windshield glass for any cracks, chips, or discoloration (haziness). Also check the sun visor for proper operation.

Shop Talk: Some small chips and cracks can be repaired by specialized windshield installers at a fraction of the cost of a replacement windshield. This procedure must be done at the early start of the chip or crack before it propagates, becoming too large to repair.

2. Check the condition of the driver's seat. Make sure that the seat is adjustable fore and aft. Most driver seats are an air ride design to cushion the operator from the punishing ride that can be inflicted while operating over rough roads. Using the air valve, feed air into the driver seat air bag, and the seat should rise. Stop and listen for any audible air leaks. Again, using the air valve, release the air from the seat's air bag, and the seat should lower itself.

3. Inspect the condition of the glass in both the trucks doors. Make sure that the windows operate properly, moving freely from closed to open and back.

4. Check the cab steps and grab handles. Make sure they are in good shape and they are securely fastened to the vehicle.

5. Inspect the mirrors of the vehicle for condition and cleanliness. Make sure they are securely fastened to the vehicle with all mounting brackets intact. Check mirror heater, motor operation, and lights if equipped.

6. Inspect cab body work and note any defects or physical damage.

7. Lubricate all cab and hood grease fittings.

8. Inspect and lubricate all door and hood hinges. If the rubber of the hood latches are cracked or weather checked, replace as necessary. Lubricate strikers, lock cylinders, and linkages. Check the hood cables for both mounting and integrity.

Note: Physical injury to persons as well as damage to the hood and truck bumper can be avoided by maintaining the condition of the hood cables.

9. Inspect the cab mounting hinges and lubricate. Check cab mounting latches and linkages and service as needed.

10. Inspect the hydraulic pump for tilting the cab. Check for fluid leaks at the pump, hydraulic lines, and the cylinders. Inspect the cab safety locking device, that it is operational and will not allow the cab to move downward in the event of a hydraulic failure. Service all components as needed.

11. Inspect the accelerator, brake, and clutch pedal for smooth non-binding operation. Check the condition of the pedals' rubber contact patch. Replace as necessary.

12. Using the manufacturer's specifications (maintenance manual), check the cab ride height. The ride height specification will indicate where the measurement must be taken. Visually inspect the condition of the cab air suspension springs (air bags). Look for any abrasions, tears, or deterioration of the bag. Listen for any air leaks and repair as necessary. Also check bag mounting, air lines, height control valve, and linkage as well as shock absorbers for any signs of leakage or damage.

13. Inspect the condition of the front bumper and the mounting of fairings and/or any other equipment/accessories mounted to the cab of the vehicle.

CAB AIR SUSPENSIONS AND DRIVER SEATS

Cab air suspensions systems have become the common method of mounting a cab because of the level of driver comfort it provides. A typical system consists of two air springs, two shock absorbers, and a leveling valve to maintain ride height. These components are shown in **Figure 10-1**. In some cases, a transverse rod is fitted for lateral stability. Suspension travel is not great, usually in the range of 1 to 3 inches, but the difference to driver comfort is considerable, especially on vehicles spec'd with unforgiving heavy-duty suspensions.

Because no two trucks cabs are alike, there are many variations in size and mounting brackets. Some systems use single-point mounts, featuring a single air bag and shock absorber, but the majority are dual point, similar to the systems shown in **Figure 10-2**.

Cab Suspension Maintenance

Moving components should be routinely lubricated and the air system checked for oil admission. Periodic shock absorber replacement is required on most systems.

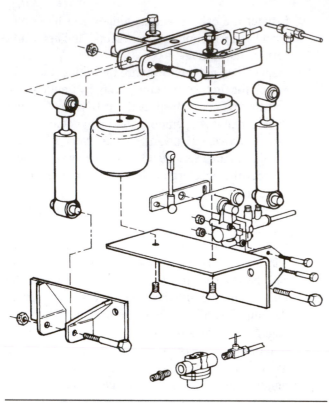

Figure 10-1 Exploded view of two-point cab air suspension.

by cab suspension manufacturers. It is also possible to use electronically controlled air valving on a more conventional system. In these new technology systems, a cab suspension ECU can be networked to the chassis data bus and coordinate cab ride improvement depending on the terrain and conditions the vehicle encounters.

Driver's Seats

A driver's comfort is usually determined by the type and quality of the driver's seat. There are three types: mechanically suspended, pneumatically (air) suspended, and solid mounted. A fleet using the latter may have problems retaining drivers. Mechanical suspension seats use a mechanical, free-moving support system to somewhat isolate the driver from suspension jolts transferred from the chassis to the cab.

Air suspension seats do a better job of isolating the driver from suspension and frame jolts. They consist of an air bag, jounce cushion, and shock absorber assembly. The air bag is charged with compressed air from the vehicle pneumatic circuit. **Figure 10-3** shows a typical air-suspended driver's seat. This type of seat has become commonplace on trucks used today, especially by fleets that want to retain their drivers.

The need for this type of service will be identified by drivers who complain about excessive bouncing. The actual service life of cab suspension shock absorbers vary with the application. They can outlast the vehicle in a linehaul application.

Electronically managed active cab suspensions that are capable of reading vibrations and activating solenoids to cancel or minimize them have been introduced

Mechanical suspension seats are the best choice if the truck does not have an air system, although at least one manufacturer offers a completely self-contained gas suspension seat that requires no air supply. It works on the same principles used in many office chairs. Solid-mounted seats are the most economical, and, as the name implies, provide the least amount of isolation from the road. (see **Figure 10-4**).

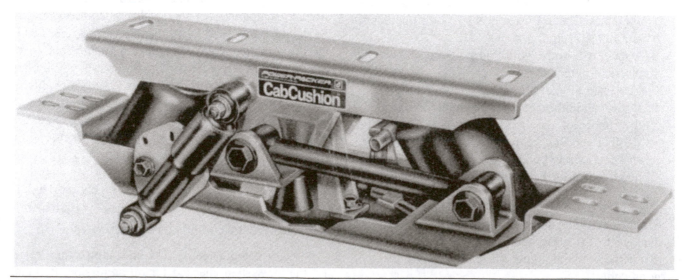

Figure 10-2 Cab air suspension system with a transverse rod, two air springs, one shock absorber, and a leveling valve. *(Courtesy of APA Engineered Solutions [Power-Packer])*

Figure 10-3 Air-suspended driver's seat. *(Courtesy of Bostrom Seating, Inc.)*

Figure 10-4 Solid-mounted seat. *(Courtesy of Bostrom Seating, Inc)*

OTHER BODY MAINTENANCE

Rust Inhibitors

Many commercial vehicles can benefit by implementing a body maintenance program. Premature failure in electrical and mechanical systems is often induced by the environment in which the vehicle is operated.

A rust inhibitor can make a big difference in maintenance costs. The maker of one such product known as KROWN claims that a preventive maintenance program will:

- Lubricate many moving parts that are normally inaccessible during a vehicle's routine maintenance, such as door locks, windows, and power antennas.

- Electrical connections and wiring also benefit by remaining virtually corrosion free.
- Potential corrosion areas (spot welds, etc.) are also protected. Welded areas are usually more susceptible to corrosion.

The maintenance for the protection offered by this and many other products is that the body is inspected and product reapplied annually where necessary.

Electronic Rust Protection

Another method of preventing rust on commercial vehicles as well as marine applications is the use of a **sacrificial anode.** The vehicle will use multiple electrodes termed "sacrificial anodes" are strategically located and attached to the vehicle where water will collect or run. When the system is turned on, a positive electrical charge is applied to these sacrificial anodes by a control module. This causes the iron atoms and impurity atoms to have a negative charge compared to the sacrificial anodes. The free electrons are now pulled through the steel to the sacrificial anodes instead of to the negatively charged iron atoms. When these free electrons arrive at the sacrificial anodes, the sacrificial anodes corrode (sacrifice). This means that corrosion that forms on the anodes would have been corrosion or rust developing elsewhere on the vehicle.

AIR CONDITIONING

1. Starting at the condenser, look for any signs of damage and inspect the mounting of the condenser. Look for any buildup of dirt, insects, or road debris that can impede heat transfer. If needed, clean the condenser with mild soap and water and a soft nylon brush. Rinse the condenser with a garden hose.

Note: Never use a pressure washer to clean the condenser, as it will damage the fragile cooling fins.

2. Inspect the refrigerant lines and connections for secure mounting and signs of leakage or abrasions. A refrigerant leak will sometimes push a little compressor oil out with it, leaving a wet oily stain on the line. Air conditioning leak checks will be discussed later in this chapter.
3. Inspect the compressor for mounting and any signs of refrigerant leakage. Check the compressor

clutch gap as well. The gap for the electromagnetic clutch can be checked with a standard business card. Just insert the card between the contact plates of the clutch assembly. The business card is thin and used like a feeler gauge. It should be able to be inserted easily but still have some friction. The gap can be adjusted by the use of shims. See the vehicle manual for complete instructions for the vehicle you are working on. Note that if the gap is too large, the clutch plates will slip, causing premature wear and eventual failure. If the gap is too tight, the plates can be in contact even when the compressor's clutch is de-energized and free wheeling, also causing frictional damage to the clutch plates and causing premature wear and eventual failure.

4. Check the air-conditioning system condition and operation. This can be accomplished by following the air-conditioning performance test as follows:

Air Conditioning Performance Test

During a PM service, the technician should performance test the air-conditioning system to verify that the system is operating properly:

- Start by running the engine at a speed of 1,500 rpm. Close all cab doors and windows.
- Turn the air-conditioning system on, setting the system for maximum cool.
- A thermometer should be inserted into the center duct and the system should be run for approximately 10 minutes to allow the system and thermometer to stabilize (to keep from fluctuating).
- Check that the air flow is coming from the appropriate vent for the position of the air selection switch.
- If fan is operating but air flow is low, check for leaves or debris that may be obstructing the air inlet.
- During the performance check, visually inspect the compressor drive belt for signs of slippage.
- Check the compressor clutch to make sure it is engaged, paying attention to any abnormal compressor noises.
- Feel the discharge line at the compressor; it should be hot to the touch. If warm or cold, it may indicate a faulty compressor or low refrigerant charge.

- Next, feel the inlet and outlet of the condenser. You should be able to feel a large change of temperature between the inlet and outlet.
- Feel the liquid line from the condenser outlet to the receiver tank; it should be warm. Cool spots in the line indicate restrictions. If the line is hot, it may indicate that the condenser if not transferring enough heat to the ambient air (check condenser fan).
- Check the inlet and outlet lines at the receiver-drier; they should be the same temperature. If they are not, it indicates that the filter-drier may be blocked or restricted.
- Feel the liquid line from the outlet of the receiver-drier to the inlet of the TXV. It should feel warm to the touch. Any cold spots in the liquid line indicate a restriction.
- Check the suction line from the evaporator outlet to the compressor inlet. It should feel cold. If humidity conditions are right, this line may be sweating and may even have some frost covering it. If the line is warm, it may indicate that the expansion valve is not metering enough refrigerant into the evaporator or that the refrigerant charge is insufficient. If the line is covered in ice, it can indicate an overcharged system or a TXV that is metering too much refrigerant into the evaporator.
- If the engine is running hot, the system will not operate as efficiently. If the water valve does not close completely, the heat from the heater core will enter the cab of the truck.
- Check the temperature of the discharge air at the dash vent to make sure the air is cool. The discharge air temperature will vary according to ambient temperature and humidity conditions.
- Now shut the unit down.
- After engine is shut down, feel across the surface of the condenser; any cold spots indicate crimped or restricted coils.

5. Belt tension should always be checked during routine maintenance for an air-conditioning system. Since the compressor is belt driven, proper belt tension is important for proper air-conditioning operation. Failure to do so can cause the compressor drive belt to slip, causing reduced air-conditioning capacity and **glazing** (belt gets hard and slippery) of the drive belt.

6. Many trucks today use an inlet filter or filters for the HVAC system. See the service manual for compliance to specified filter change intervals.

REFRIGERANT LEAK DETECTION

All air-conditioning systems leak refrigerant over time, and the system will eventually become depleted of refrigerant. Systems that are in good condition may lose up to half a pound per year, and this is considered normal. If refrigerant losses are more than this, the technician must locate the refrigerant leak and repair it as necessary.

Refrigerant leaks can occur anywhere throughout the system. This being said, there are a few locations that seem to cause more problems than others:

- Leaks can be around refrigerant lines at the compressor and at the compressor shaft seal.
- Leaks can be around various hose fittings and joints throughout the system.
- Condensers are notoriously susceptible to leaks, as road debris is often driven through it at high speed, nicking or puncturing the coil.
- Refrigerant can also be lost through hose **permeation.**
- Also, don't rule out the evaporator coil itself.

Before performing one of the many leak detection methods, first perform a visual inspection by looking for signs of compressor oil. Whenever refrigerant is lost, some of the compressor lubrication oil will also be pushed out at the site of the leak. Look for wet oily spots along refrigerant line hoses and components. Look for damage and corrosion of the refrigerant lines, hoses, and components. If a leak is suspected, it may be confirmed by using one of the following methods.

Leak Detection Methods

1. One of the simplest methods of finding leaks in an air-conditioning system is with a soapsuds solution. The solution is sprayed or applied to the suspected leak location or fitting, and if the leak is large enough, it will cause bubbles to form at the point of the leak. This type of leak detection method is used when no other form of detection equipment is available. It may also be used to pinpoint a large leak that makes a sensitive electronic leak detector activate anywhere in the vicinity of the leak.

2. Another method along the same line as the soapsuds solution is a commercially available **fluid-type leak detector.** This solution can find smaller leaks than the soapsuds solution. The fluid is sprayed or applied to the suspected leak area to be tested. If a leak is present, the liquid will form clusters of bubbles or large bubbles (depending on the size of the leak) at the site of the leak. This fluid-type leak detector may also be

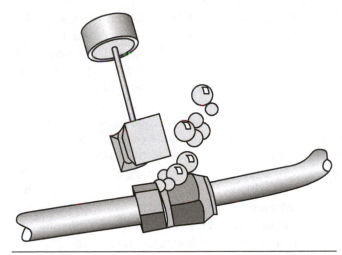

Figure 10-5 A fluid-type leak detector is applied to a suspected leakage point.

used to pinpoint a large leak that makes a sensitive electronic leak detector activate anywhere in the vicinity of the leak. (see **Figure 10-5**).

3. **Flame-type leak detectors,** although not commonly used today, were very common for finding CFC refrigerant leaks. Another name for this device is a halide torch. The device is threaded on top of a disposable propane tank. The propane is burned in the presence of a copper element. The flame will change colors (green or turquoise) when it comes in contact with a CFC refrigerant, depending upon the size of the leak. The technician runs an inlet hose over any suspected areas and watches for any change in color indicating a leak. These leak detectors are a potential fire hazard, and when the refrigerant is burned, it can produce poisonous phosgene gas. (see **Figure 10-6**).

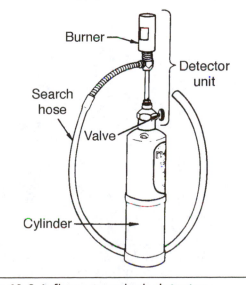

Figure 10-6 A flame-type leak detector.

4. **Ultraviolet (UV) leak detectors,** also called black light leak detectors, are used to locate very small and troublesome leaks. Some air-conditioning systems have phosphorous dye installed during production and are labeled on the compressor or receiver drier if so equipped. The phosphorous dye may also be installed in the low side of an air-conditioning system in a small specific quantity. The dye will mix with the compressor oil and will be pushed out of the system with the escaping refrigerant. The technician can then broadcast a standard black light over the system, and any leak found will fluoresce bright yellow. This method of leak detection is used to pinpoint hard-to-find leaks, especially cold or hot leaks (see **Figure 10-7**).

5. **Electronic leak detectors** should be used by any shop regularly servicing air-conditioning systems. It is the most sensitive of all the leak detection methods. Some can find a leak of one half of an ounce per year. The sensitivity of this device can make pinpointing leaks in confined spaces very difficult. Electronic leak detectors are small handheld instruments with a flexible probe that the technician runs over the refrigerant lines and fittings but without actually making contact with the lines. The end of the probe should be moved at a rate of 1 to 2 inches per second (25–50 mm). When the probe senses a leak, the detector will set off an audible alarm or a visual light depending which option is selected by the operator. The area being tested should be free of oil, grease, and residual refrigerant before starting the leak detection process. Suspected leakage areas

should be cleaned using soap and water, not a solvent. A detected leak should be an active flow leak, not a residual condition caused by refrigerant trapped under a film of oil. Electronic leak detectors can be refrigerant specific as to the type of refrigerant they can detect. (see **Figure 10-8, Figure 10-9, and Figure 10-10**).

6. **Ultrasonic leak detectors** are another way of finding refrigerant leaks. This type of leak detector actually listens for them. This is

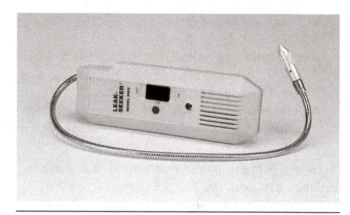

Figure 10-8 Electronic leak detectors are extremely accurate in finding refrigerant leaks. *(Courtesy of John Dixon)*

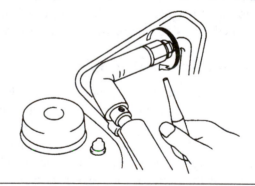

Figure 10-9 Tracing a suspected leak with an electronic leak detector.

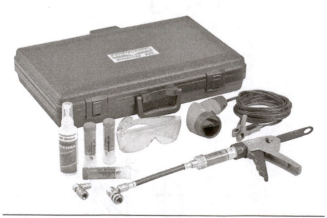

Figure 10-7 An ultraviolet (UV) leak detector kit, used to pinpoint hard-to-find leaks. *(Courtesy of Robinair, A Business Unit of Service Solutions)*

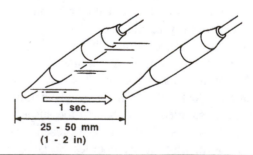

Figure 10-10 The rate of speed at which the probe of an electronic leak detector should be moved.

accomplished with the use of an ultrasonic tester. This device is able to detect sounds in the ultrasonic frequency that can't be detected by the human ear. The detector then converts and amplifies the sound so that the technician can hear it using a headset. Some detectors will also display the sound/leak rate. Ultrasonic leak detectors are fast and accurate, as they are not falsely triggered like electronic detectors. These detectors don't usually need to be re-calibrated to compensate for background noise. Ultrasonic leak detectors can only be used on systems that are pressurized or in a vacuum because they sense the frequency of escaping gases or air entering the system. An ultraviolet detector can be used when the system is empty, as the dye will already have been pushed out of the leak site. Ultrasonic leak detectors can also be used to find vacuum leaks on vehicles or can be used in a preventive maintenance pro-gram to detect bearings or solenoids that are starting to fail. In order to use the ultrasonic leak detector, the technician puts on the head-set and turns the unit on. The technician can then adjust the sensitivity level and begin running the probe over any suspect leak areas. The sound of the leak will be amplified in the headset. The sound will get louder as the probe is positioned closer to the leak, allowing the technician to actually pinpoint it by the loud-ness of the sound. Most units automatically block out ambient noise, wind, stray gases, or other contaminants so that it will not detect a false leak. (see **Figure 10-11**).

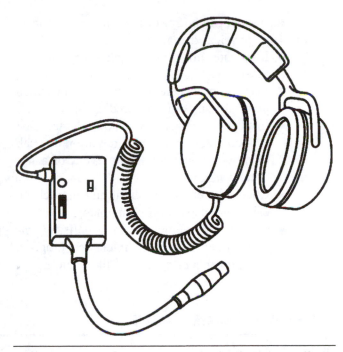

Figure 10-11 An ultrasonic leak detector allows technicians to hear refrigerant or vacuum leaks in the ultrasonic range. *(Courtesy of SPX Corp.)*

CAUTION *Air-conditioning systems are con-stantly under pressure. Before internal repairs are made to any air-conditioning system, refrigerant must be **recovered** (removed).*

At this point only technicians certified with the Environmental Protection Agency (EPA) under Section 609 of the Clean Air Act are able to make any internal repairs or maintenance to the air-conditioning system.

Summary

- Inspect keys for any cracks that could al-low the key to break inside the ignition or lock cylinder.
- With the vehicle running, check the opera-tion of the climate control system, and check the operation of all instrument/gauges and panel lights. Record oil pressure and charg-ing voltage.
- With the vehicle running, check the oper-ation of the climate control system.
- Check the operation of any accessories that the vehicle may have.
- Using the onboard diagnostic system, ex-tract and record engine, transmission, and brake monitoring information and codes.

- Inspect the condition of all onboard safety equipment.
- Inspect seat belts and sleeper restraints for proper operation.
- Inspect operation of windshield wipers and inspect wiper blades and arms.
- Check the vehicle for all required vehicle permits.
- Inspect the windshield glass for any cracks, chips, and discoloration.
- Check the condition of the driver's seat.
- Inspect the cab steps and grab handles.
- Inspect all mirrors on the vehicle for good visibility and cleanliness.

- Inspect cab bodywork and note any defects or physical damage.
- Lubricate all moving parts on the cab assembly.
- Inspect the accelerator, brake, and clutch pedal for smooth non-binding operation.
- Inspect the cab safety locking device, that it is operational and will not allow the cab to move downward in the event of a hydraulic failure.
- Check the cab ride height and inspect the cab's air suspension components.
- Commercial vehicles can benefit by implementing a body maintenance program that prevents corrosion of the vehicle.

- When inspecting the air-conditioning system, check all lines and components for secure mounting, possible leaks, and cleanliness.
- The proper gap in the compressor clutch plates can be checked with a standard business card.
- An air-conditioning performance test confirms that the system is functioning properly.
- Air-conditioning systems leak refrigerant over time and will eventually become depleted of refrigerant.
- To test for refrigerant leaks, use one of the methods that are appropriate for the vehicle.

Review Questions

1. Before testing the operation of the air horn, the technician should ensure that:

   A. the vehicle air brakes are released.

   B. the ignition key is in the on position.

   C. there is sufficient air in the reservoirs.

   D. the electric horn operates correctly.

2. While performing a cab inspection, you find you can't tell if dash instrument lighting is operating. What should you do?

   A. Turn the switch till the dash lights are at their lowest intensity.

   B. Move the vehicle into direct sunlight so the dash shows up better.

   C. Move the vehicle into non-direct sunlight so the dash lights will show up better.

   D. None of the above

3. What item is a required safety item that must be in the cab and in good working order?

   A. Reflective triangles

   B. First-aid kit

   C. Latex gloves

   D. Spare fuses

   E. All of the above

4. When a truck's windshield receives a small stone chip, the windshield is non-repairable and must be replaced immediately.

   A. True

   B. False

5. To check the ride height of a cab with an air ride suspension, the technician is first required to know what the measurement is and where to measure it. Where will the technician find this information?

   A. On the vehicle firewall

   B. On the vehicle door frame

   C. In the driver's manual

   D. In the service manual

6. What is the name of the device designed to corrode away instead of the vehicle body as a form of preventive body rust protection?

    A. Sacrificial electron   C. Sacrificial anode

    B. Melting rod   D. Magnetic coupler

7. When an air-conditioning system is performance tested, the truck's engine should be run at _____ rpm.

    A. 1,000   C. 1,500

    B. 1,250   D. 1,750

8. When an air-conditioning system is performance tested, the vehicle should be run for around 10 minutes to _____.

    A. pre-cool the system   C. subcool the refrigerant

    B. bring the truck's engine up to D. stabilize the air conditioning system
       operating temperature

9. When an air-conditioning system is performance tested, what should the suction line feel like to your hand?

    A. Hot   C. Cool

    B. Warm   D. Cold

10. What type of refrigerant leak detector is the most sensitive?

    A. Fluid-type leak detectors   C. Ultraviolet (UV) leak detectors

    B. Flame-type leak detectors   D. Electronic leak detectors

11 Steering and Suspension

Objectives

Upon completion and review of this chapter, the student should be able to:

- List the components of the steering system and explain their function.
- Perform a steering axle inspection.
- Verify a steering complaint.
- Perform a complete steering knuckle inspection.
- Explain how to perform a tie-rod inspection.
- Perform a wheel bearing inspection.
- Explain some of the basic steering geometry.
- List the importance of a suspension system.
- Explain some of the basic terminology used when discussing suspensions.
- Explain the different types of suspension systems used in the heavy-duty truck industry.
- Perform service inspections on the various types of suspensions.
- Explain how to identify and maintain U-bolts.
- Perform inspection procedures for air spring suspensions.
- Explain and perform servicing procedures for height control valves.

Key Terms

Ackerman arm

air spring suspensions

ball joint

camber

caster

convoluted

dampening

drag link

height control valve

jounce

kingpin

leaf spring

Pitman arm

rebound

reversible sleeve

spring pack

steering control arm

steering gear

steering knuckle

toe

toe-in

toe-out

unsprung weight

INTRODUCTION

As with the braking system, there is absolutely no room for error in regard to steering system maintenance. These two systems are the only things the driver has to control the path of the vehicle on public roads.

While performing a PM inspection, a technician must be able to find any faults that may be present in the steering system. Often, improper tire wear can be the first indication that there is a problem with the system.

The suspension is another area that must be maintained and checked during a service to ensure proper ride characteristics as well as to maintain axle alignment.

SYSTEM OVERVIEW

This chapter starts by explaining the steering system components and their functions as well as how to check them for wear during routine inspections. The chapter also explains some of the basic steering geometry so that you will understand the effects or consequences of adjusting any of the steering components incorrectly.

We then discuss some of the more common types of suspension systems used by the heavy-duty trucking industry as well as some of the maintenance checks and adjustments that must be performed to keep the suspension system in top operating condition.

STEERING SYSTEM COMPONENTS

The steering system of the truck is made up of the steering wheel, steering column, steering gear, and steering linkages that actually move the steering tires. All of these components must be inspected and maintained on PM service.

Steering Wheel

The steering wheel is the driver's means to control the direction that the vehicle travels. It is therefore the primary input to the steering system. The steering wheel is formed from a strong steel rod that is formed into the shape of a wheel and supported by spokes.

When the driver applies a turning effort to the wheel, it becomes a turning force on the steering column. The larger the steering wheel diameter, the more torque is generated from the same amount of driver effort. Steering wheels on heavy-duty trucks are typically 22 inches in diameter. This large diameter can help a driver control a vehicle equipped with power-assisted steering when the power assist fails.

Many trucks also incorporate an electric horn at the center hub of the steering wheel.

Steering Column

The steering column connects the steering wheel to the steering gear. **Figure 11-1** shows a complete steering column assembly. The major components of the steering column assembly are a jacket, bearing assemblies, a steering column shaft, and wiring and contact assemblies for the electric horn. At the upper end of the steering column shaft are threads machined for a nut that holds the steering wheel in place. Straight external splines on the column shaft match internal splies on the steering wheel hub. This allows the two components to be coupled with zero axial play. The lower end of the column shaft has external splines that mate to internal splines on the column yoke.

The steering column is mounted to the firewall and dash by brackets. Coupled to the steering column upper shaft by a pair of yokes and the U-joint assembly is the lower shaft assembly. The U-joint permits some angular deviation between the upper and lower column shafts. The lower column shaft assembly connects to the steering gear.

STEERING COLUMN MAINTENANCE

The steering column should be routinely lubricated, to keep the component moving freely. The U-joints of the steering column are phased like the U-joints of a driveshaft and this must be retained, or the result will be binding in the assembly. The U-joints in a steering shaft should be lubed in the same manner as those in a driveshaft. Many steering-related problems as well as hard steering complaints are the result of problems caused by U-joint failures.

Steering Gear

The **steering gear,** sometimes referred to as the steering box, is used to multiply the driver's efforts at the steering wheel as well as change its direction. There are two general categories of heavy-duty steering gears: worm and sector shaft and reticulating ball. **Figure 11-2** shows a steering knuckle assembly.

Pitman Arm

A **Pitman arm** is a steel lever, spline-attached to the sector (output) shaft of the steering gear. The end of the Pitman arm moves through an arc with the sector shaft center, forming its center. The Pitman arm functions to change the rotary motion of the steering gear sector shaft to linear motion. The length of the Pitman arm affects

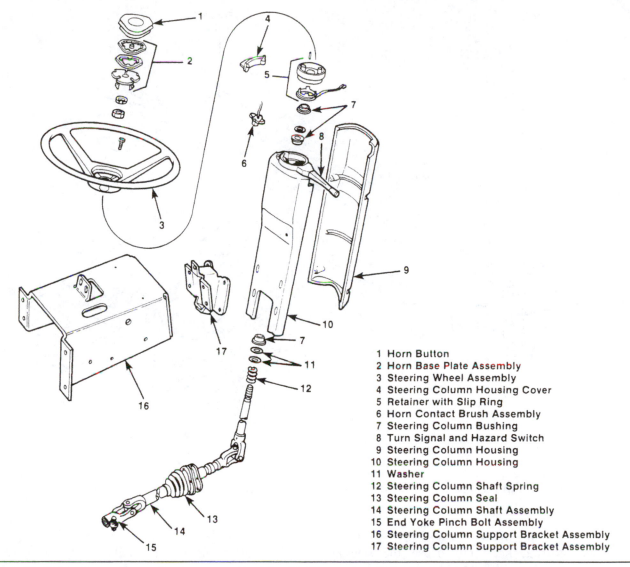

1 Horn Button
2 Horn Base Plate Assembly
3 Steering Wheel Assembly
4 Steering Column Housing Cover
5 Retainer with Slip Ring
6 Horn Contact Brush Assembly
7 Steering Column Bushing
8 Turn Signal and Hazard Switch
9 Steering Column Housing
10 Steering Column Housing
11 Washer
12 Steering Column Shaft Spring
13 Steering Column Seal
14 Steering Column Shaft Assembly
15 End Yoke Pinch Bolt Assembly
16 Steering Column Support Bracket Assembly
17 Steering Column Support Bracket Assembly

Figure 11-1 Steering column for a conventional cab. *(Courtesy of Navistar International Corp.)*

leverage and therefore steering response. A longer Pitman arm will generate more steering motion at the front wheels for a given amount of steering wheel movement. The Pitman arm is shown in **Figure 11-3.**

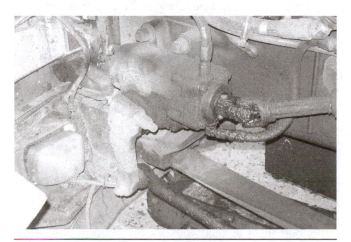

Figure 11-2 Steering gear. *(Courtesy of John Dixon)*

Drag Link

The **drag link** is a forged rod that connects the Pitman arm to the steering control arm. The drag link can be a one-piece or two-piece component. The two-piece design allows the technician to adjust its length, which makes it easy to center the steering gear with the wheels straight ahead. One-piece drag links are used on systems with very close tolerances. Other components are used to make adjustments to the system when a one-piece drag link is connected at each end by ball joints. These ball joints help isolate the steering gear and Pitman arm from axle motion. The drag link can be seen in **Figure 11-3.**

Steering Control Arm

The **steering control arm** (steering arm or lever) connects the steering knuckle to the drag link on the driver's side of the vehicle. It is usually a drop-forged,

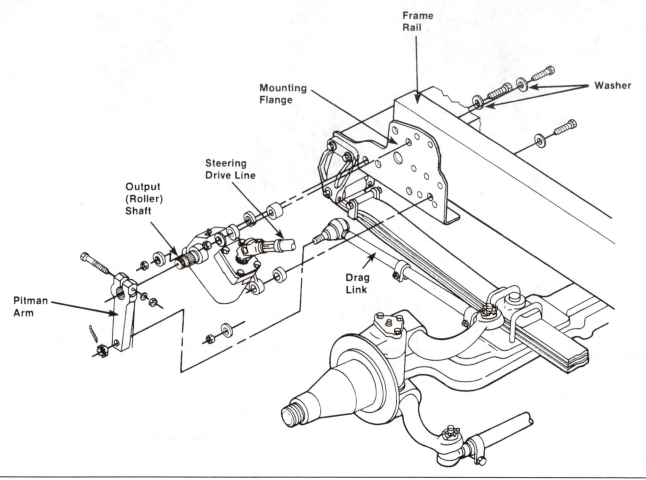

Figure 11-3 Typical manual steering gear installation. *(Courtesy of Daimler Trucks North America)*

tempered steel component. When the drag link is moved, the steering arm will cause the steering knuckle to move, changing the direction of the front tire. See steering components in **Figure 11-4.**

Steering Knuckle

The **steering knuckle** in **Figure 11-5** is mounted by **kingpins** (knuckle pins) to the solidly mounted front axle. Kingpins allow for the pivoting action of the steering knuckle required to control the direction of the vehicle. The steering knuckles incorporate the spindle onto which wheel bearings and wheel hubs are mounted, plus a flange to which the brake spider is mounted. A steering control arm is attached to the top of the knuckle assembly on the left side steering knuckle, and Ackerman (tie-rod) arms are attached to the bottom of both the left and right steering knuckles.

Kingpins

Kingpins can be either tapered or straight. Tapered pins are drawn into the axle center and secured by tightening a nut at the upper pin end. Straight kingpins

are secured to the axle with tapered draw keys that bear against flats on the pins. Tapered pins are usually sealed and may not have a grease nipple, so they cannot be lubricated on a service. Straight pins have a cap on either end (top and bottom) to retain grease. Grease fittings are used to lubricate the steering knuckles. (see **Figure 11-6**).

KINGPIN MAINTENANCE

The kingpins must be lubricated on all PM services. A good-quality chassis grease can be used to lubricate most steering knuckles, but always consult the OEM. The kingpin is held securely to the front axle, while the steering knuckle pivots on the kingpins. The steering knuckles should be greased by raising the steering axle off the ground. This removes the weight from the wheels and helps properly distribute grease through the kingpins. Some new vehicles use unitized steering axles; the steering knuckle on these is lubed-for-life, and no attempt should be made to lubricate these.

Ball Joints

A **ball joint** is a ball-and-socket assembly made up of a forged steel ball with a threaded stud attached to

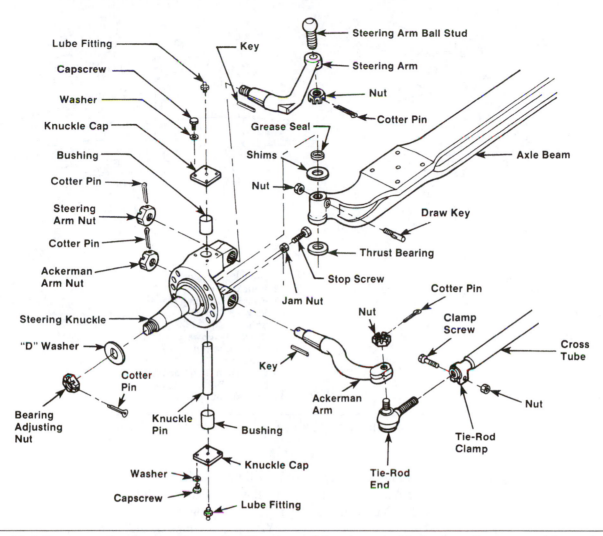

Figure 11-4 Components of a steering axle. *(Courtesy of Roadranger Marketing. One great drive train, two great companies—Eaton and Dana Corporations.)*

it. A socket shell grips the ball. This assembly will allow movement in any direction from the fixed ball. Ball joints are necessary to compensate for movement between the steering linkage and the frame of the vehicle that results when the front axle springs flex, and do so without affecting steering. Ball joints are used in the ends of steering control arms, idler arms, and tie-rod ends.

Ackerman Arm

The **Ackerman arm** or tie-rod arm is used to transfer and synchronize steering action on both steer wheels on a steering axle. Ackerman arms are drop-forged, tempered steel levers that are angled to allow the outer wheel to turn at less of an angle that the inner wheel (toe-out on turn). One end of the Ackerman arm is keyed and bolted to the lower portion of the steering knuckle. The other end is taper bored to mount the tapered stud of the tie-rod ball joint.

Tie-Rod Assembly

The tie-rod assembly consists of two tie-rod ends fastened to a cross tube. The assembly transfers any

Figure 11-5 Steering knuckle. *(Courtesy of John Dixon)*

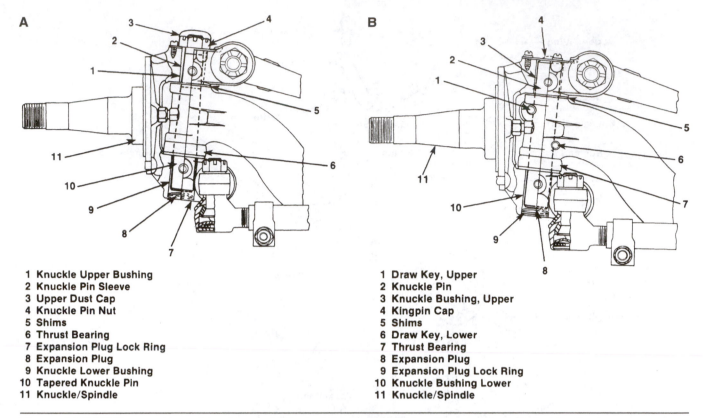

A

1 Knuckle Upper Bushing
2 Knuckle Pin Sleeve
3 Upper Dust Cap
4 Knuckle Pin Nut
5 Shims
6 Thrust Bearing
7 Expansion Plug Lock Ring
8 Expansion Plug
9 Knuckle Lower Bushing
10 Tapered Knuckle Pin
11 Knuckle/Spindle

B

1 Draw Key, Upper
2 Knuckle Pin
3 Knuckle Bushing, Upper
4 Kingpin Cap
5 Shims
6 Draw Key, Lower
7 Thrust Bearing
8 Expansion Plug
9 Expansion Plug Lock Ring
10 Knuckle Bushing Lower
11 Knuckle/Spindle

Figure 11-6 (A) Tapered knuckle pin; and (B) straight knuckle pin. *(Courtesy of Arvin Meritor)*

movement from the right side steering knuckle to the right side Ackerman arm to the left side Ackerman arm and on to the left side steering knuckles. You can see the tie-rod assembly in **Figure 11-7.**

The tie-rod ends are ball joints that connect to the Ackerman arms on each steering knuckle. A tie-rod cross tube is a steel rod that runs parallel to the front axle, extending from one side to the other. Each tie-rod end assembly is threaded onto either end of the tie-rod tube and the ball joints secured to the Ackerman arms. The threaded ends of the tie-rod have opposite threads,

one will be right hand thread and the opposite side left hand threads. The will allow the tie-rod assembly to be shortened or lengthened.

When the clamps of the tie-rod ends are loosened, the tie-rod tube can be turned to adjust the toe setting of the vehicle. Lengthening the assembly will increase toe setting, and shortening will reduce toe setting. This is the method used to set steering toe, and it will be discussed later in this chapter.

STEERING AXLE INSPECTION

Always consult specific service procedures and maintenance intervals prescribed by the OEM. The following procedures are general guidelines for periodic steering axle inspection:

- General Inspection: A thorough visual inspection for proper assembly, broken parts, and looseness should be performed each time the vehicle is lubricated. In addition, ensure that the springs to axle mounting nuts and steering connection fasteners are secure.
- Wheel Alignment: Front steering wheel alignment should be checked periodically according to OEM recommended service intervals (typically 3 months or 25,000 miles/40,000 km), If

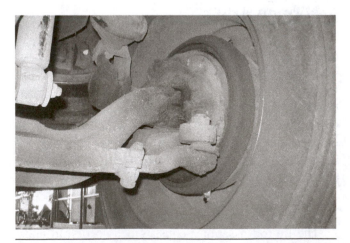

Figure 11-7 Tie-rod ends. *(Courtesy of John Dixon)*

excessive steering effort, vehicle wander, or even uneven excessive tire wear is evident, the wheel alignment should be checked immediately.

- Steering Axle Stops: Although steering axle stops should be checked periodically, they seldom need adjustment. However, if the steering turning radius is insufficient or excessive (resulting in tire or wheel contact with the frame), the stops should be adjusted.
- Tie-Rod Ends: Tie-rod ends should be inspected each time the axle is lubricated; check for torn or cracked seals and boots, worn ball sockets, or loose fasteners. Have an assistant move the steering wheel back and forth while checking for excessive movement of the ball joint.
- Steering Knuckle Thrust Bearing: The knuckle thrust bearings should be checked each time the hub/drums are removed. Knuckle vertical play should be adjusted each time the kingpins are removed for service, at each axle overhaul, or whenever excessive knuckle vertical movement is noted.
- Kingpins: The kingpin and its bushing should be inspected whenever the pins are removed for service, at axle overhaul, or when looseness is noted.
- Wheel Bearings: The wheel bearings should be inspected for damage or wear each time the hub or drum is removed. Check for signs of wear or distress.
- Lubrication: Steering axle components should be lubricated at least every 25,000 miles (40,000 km). Good-quality chassis grease should be used. Kingpins, thrust bearings, tie-rod ends, and any other steering linkage should be lubricated at each PM service interval. Those without grease fittings are usually permanently lubricated. Consult the OEM service literature for the recommended steering gear lubricant. This may be gear lube, engine oil, or hydraulic transmission fluids. Because most steering gears have a number of lubricant options, you should try to top with the same lubricant that is already in the gear or reservoir.

Inspection Procedure

When a steering problem is reported, systematically inspect the vehicle steering system, front and rear suspensions, and trailer suspensions. In most cases, a road test will be required, but never take a truck out onto a road until you are sure it is roadworthy. If a reported problem occurs only when the vehicle is loaded, you should test drive the vehicle loaded, and when you check out steering systems, remember that other chassis systems can cause steering problems.

CAUTION *All steering mechanisms are critical safety items. A vehicle should be deadlined (OOS report) when a defect is reported. It is essential that instructions in the service literature be adhered to. Failure to observe these procedures may result in loss of steering with life-threatening results.*

Perform the following sequence of checks when symptoms such as rapid tire wear, hard steering, or erratic steering indicate a problem in the steering system:

1. Check that the front tires are the same size and model. Ensure they are equally and adequately inflated. Underinflated tires cause hard steering. Overinflated tires reduce the road contact footprint, reducing control.
2. If the steering problem occurs only when the vehicle is loaded, make sure that the fifth wheel is adequately lubricated.
3. Inspect the steering linkages for loose, damaged, or worn parts. Steering linkage components include the tie-rod assembly, steering arms, bushings, and other components that carry movement of the Pitman arm to the steering knuckles. The wheels should turn smoothly from stop to stop through a full turn cycle.
4. Inspect the drag link, steering driveshaft(s), and upper steering column for worn or damaged parts.
5. Ensure that the steering column components, especially the U-joints, are adequately lubricated.
6. Check front axle wheel alignment, including wheel bearing adjustment, caster, camber angles, and toe-in.
7. Check the rear axle alignment. Rear axle misalignment can cause hard or erratic steering. If needed, align the rear axle(s).
8. Inspect the front axle suspension for worn or damaged parts.
9. With the front wheels straight ahead, turn the steering wheel until motion is seen at the wheels. Align a reference mark on the steering wheel with a mark on a ruler and then slowly turn the steering wheel in the opposite direction until motion is again observed at the wheels. Measure the lash (freeplay) at the steering

wheel. Too much lash exists if the steering wheel movement exceeds:

- 2-3/4 inches (70 mm) for a 22-inch (0.56 m) steering wheel
- 2-1/4 inches (57 mm) for 20-inch (0.51 m) steering wheel

10. Turn the steering wheel through a full right and full left turn. If the front wheels cannot be turned to the right and left axle steering stops without binding or noticeable interference, the problem is likely in the steering gear.
11. Secure the steering wheel in the straight ahead driving position. Move the front wheels from side to side. Any play in the steering gear bearings will be felt in the drag link ball joint at the Pitman arm. If any bearing play exists, adjustments to the steering gear may be in order.

CAUTION *Do not drive a vehicle with too much lash in the steering gear. Excessive lash is a sign of an improperly adjusted steering gear or worn or otherwise damaged steering gear components. Driving the vehicle in this condition could result in a loss of steering control.*

12. Check the tractor fifth wheel. Lack of lubrication between the fifth wheel plate and trailer upper coupler can cause serious steering problems. This is most common in tractor-trailer combinations that seldom uncouple. With most fifth wheels, the tractor and trailer must be split to apply grease directly to the fifth wheel top plate.

CAUTION *Before performing any servicing procedure on the front suspension, set the parking brake and chock the drive wheels to prevent the vehicle from moving. After jacking the truck up until the steering axle wheels are raised off the ground, support the chassis with safety stands. Never work under a vehicle supported only by a jack.*

STEERING KNUCKLE INSPECTION

Steering knuckles can be the source of a number of steering-related problems and many of these originate from lack of lubrication. The following tests should be performed to check out steering knuckles and kingpins (see **Photo Sequence 9**).

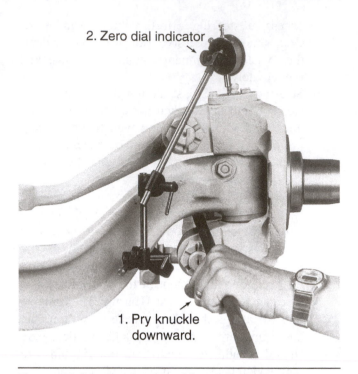

Figure 11-8 Inspecting knuckle vertical play. *(Courtesy of Roadranger Marketing. One great drive train, two great companies—Eaton and Dana Corporations.)*

Steering Knuckle Vertical Play

1. Mount a dial indicator on the axle beam. Position the indicator plunger on the knuckle cap. (see **Figure 11-8**).
2. Pry the steering knuckle downward.
3. Zero the dial indicator.
4. Lower the front axle to obtain the dial indicator reading. If the reading exceeds the OEM specification (typically 0.04 inch or 1.0 mm), inspect the thrust bearing. Replace them if necessary.

Kingpin Upper Bushing Freeplay

1. Mount a dial indicator on the axle, as shown in **Figure 11-9.** Place the indicator plunger on the upper part of the knuckle, as shown.

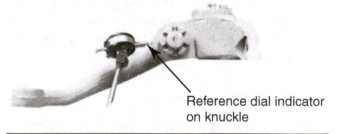

Figure 11-9 Inspecting freeplay in upper bushing. *(Courtesy of Roadranger Marketing. One great drive train, two great companies—Eaton and Dana Corporations.)*

PHOTO SEQUENCE 9 — Measuring Steering Kingpin Wear and Vertical End Play

P9-1 Apply the parking brakes, block the rear wheels, and use a floor jack to raise the front axle until the tires are off the shop floor.

P9-2 Lower the vehicle so that it is supported securely on safety stands with the front tires still off the shop floor.

P9-3 Mount a dial indicator on the front axle I-beam and position the dial indicator plunger on the inner side of the upper end of the steering knuckle. Zero the dial indicator.

P9-4 While a helper moves the top of the wheel and tire inward and outward, observe the dial indicator reading. If the total movement on the dial indicator exceeds the specified kingpin bushing movement, the kingpin bushing must be replaced.

P9-5 Mount the dial indicator on the front axle I-beam with the dial indicator plunger touching the inner side of the lower end of the knuckle. Zero the dial indicator. While a helper moves the bottom of the tire inward and outward, observe the dial indicator reading. If the dial indicator reading exceeds the specified kingpin bushing movement, replace the lower kingpin bushing.

P9-6 Mount the dial indicator on the axle I-beam and position the dial indicator plunger on top of the upper knuckle joint cap.

P9-7 Use a pry bar to force the steering knuckle downward. Check the reading on the dial indicator.

P9-8 Next, observe the dial indicator reading while a helper uses a large pry bar to lift upward on the tire and wheel. If the dial indicator reading exceeds the truck manufacturer's specifications, remove the steering knuckle and inspect the thrust bearing. Replace this bearing if necessary and install the required shim thickness.

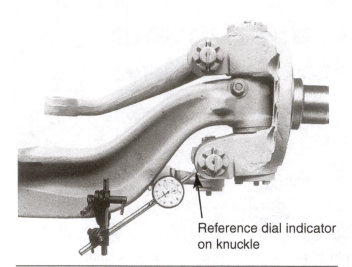

Reference dial indicator on knuckle

Figure 11-10 Measuring freeplay in lower bushing. *(Courtesy of Roadranger Marketing. One great drive train, two great companies—Eaton and Dana Corporations.)*

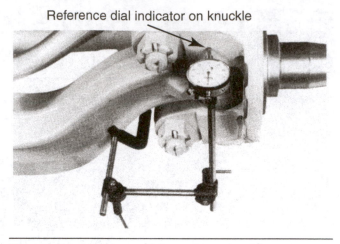

Reference dial indicator on knuckle

Figure 11-11 Measuring upper bushing torque deflection. *(Courtesy of Roadranger Marketing. One great drive train, two great companies—Eaton and Dana Corporations.)*

2. Move the top of the wheel in and out with a push/pull motion. Have someone monitor the dial indicator readings.
3. Readings that exceed the OEM specifications (typically 0.015 inch or 0.38 mm) indicate the need for bushing replacement.

Kingpin Lower Bushing Freeplay

1. Mount the dial indicator on the axle. Reference the plunger on the lower tie-rod end socket of the steering knuckle, as shown in **Figure 11-10.**
2. Move the bottom of the wheel in and out with a push/pull motion. Have an assistant read the dial indicator.
3. A dial indicator reading that exceeds the OEM specifications (typically 0.015 inch or 0.38 mm) indicates that the lower bushing should be replaced.

Kingpin Upper Bushing Torque Deflection

1. Mount the dial indicator to the axle, referencing the upper knuckle steering arm socket area. (see **Figure 11-11**).
2. Have someone apply the foot brake. Try to roll the wheel forward and backward and note the deflection.
3. Readings in excess of the manufacturer's specifications (typically 0.015 inch or 0.38 mm) indicate that the top bushings should be replaced.

Lower Bushing Torque Deflection Test

1. Mount the dial indicator on the axle and the plunger on the lower bushing area.
2. Have someone apply the foot brake. Try to roll the wheel forward and backward and note the defection.
3. Readings that exceed the OEM specifications (typically 0.015 inch or 0.38 mm) indicate that the lower bushing should be replaced.

TIE-ROD INSPECTION

1. Shake the tie-rod or cross tube. Movement or looseness between the tapered shaft of the ball and the cross-tube socket members indicates that the tie-rod end assembly should be replaced.
2. The threaded portion of both tie-rod ends must be inserted completely beyond the tie-rod split recess, as shown in **Figure 11-12.** This is important for adequate clamping force. If this is not possible, components will have to be replaced. Check to see if the cross tube or the tie-rod end is at fault.
3. Ball socket torque (without the boot) should be 5 inch-pounds or more on a disconnected tie-rod end. To perform this test, install the nut on the tapered ball joint shaft and install an inch-pound torque wrench. A flexible beam type torque wrench works best for this test. Rotate the torque wrench and note the torque required to rotate the socket. Loose assemblies will adversely affect the steering system performance and might prevent adjustment of the steering assembly to the vehicle OEM alignment specifications.

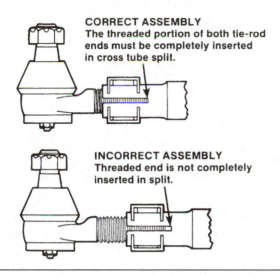

CORRECT ASSEMBLY
The threaded portion of both tie-rod ends must be completely inserted in cross tube split.

INCORRECT ASSEMBLY
Threaded end is not completely inserted in split.

Figure 11-12 Correct and incorrect tie-rod installation. *(Courtesy of Roadranger Marketing. One great drive train, two great companies—Eaton and Dana Corporations.)*

4. If the tapered shank-to-tie-rod arm connection is loose or the cotter pin is missing, disconnect and inspect these components for worn contact surfaces. If either one is worn, replace it.

WHEEL BEARING INSPECTION

The following procedure can be used on steering axles that do not use preset wheel bearing assemblies:

1. Remove the wheel bearings from the wheel hub.
2. Thoroughly clean the bearings, spindles, hubcap, and hub cavity. Parts can be washed in solvent and rinsed completely and thoroughly dried off.
3. Clean hands and tools to be used before assembly.
4. Closely examine the bearing cups, rollers, and cage assembly. Don't forget to examine the inner race by holding the bearing assembly up to the light.
5. Replace any damaged or distressed bearings with mated bearing assemblies. (Use new cups and cones from the same manufacturer.)
6. Lubricate bearings, after making certain they are free of moisture or other contaminants. Refer to Chapter 8 for additional information on installing bearings and the procedure used for pre-set hubs.

STEERING GEOMETRY

The major front end alignment settings are **toe, camber,** and **caster.**

Of these alignment angles, toe if set incorrectly will cause the most amount of tire wear, followed by camber. Improper caster by itself generally does not cause a tire wear issue.

Toe

Toe is defined as the difference in distance that the front if the tire is compared to the back of the tire on the steering axle as viewed from above. **Toe-in** exists when the tires are closer together at the front than at the rear. If toe-in is excessive, the tire will scrub (wear) in a direction from the outside toward the inside of the tire, resulting in feather edge wear on the inner edge. **Toe-out** is when the tires are closer together at the rear when viewed from above. Excessive toe-out wear can be indicated by tire scrub from the inside toward the outside of the tire, resulting in feather edge wear on the outer edge. (see **Figure 11-13**).

Camber

Camber is the tilt of the tires as viewed from the front of the vehicle, Positive camber exists when the tire are

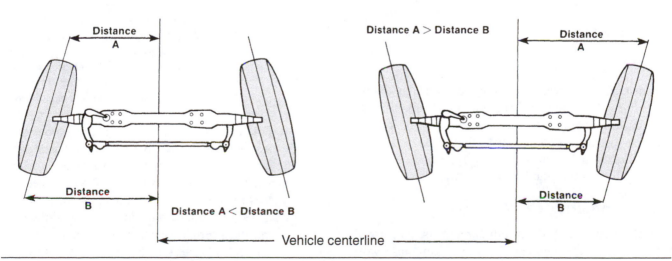

Figure 11-13 Toe-in (left) and toe-out (right). *(Courtesy of Navistar International Corp.)*

POSITIVE CAMBER

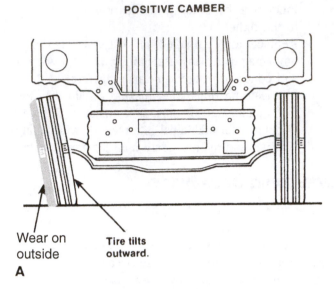

Wear on
outside

**Tire tilts
outward.**

A

NEGATIVE CAMBER

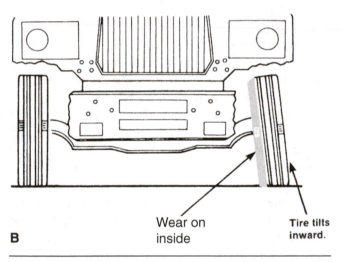

Wear on
inside

**Tire tilts
inward.**

B

Figure 11-14 (A) Positive camber results in wear on the outer edge of the tire; (B) negative camber results in tire wear on the inner edge.

tire are closer together at the bottom. Negative camber exists when the tires are closer together at the top. (see **Figure 11-14**).

Caster

Caster is the forward or rearward tilt of the kingpin centerline when viewed from the side of the vehicle. Zero caster occurs when the centerline of the kingpin is exactly vertical. Positive caster indicates the kingpin is tilted rearward as shown in **Figure 11-15.** Negative caster indicates that the kingpin is tilted forward.

Pre-alignment

Before any alignment adjustment is performed, always check the vehicle for loose kingpins, worn wheel bearings and tie-rod ends, and any looseness in the steering system. Adjust wheel bearing end play

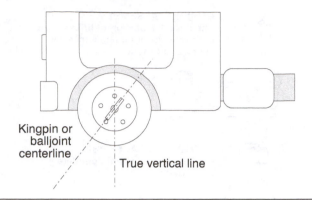

Kingpin or
balljoint
centerline

True vertical line

Figure 11-15 Caster is the forward or rearward (shown) tilt of the kingpin when viewed from the side of the vehicle.

in accordance the recommendation of the OEM. Attempts to correct alignment on a vehicle with worn or loose components are pointless.

STEERING MECHANISM OUT-OF-SERVICE CRITERIA

(Taken from the North American Standard Out-of-Service Criteria by the Commercial Vehicle Safety Alliance, April 1, 2006)

1. **Steering Wheel Free Play**
   In reference to the chart below, if any of these values, inch movement or degrees, are met or exceeded, the vehicle shall be placed out of service.

....

Note: For power steering systems, engine must be running.

....

Steering Wheel Diameter	Manual System Movement 30°	Power System Movement *45°
16" (41 cm)	4-1/2" (11.5 cm) (or more)	6-3/4" (17 cm) (or more)
18" (46 cm)	4-3/4" (12 cm) (or more)	7-1/8" (18 cm) (or more)
19" (48 cm)	5" (13 cm) (or more)	7-1/2" (19 cm) (or more)
20" (51 cm)	5-1/4" (13 cm) (or more)	7-7/8" (20 cm) (or more)
21" (53 cm)	5-1/2" (14 cm) (or more)	8-1/4" (21 cm) (or more)
22" (56 cm)	5-3/4" (15 cm) (or more)	8-5/8" (22 cm) (or more)

- For power systems, if steering wheel movement exceeds 45 degrees before steering axle tires move, proceed as follows: Rock steering wheel left to right between points of power steering valve resistance. If that motion exceeds 30 degrees (or the inch movement values shown for manual steering) vehicle shall be placed out-of-service. This test is to differentiate between excessive lash and power systems designed to avoid providing steering assist when the steering wheel is turned while the truck is motionless (not moving forward or backward).

2. **Steering Column**
   - Any absence or looseness of U-bolt(s) or positioning part(s)
   - Obvious repair-welded universal joint(s)
   - Steering wheel not properly secured

3. **Front Axle Beam and All Steering Components Other Than Steering Column (Including Hub)**
   - Any crack(s)
   - Any obvious welded repair(s)

4. **Steering Gear Box**
   - Any mounting bolt(s) loose or missing
   - Any crack(s) in gear box or mounting brackets
   - Any obvious weld repairs
   - Any looseness of the yoke-coupling to the steering gear input shaft

5. **Pitman Arm**
   - Any looseness of the Pitman arm on the steering gear output shaft
   - Any obvious welded repair(s)

6. **Power Steering**
   - Auxiliary power assist cylinder loose

7. **Ball and Socket Joints**
   - Any movement under steering load of a stud nut
   - Any motion, other that rotational, between any linkage member and its attachment point of more than 1/8 inch (3 mm) measured with hand pressure only
   - Any obvious welded repair(s)

8. **Tie-Rods and Drag Links**
   - Loose clamp(s) or clam bolt(s) on tie-rods or drag links
   - Any looseness in any threaded joint

9. **Nuts**
   - Nuts that are loose or missing from tie-rod, Pitman arms, drag link, steering arm, or tie-rod arms

10. **Steering System**
    - Any modification or other condition that interferes with free movement of any steering component

SUSPENSIONS

The purpose of the suspension system is to support the vehicle and its intended load. It also isolates the vehicle's frame from shock loads encountered when the vehicle passes over the surface of the road. With no suspension, these forces would be transferred directly to the truck frame. Without the suspension performing its job, the harsh pounding would prevent the truck frame, its load, and the driver from lasting very long.

A properly engineered suspension system plays a number of roles:

- It stabilizes the truck when traveling over smooth highway as well as over rough terrain.
- It cushions the chassis from road shock and enables the driver the steer the truck.
- It maintains the proper axle spacing and alignment.
- It keeps the wheels in contact with the ground after going over road irregularities.
- It provides a smooth ride when both loaded and unloaded.

Since trucks are specified by the heaviest load they are expected to haul, the suspension must be able to support that maximum load. However, heavy-duty suspensions that perform comfortably when fully loaded can be harsh and unforgiving when not loaded down. A really good suspension should perform well under loaded, unloaded, highway, and off-road conditions.

For the purpose of this chapter we will be looking at the maintenance issues related to common style truck suspensions, those being:

- Leaf springs
- Equalizer beam: leaf spring and solid rubber spring
- Air spring: pneumatic-only and combination air/leaf spring

Before we begin the maintenance issues related to the various suspensions, there are some suspension terms that will be used:

- **Jounce** literally means "bump." The suspension is moving upward. In suspension terminology, it is the most compressed condition of a spring. For instance, many suspensions use jounce blocks to prevent frame-to-axle contact known as suspension slam.
- **Rebound** is the suspension moving in a downward direction. This occurs after jounce; the spring kicks back.
- **Unsprung weight,** an important factor in a suspension, means the weight of any chassis components not supported by the suspension, for instance, the axles and wheels. Ideally it is kept as low as possible because of the reaction effect, which is one of the reasons for spec'ing aluminum wheels: rebound is quicker.

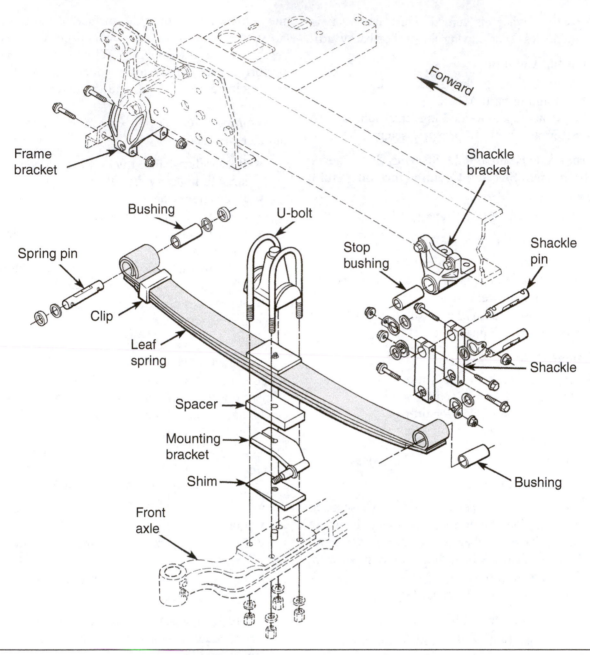

Figure 11-16 Front suspension, multileaf shackle spring. *(Courtesy of Daimler Trucks North America)*

- Oscillation is either rhythmic or irregular vibrations or movements in a suspension. For instance, a good suspension will minimize jounce/rebound oscillations by using dampening devices such as shock absorbers and multileaf spring packs.
- Leaf Spring Suspension: A leaf spring is a steel plate or stack of clamped steel plates. Most leaf springs used in trucks today are manufactured from spring steel. Spring steel is mild alloy steel that has been tempered, that is, heat treated. The result is to provide a leaf spring plate with considerable ability to flex without permanent deforming, Leaf springs may consist of a single

leaf or a series of leaves clamped together known as a **spring pack** (see **Figure 11-16** and **Figure 11-17**).

Servicing Leaf Springs

Steel spring suspensions tend to be low maintenance over the lifetime of the vehicle. The U-bolts should be tightened on a PM service along with lubricating pin bushings and visually inspecting the spring clearance. On new trucks and trailers, it is good practice to retorque U-bolt nuts after the first month or 1,000 miles of operation. Thereafter, tightening of U-bolts should be built into a PM schedule C type

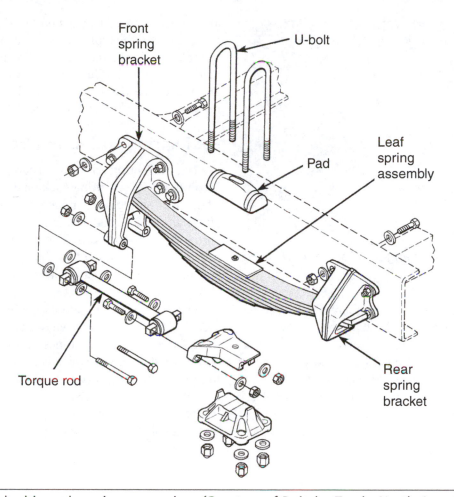

Figure 11-17 Single drive axle spring suspension. *(Courtesy of Daimler Trucks North America)*

service occurring typically between 35,000 and 50,000 linehaul miles, or every 6 months, depending on the application. If the vehicle is operated in an environment with a lot of corrosion (Northern States and Canada), retorquing cannot be accurately performed. In this case, tap the U-bolt with a hammer and look for any signs of movement. This should give you a good indication if it has lost its tension.

If spring pins fail to take grease, the first thing you should do is jack up the frame removing the weight from the spring and try again. If this does not work, try using a hand grease gun with the weight off the spring. A hand grease gun develops more pressure than an air-actuated grease gun. If this fails, you can try heat using a rosebud (make sure the bracket/shackle is not made out of aluminum alloy). But when you spend extra time properly lubing a spring pin, remind yourself that without grease, the pin/bushing will fail if not lubricated. It is only a question of when the failure will occur.

Inspect spring ends to ensure that they have not shifted and come into contact with the sides of equalizer or hanger brackets, as shown in **Figure 11-18**. Inspect spring hangers for cracks (see **Figure 11-19**). Spring end contact indicates that the spring assemblies are not

seated on the axle housing or there is a need for suspension alignment. Check U-bolt seats, U-bolts, and spring assembled for integrity.

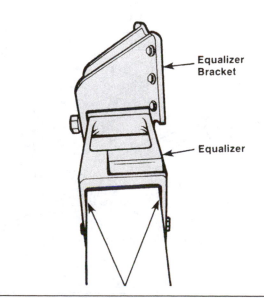

Figure 11-18 Spring end clearance. *(Courtesy of Navistar International Corp.)*

Figure 11-19 Cracked hanger. *(Courtesy of John Dixon)*

A wire brush and solvent or pressure washer can be used to clean the suspension system and remove dirt, grease, and scale. When using solvents, follow the safety precautions recommended by the solvent manufacturer. Inspect the shackle brackets and the spring shackles for cracks, wear, and other damage. Replace any damaged components. Inspect spring leaves for cracks and corrosion. If any cracked or broken leaves are observed, most suspension OEMs recommend that the spring pack be replaced as a unit. The other option is to replace only the damaged leaf(s) in the pack. Never paint leaf springs. This would lower interleaf friction and reduce the dampening capability. If severe rusting or corrosion is observed, replace the spring assembly.

Leaf Spring Equalizing Beam

This leaf spring type suspension (see **Figure 11-20**). incorporates a leaf spring pack on each equalizer beam. These leaf springs are mounted on saddle assemblies above the equalizing beams and pivot at the front on spring pins and brackets. The rear of each spring pack has no rigid attachment to the spring brackets but is free to move forward and backward to compensate for spring deflection.

This type of suspension uses a lever principle to divide the load equally between two axles as well as reduce the effect of road irregularities. Torque rods are used to absorb windup torque, which is the tendency of the axles to turn backward or forward on their axis due to starting or stopping forces. The cross-tube connects the equalizing beams, ensuring correct alignment of the tandem, and prevents damaging load transfer.

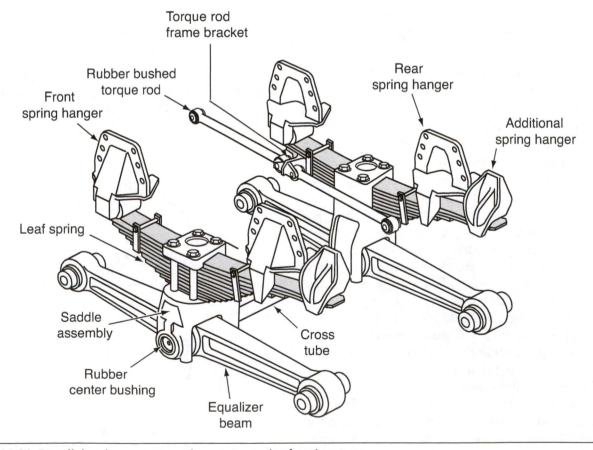

Figure 11-20 Equalizing beam suspension system—leaf spring-type.

Figure 11-21 Solid rubber suspension. *(Courtesy of John Dixon)*

Solid Rubber Spring Equalizing Beam

Solid rubber suspensions have been around for years. Unlike leaf spring suspensions, these use hard rubber cushions to absorb the shock instead of a steel leaf. (see **Figure 11-21**). On these units, rubber load cushions are mounted on a saddle assembly at each side. Mounted between the frame brackets and the suspension, each rubber block unit is secured by four rubber-bushed drive pins, each of which passes through the rubber cushion. All driving, braking, and cornering forces are transmitted through these pins. (see **Figure 11-22**).

When the vehicle is unloaded, the spring rides on the outer edges of the spring load cushions. As the load increases, the crossbars of the cushions are progressively brought into contact to absorb the additional load. Cushioning and alignment are accomplished by the four drive pins incased in rubber bushings. The bushings permit the drive pins to move up and down in direct relation to movement of the load cushion.

Another style of solid rubber suspension is one manufactured by the Chalmers Company. Again, it is a walking beam style suspension but uses only a single circular rubber spring surrounded by a steel retaining canister. As load is applied to the spring, the canister contains the rubber spring, allowing it to become stiffer. This equalizing beam and rubber cushion only support and cushion the load and also provide oscillation. Locating and guiding the axles is done by torque rods.

Servicing the Equalizing Beam

Power wash the suspension and inspect the components for cracks or damage. Inspect rubber bushing for damage or deterioration and plan to replace them if they show any indication of fatigue.

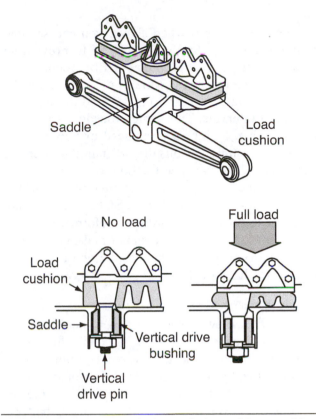

Figure 11-22 Sectional view of the rubber cushion-type equalizing beam suspension.

Equalizing beam bushings can be replaced using special service tools, and it is recommended that you use them. There are also many service facilities that specialize in the replacement of suspension system components. The bushings can be removed with standard shop tools, but it can be a tough job. If a portable press is available, this can be used in conjunction with steel pipe sections with diameters equivalent to the bushing sleeves as drivers for both removal and installation.

CAUTION *It is not recommended to remove bushings by burning them out; once alight, they burn for a long period of time, producing high heat and noxious fumes.*

U-BOLTS

U-bolts perform three very important functions:

1. Provide a positive clamping force between top plates, spring seats, and axles.
2. Provide a flex point away from the spring center bolt area (weakest point of the spring).
3. Keep spring pack together, eliminating shearing of spring center bolts.

From the functions listed above, you can see that maintaining proper torque is the key to preventing U-bolt and suspension problems. OEMs recommend that U-bolts be torqued at time of delivery, after the first 1,000 miles, after 3,000 miles, and thereafter at 5,000-mile intervals. This same procedure should be followed when springs are replaced or new U-bolts have been installed. Consult the manufacturer for recommended torque specification.

If the vehicle is operated in an environment with a lot of corrosion (Northern United States and Canada), retorquing cannot be accurately preformed. In this case, tap the U-bolt with a hammer and look for any signs of movement. This should give you a good indication if it has lost its tension.

Tech Tip: Never heat a U-bolt to achieve a retorque. Never reuse U-bolts, as some elongation (stretching) occurs when U-bolts are installed and properly torqued. This elongation is used to maintain the clamping force on the springs and axle. Because of this, a used U-bolt will not be able to produce the clamping force required.

U-bolt Identification

To identify a U-bolt, there are four specifications that are required (see **Figure 11-23**).

1. Type of bend (round, semi-round, square, etc.)
2. Thread size and pitch (7/8"-14, 1 1/8"-12, etc.)
3. Length of legs (from inside highest point to end of thread)
4. Width (inside legs)

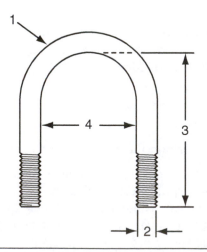

Figure 11-23 U-bolt identification.

AIR SPRING SUSPENSIONS

Air spring suspensions, also known as air ride suspensions, are now extremely common in the heavy-duty trucking industry. These systems may be fully pneumatic (all air springs) or a combination air/leaf spring suspension as in **Figure 11-24** and **Figure 11-25**.

The advantages of air bag or air spring suspensions offer the ultimate in smooth shock and vibration-free

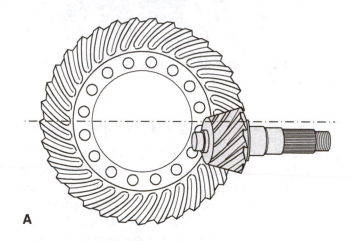

A

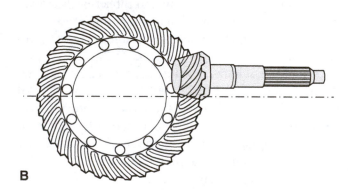

B

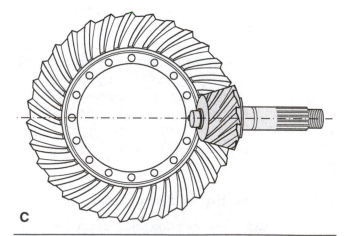

C

Figure 11-24 (A) Hypoid gearing; (B) amboid gearing; and (C) spiral bevel gearing arrangements. *(Courtesy of Arvin Meritor)*

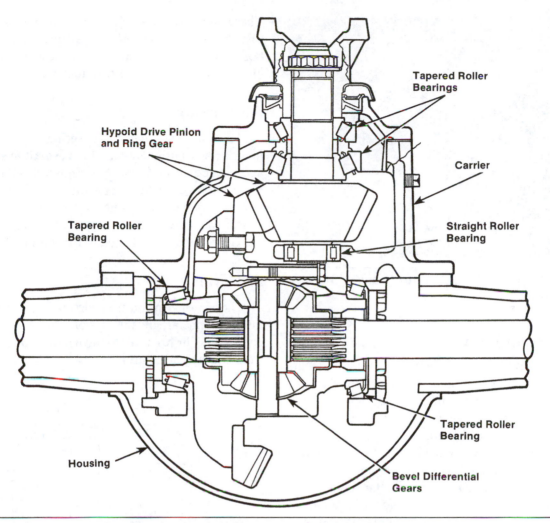

Figure 11-25 Standard single-reduction carrier. (*Courtesy of Arvin Meritor*)

ride. They also automatically maintain a constant ride height under all conditions, regardless of load. In addition, they do a better job of isolating the vehicle from shock than do leaf springs or solid rubber suspensions. This being said, the air ride suspension has no ability to dampen suspension oscillations. For this reason, they use auxiliary dampening mechanisms such as shock absorbers. Higher suspension oscillation requires the use of axles using thicker walls, compared to other types of suspensions. One other disadvantage that some manufacturers have been able to eliminate is called dock walk. This occurs when a suspension is repeatedly loaded and unloaded and literally creeps or walks away from the loading dock. This occurs because the brakes are set on the vehicle and when the suspension is jounced, it will push the vehicle forward. In some cases when trailers are left with the suspension aired up and air pressure leaks down, over time the suspension lowers and may even collapse the landing gear of the trailer if the trailer is in a loaded state and the landing gear digs into the ground instead of sliding across it.

The major components of an air suspension system:

- Height control valve
- Pressure regulator
- Air lines
- Air bags or spring
- Shock absorbers

Height Control Valve

The **height control valve** is really the brains of the air ride system. Its function is to maintain the chassis ride height. The height control valve is usually mounted at the rear of the truck frame, but other arrangements are also used. The valve has a lever rigidly connected to the rear axle assembly by means of a linkage rod. There are three air ports as well on the valve. One port is the supply air port to the valve. The second port delivers air from the valve to the air springs. The third port is the exhaust port (see **Figure 11-26**).

Figure 11-26 Typical height control valve. *(Courtesy of John Dixon)*

HEIGHT CONTROL VALVE OPERATION

With the lever of the height control valve in the neutral (middle) position, there is no air flow from the height control valve. When the suspension is loaded, it will move downward, causing the lever of the height control valve to move upward. In this state, air from the supply port of the valve is allowed to flow into the air springs, causing the suspension to move upward. Once the suspension moves upward far enough, the linkage will pull the lever down into the neutral position, causing the air from the supply port to stop flowing. When the load is removed, the suspension will move upward, causing the linkage to pull the lever downward. In this position, the valve will open a port between the air springs and the exhaust port. With the air exhausting from the air springs, the suspension will lower. As the suspension lowers, the linkage will again position the lever into the neutral position, stopping the air flow from the air springs to the exhaust port of the height control valve.

Pressure Regulator

In some trailer applications like trailer lift axles, a height control valve may not be used, but instead a pressure regulator is used. With the lift axle in the down position, the operator can increase or decrease the air pressure to the lift axle down bags (air springs). By adjusting the air pressure, the operator also controls the amount of weight the axle will be supporting.

Most OEMs will also use some type of pressure-regulating device to control the amount of air pressure that is being supplied to the height control valve. These will sometimes be called pressure protection valves.

These valves require 90 psi before they open, ensuring that the reservoir always has 90 psi for braking purposes and that the suspension system will not siphon all the air in the case of a serious air leak.

Air Springs

There are two types of air spring that may be used in an air suspension system: the **reversible sleeve** type and the **convoluted** type. See **Figure 11-27(A)** for the reversible sleeve and the convoluted style of air spring. The convoluted air spring is also available in three styles: single, double, and triple convolutions. A double convoluted type is pictured in **Figure 11-27(B).**

Air Lines

The air suspension uses air lines to connect all of the components of the system together. Air lines connect the air reservoir to the height control valve and from the height control valve. Also, air lines located

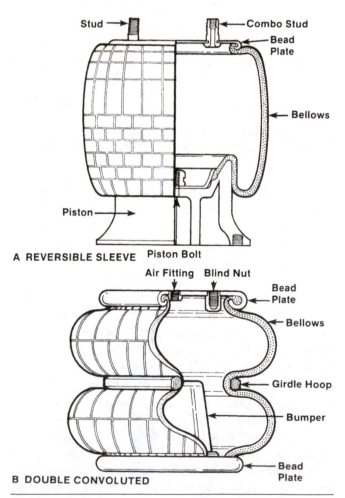

Figure 11-27 Components of reversible sleeve and convoluted air springs. *(Courtesy of Firestone Industrial Products Co.)*

between the front and rear axle air springs transfer air pressure between the axles. In this way, the height control valve maintains a consistent frame rail height and equal load distribution between the axles.

Shock Absorbers

Shock absorbers are necessary on many types of suspensions to reduce potentially damaging shock and vibration on heavy-duty trucks. There are three advantages of reducing or eliminating shock and vibration:

1. Reduced wear to high-maintenance areas such as electrical systems, suspension parts, and cooling systems as well as premature wear to other components on the vehicle, with resulting downtime
2. Reduced driver fatigue
3. Reduced cargo damage

All of this is accomplished by stopping the vibration at its source: road shock transferred through the axle.

As a wheel passes over an imperfection in the road surface, the vehicle's springs deflect (jounce) and respond by over returning (rebound). This causes the suspension to oscillate and will continue, slightly reduced, each time the spring deflects. These oscillations, coupled with road-induced shock, can cause severe damage to truck components. The jounce and rebound force **dampening** (resistance to movement) built into shock absorbers reduces component failures.

SHOCK ABSORBER MAINTENANCE

Shock absorbers should be checked at every PM service. OEMs recommend checking shock absorbers at intervals of 12,000 miles. To check shock absorbers on a truck, you can use the following sequence:

1. To check for noise, make sure it is not coming from some other part of the vehicle, then check shock absorber brackets and mounting stud nuts to ensure that they are tight and that the shock absorber is not striking or rubbing on the frame or some other part of the chassis.
2. Check the rubber mounting bushings and replace if worn. **Figure 11-28** shows poor shock absorber bushings.
3. Disconnect the bottom mount and work the shock absorber by hand to check that the outer tube is not contacting or rubbing against the fluid reservoir tube.
4. When the shock absorber is disconnected, check the piston movement by pulling and pushing the

Figure 11-28 Poor shock absorber bushings. *(Courtesy of John Dixon)*

absorber down and up slowly through its travel stroke, making sure the piston does not bind in the pressure tube. Typically, there should be considerable resistance when extending the absorber but only slight resistance when collapsing it. Also note the rate of effort for distance of travel. This rate should not vary.
5. If it is determined that the shock absorber is defective, replace it.

To tell if the shock absorber is operating properly, you can check it immediately when the truck comes off the road. When a shock absorber is performing its function of resisting movement, it is absorbing this energy. The energy is absorbed in the form of heat transferred to the fluid contained within the shock absorber. Therefore, when a vehicle comes in from the road, you can feel all shock absorbers being warm with the palm of your hand. A cold shock is a nonfunctioning shock absorber.

Shock absorbers should also be inspected when a vehicle is experiencing:

- Frequent lightbulb replacements
- Excessive kingpin wear
- Premature tire wear
- Air spring damage

Shock absorbers are integral and hard-working components, especially on an air ride suspension, because air ride springs have no resistance to oscillation. As well, shock absorbers are the stroke-limiting device for air ride suspensions. A shock that is too long for the intended application or with worn or broken mounts will allow the air springs to overextend, damaging the air spring. Therefore, shock absorbers in an air ride suspension should be replaced regularly and in pairs, because if one shock has failed, the other is never very far behind.

The correct shock for each application not only increases its own life but the life of all vehicle components.

Note: Most distributors of shock absorbers store them on their side. In this position, the oil will flow to a level that hinders normal shock absorber operation. If a new shock absorber seems to have little or no resistance, it should be placed in its normal operating (vertical) position for 24 hours to allow the oil to find its proper level and then be rechecked.

SHOCK ABSORBER REPLACEMENT

The task of replacing a shock absorber is straightforward and one that a new technician should be able to accomplish as follows:

1. Remove the nuts from the top and bottom bolts/studs from the eyes of the shock absorber.
2. Remove the shock absorber from the vehicle.
3. Insert new rubber bushing into the eyes of the new shock absorber.
4. Install the top bolt/stud through the top mounting bracket on the frame rail.
5. Place a flat washer over the bolt and thread the bolt into the upper rubber bushings of the new shock absorber.
6. Place a second flat washer over the bolt and then turn the nut finger tight.
7. Extend the lower part of the shock absorber until the mounting holes line up in the lower brackets and shock absorber eye.
8. Install the bottom bolt/stud through the lower mounting bracket on the axle.
9. Place a flat washer over the bolt and thread the bolt into the lower rubber bushing of the new shock absorber.
10. Place a second flat washer over the bolt and turn the nut finger tight.
11. Check to make sure that the rubber bushings are seated properly and tighten the nuts to the specified torque.

Servicing Air Suspensions

Air suspension systems tend to require a little more attention in inspection and preventive maintenance than mechanical suspensions. Some of the commonly required service procedures will be looked at next, including repairs and height control valve adjustments.

Air Suspension Inspection

Air ride suspensions should be checked daily and incorporated into the driver pre-trip inspection and PM schedules. The truck or trailer should be observed to be level and riding at the correct height in both a loaded and unloaded condition. Any broken or loose components should be noted and repaired immediately.

Monthly or B schedule service should include inspections of clearance around the air springs, tire shock absorbers, and other moving parts in the suspension system. When air spring bellows make contact with other chassis components, the problem almost never originates in the air spring assembly.

Quarterly or on C schedule service, all welds should be visually inspected, and the frame attachment joints and fasteners, cross-members, and pivot connections inspected for wear, cracks, and corrosion. U-bolt lock nuts should be retorqued every 100,000 miles.

Air springs should last a very long time, as long as they are not rubbed, scuffed, or twisted (improper installation or suspension misalignment), and they are easily punctured by contact with sharp objects. Even a plastic air line or plastic tie will eventually puncture the air spring if left unchecked. If an air spring does fail, the chassis weight settles onto the pedestal's jounce blocks (if equipped), which have almost no ability to buffer road forces when used in this way. In this situation, the driver should be instructed to proceed with caution to the closest service facility.

Inspection Procedure for Rolling Lobe

Rolling lobe style pistons should be inspected for corrosion and cracks. If corrosion or cracks are present, replace the piston. Slight corrosion may be cleaned up as long as a smooth rolling surface is restored. Also, inspect the flexible member for damage. Any time excessive wear is found on the flexible member, the air spring must be replaced.

Inspection Procedure for Convoluted Spring

Convoluted-style air springs should be inspected for foreign material (such as stones, road dirt buildup, etc.) located in the girdle area or under the retainer plates. These materials will cause excessive wear on the flexible member, leading to failure. If excessive wear is found, the air spring must be replaced.

CAUTION *Exercise extreme caution around air springs. The forces that these springs contain when pressurized can cause serious bodily injury or death if they explode. Never use an air spring as a jack. Avoid using torches or sharp objects around loaded air springs. If service work must be performed around the air springs, remove the air from the suspension before proceeding.*

Checking the Air Ride Height Control Valve

The height control valves (also known as leveling valves) are often thought to be defective and replaced unnecessarily. A simple diagnostic routine can determine whether the valve is defective or whether proper adjustment has not been performed. The diagnostic routine outlined here references one type of height control valve, but the procedure is similar to that used on other manufacturers' valves. To verify the performance of the height control valve, perform the following service checks:

1. Remove the bolt that attaches the linkage to the lever of the height control valve.
2. Ensure that air pressure in the reservoir is above 100 psi.
3. Raise the height control valve lever 45 degrees above horizontal. Air pressure at the air springs should begin to increase within 15 seconds, which will cause the truck or trailer chassis to rise.

CAUTION *Most height control valves have a delayed reaction time that can be as long as 15 seconds. This is used to prevent continuous correction cycling; remember this when diagnosing height control valve problems. Also, before condemning a valve that does not allow the springs to inflate, make sure the height control valve is receiving air pressure from the reservoir. Often, a pressure protection valve can be the culprit.*

4. Lower the height control valve lever 45 degrees below horizontal. Air should exhaust from springs through the valve's exhaust port within

15 seconds. Let air exhaust until air spring height is approximately 10 inches.
5. Raise the valve lever to 45 degrees above horizontal again, until air spring height is approximately 12.5 inches. Release valve lever.
6. Check the valve body, all tubing connections, and air springs for leaks with soapy water.
7. Check air spring height 15 minutes after performing step 5.
8. If no leaks are found, the valve functions as just described, and air spring height has not changed in 15 minutes, the valve is functioning properly.
9. Install the bolt that attaches the linkage to the height control valve lever. Torque nut to specification.

Ride Height Adjustment

Checking and adjusting ride height is another fairly simple task that an entry-level technician should be able to perform. Some versions of height control valves have a centering pin and bosses. The pin is positioned in the bosses after setting the height. The pin is used to lock the height control lever in the neutral position (no air in or out of the valve). For this example we will be taking our measurements from rear axle of the tandem, but always follow the OEM literature for measuring locations and distances. To adjust the height control valve proceed as follows:

1. Make sure the centering pin is removed from the bosses as indicated in **Figure 11-29.**
2. Make sure that the reservoir has at least 100 psi air pressure, as indicated on the dash air pressure gauge of the vehicle.

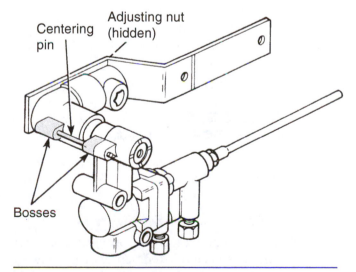

Figure 11-29 Height control valve centering pin. *(Courtesy of Navistar International Corp.)*

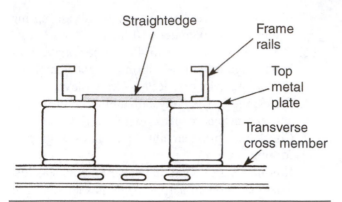

Figure 11-30 Ride height adjustment. *(Courtesy of Navistar International Corp.)*

3. Disconnect the linkage connected the lever of the height control valve.

4. The height adjustment is checked (for this example) at the rear axle. Place a straightedge on the centerline of the top metal plate, on the air springs, as shown in **Figure 11-30.** Do not position the straight edge on the frame rails. Measure the distance from the bottom edge of the straightedge to the top edge of the transverse cross-member.

5. With no air in the suspension system, jack the vehicle frame until the specified ride height distance is obtained.

6. With the ride height at the correct level, lock the valve lever with the centering dowel (pin).

7. Adjust the linkage so that it can be installed between the vehicle frame mount and the height control lever.

8. Tighten the fine adjustment nut on the lever to 115 to 130 inch-pounds (see OEM specs).

9. Remove the centering dowel and lower the vehicle and remove the jack.

10. Air up the suspension system and re-measure to ensure ride height is correct.

SUSPENSION SYSTEM OUT-OF-SERVICE CRITERIA

(Taken from the North American Standard Out-of-Service Criteria by the Commercial Vehicle Safety Alliance, April 1, 2006)

1. **Axle Parts/Members See Figure 11-31**

   ■ Any U-bolt(s) or other spring-to-axle clam bolt(s) cracked, broken, loose, or missing

   ■ Any spring hanger(s), or other axle positioning part(s) cracked, broken, loose, or

missing resulting in shifting of an axle from its normal position

Note: After a turn, lateral axle displacement is normal with some suspensions, including composite springs mounted on steering axles.

2. **Spring Assembly**

   ■ One fourth or more of the leaves in any spring assembly broken

   ■ Any leaf or portion of any leaf in any spring assembly missing or separated

   ■ Any broken main leaf in a leaf spring

   1. Any leaf of a leaf spring assembly is a main leaf if it extends, at both ends, to or beyond:

      a. The load bearing surface of a spring hanger or equalizer

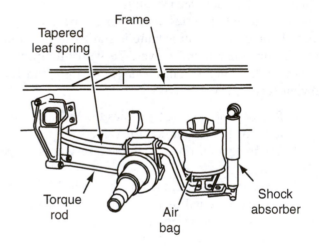

AIR SUSPENSION

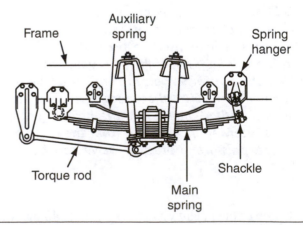

Figure 11-31 Suspension components that need to be inspected.

b. The spring end cap or insulator box mounted on the axle

c. A spring eye

2. Any leaf or a helper spring assembly is a helper main leaf if it extends, at both ends, to or beyond the load bearing surface of its contact pad, hanger, or equalizer.

3. The radius rod leaf, in springs having such a leaf, has the same function as the torque or radius components referenced in "Torque Radius, Tracking, or Sway Bar Components" and should be treated as such a component for purposes of Out of Service.

- Coil spring broken
- Rubber spring missing

- One or more leaves displaced in a manner that could result in contact with a tire, rim, brake drum, or frame
- Broken torsion bar spring in torsion bar suspension
- Deflated air suspension (i.e., system failure, leak, etc.)

3. **Torque Rods, Tracking, or Swaybar Components**

Any part of a torque, radius, or tracking component assembly or any part used for attaching some to the vehicle frame or axle that is cracked, loose, broken, or missing (including spring leaves used as a radius or torque rod, missing bushings but not loose bushings, in torque, track rods or sway bars) (see **Figure 11-31**).

Summary

- The steering wheel is the driver's means to control the direction that the vehicle travels.

- The larger the steering wheel diameter, the more torque is generated from the same amount of driver effort.

- The steering column connects the steering wheel to the steering gear.

- The steering gear, sometimes referred to as the steering box, is used to multiply the driver's efforts at the steering wheel.

- The steering knuckles should be greased by raising the steering axle off the ground.

- Ball joints are necessary to compensate for movement between the steering linkage and the frame of the vehicle.

- Ackerman arms or tie-rod arms are used to transfer and synchronize steering action on both steer wheels on a steering axle.

- The tie-rod ends are ball joints that connect to the Ackerman arms on each steering knuckle.

- When a steering problem is reported, systematically inspect the vehicle steering system, front and rear suspensions, and trailer suspensions.

- All steering mechanisms are critical safety items. A vehicle should be deadlined (OOS report) when a defect is reported.

- Do not drive a vehicle with too much lash in the steering gear.

- Steering knuckles can be the source of a number of steering-related problems, and many of these originate from lack of lubrication.

- The major front end alignment settings are toe, camber, and caster.

- The purpose of the suspension system is to support the vehicle and its intended load.

- A leaf spring is a steel plate or stack of clamped steel plates.

- Equalizing beam suspension uses a lever principle to divide the load equally between two axles as well as reduce the effect of road irregularities.

- Solid rubber suspensions use hard rubber cushions to absorb the shock instead of steel leaves or air.

- Maintaining proper torque is the key to preventing U-bolt and suspension problems.

- Air ride suspensions do a better job of isolating the vehicle from shock than do leaf springs or solid rubber suspensions.

- The height control valve's function is to maintain the chassis ride height.

- Two types of air spring may be used in an air suspension system: the reversible sleeve type and the convoluted type.

- Air lines connect all of the components of the system together.

- Shock absorbers reduce potentially damaging shock and vibration.

Review Questions

1. Which of these front suspension angles can cause the greatest tire wear?

   A. Incorrect kingpin inclination

   B. Incorrect toe

   C. Incorrect caster

   D. Incorrect camber

2. To properly inspect upper and lower kingpin bushing wear, a(n)_____ should be used.

   A. vernier gauge

   B. bushing gauge

   C. inclinometer gauge

   D. dial indicator

3. All the following are parts of the steering column except:

   A. U-joint.

   B. turn signal switch.

   C. drag link.

   D. boot seal.

4. The Pitman arm connects the:

   A. steering gear to the drag link.

   B. steering gear to the left upper steering arm.

   C. steering arm to the tie-rod.

   D. drag link to the tie-rod.

5. When replacing a steering or Pitman arm, you should perform all of the following except:

   A. road test when repairs are completed.

   B. replace both outer tie-rod ends.

   C. check and correct for changes in wheel alignment.

   D. lube the replacement part after installation.

6. Which component must you remove to access the kingpin?

   A. Knuckle cap

   B. Spindle

   C. Ackerman arm

   D. Axle

7. One of the following examples will not cause a hard steering complaint?

   A. Underinflated tires

   B. Fifth wheel not lubricated

   C. Universal joints not lubricated

   D. Overinflated tires

   E. None of the above

8. Which of the following components uses universal joints?

   A. Steering gear

   B. Steering shaft

   C. Pitman arm

   D. Tie-rod end

9. Which of the following components contains ball joints?

   A. Steering gear

   B. Pitman arm

   C. Steering knuckle

   D. Tie-rod end

10. Which of the following components connects the Pitman arm to the steering arm?

    A. Drag link   C. Cross tube

    B. Tie-rod assembly   D. Steering column

11. Straight kingpins are secured to the axle with:

    A. nut.   C. tapered draw keys.

    B. cotter pins.   D. tapered bearings.

12. U-bolts are generally reusable.

    A. True   B. False

13. Rebound refers to spring loading or suspension moving up.

    A. True   B. False

14. Shock absorbers:

    A. reduce suspension spring   D. reduce suspension spring and linkage wear.
       oscillation.
                                    E. All of the above
    B. resist road impact.

    C. resist vehicle sway.

15. Which component uses a lever principle to distribute a load equally between the axles to reduce the effects of road irregularities?

    A. Jounce block   C. Shock absorber

    B. Equalizer   D. Rigid torque arm

16. A shock absorber that is ineffective will be warmer than the rest of the shocks when the unit comes in after a long run.

    A. True   B. False

17. What is the function of the height control valve?

    A. Maintain the desired air   C. All of the above
       pressure in the air springs
                                    D. None of the above
    B. Maintain a constant ride
       height under all conditions

18. Leaf springs have the ability to isolate the vehicle from road shock better than air suspensions.

    A. True   B. False

19. With the arm of the height control valve in the up position (after time delay), air will flow from the springs to the exhaust port of the height control valve.

    A. True   B. False

20. Reversible sleeve air springs use only one girdle hoop.

    A. True   B. False

21. When an air spring has been overextended, the bellow can be torn away from the bead plate or piston or the bead plate can be bent or cracked around fittings. Shock absorbers can be the main cause of this damage if shock absorbers are too short for their intended application.

    A. True   B. False

22. A triple convoluted air spring will use two girdle hoops.

    A. True   B. False

23. When a height control valve is not used in an air suspension system, what could be used in its place?

    A. Manual valve and pressure   C. All of the above
       gauge
                                          D. None of the above
    B. Pressure regulator and gauge

24. Through what part or component does air enter or exit a reversible sleeve air spring?

    A. Stud   C. Air fitting hole

    B. Piston   D. Combo stud

CHAPTER 12

Electrical Systems and Auxiliary Components

Objectives

Upon completion and review of this chapter, the student should be able to:

- Demonstrate safe working procedures around batteries.
- Explain the role of the battery in a truck's electrical system.
- Verify the condition of a battery using a voltmeter, hydrometer, refractometer, and carbon pile tester.
- Describe battery maintenance procedures.
- Describe and demonstrate the safe charging procedure for batteries.
- Jump start a vehicle with a flat battery.
- Explain the role of the charging system.
- Verify the performance of an alternator.
- Explain what full fielding and an alternator will accomplish.
- Demonstrate how to test a starter to ensure it is in good condition.
- Explain the purpose of a liftgate.
- Describe maintenance procedures that must be performed on a liftgate.
- Troubleshoot for problems with a hydraulic liftgate.

Key Terms

ampere-hour rating

BCI (Battery Council International)

cold cranking amps rating

deep cycle

distilled water

electrolyte

fast charging

full field test

gasification

gasses

hydrometer test

jump starting

liftgate

load test

neutralize

open circuit voltage test

refractometer

reserve capacity rating

slow charging

specific gravity

INTRODUCTION

This chapter deals with a truck's basic electrical system. This system must be kept in the best possible condition at all times. The truck technician must know how to perform periodic maintenance and inspections on the truck's electrical system and electrical components. Without proper maintenance, the vehicle may very well have a roadside breakdown, a no-start condition, or a failure of important electrical circuits, which can lead to expensive downtime and repairs.

SYSTEM OVERVIEW

We start by examining the storage source for the vehicle's electrical system, the battery, and the safety practices that must be adhered to when performing routine maintenance on the battery. We then look at the different types of batteries. The technician must know how to identify the different styles of batteries in order to know what the required maintenance is for the style of batteries he is maintaining. If the technician is required to replace a battery, he must know what ratings to look for in a replacement battery, so that the correct battery will be selected for the vehicle. These ratings include amp-hour rating, cold cranking amps, reserve capacity, and BCI dimensions. You should also be able to test a battery to determine its serviceability. The most common tests are discussed, which are open circuit voltage test, a hydrometer test for the state of the batteries charge, and a load test to determine the batteries performance level.

Safe battery charging is also discussed, as well as jump starting procedures, charging systems testing, and engine starting systems testing.

Lastly, we look at hydraulic liftgate maintenance procedures and troubleshooting procedures.

BATTERY SAFETY

When working around batteries, technicians should exercise caution to avoid personal injury and for the safety of others in the immediate area.

Figure 12-1 illustrates a typical warning label affixed to the case of a battery. Batteries contain sulphuric acid and generate hydrogen gas, which is highly flammable. Always comply with the following safety tips when working with batteries:

- Keep flames or sparks away from batteries. No smoking around batteries.
- Always wear eye protection and rubber gloves to protect yourself from chemical burns when handling batteries.

Figure 12-1 The warning label found on a battery. *(Courtesy of Battery Council International © 2005)*

- Never connect or disconnect live circuits. Always turn off the unit, battery charger, or tester when attaching or removing leads. (Sparks can be produced when making or breaking live circuits.)
- Batteries should always be installed in a vented battery box, as they emit hydrogen gas when charging.
- Work in a well-ventilated area when charging batteries.
- Always keep the battery top upright to prevent spilling of the electrolyte.
- Never work alone on batteries in case of accidents.

CAUTION *Battery **electrolyte** contains sulphuric acid, which can cause severe personal injury (burns) and damage to clothing and equipment. If electrolyte is accidentally spilled or splashed on your body or clothing, immediately **neutralize** by washing with a solution of baking soda and water. The solution should be 0.25 pounds baking soda to 1 quart water (115 grams baking soda to 1 liter of water).*

Electrolyte splashed into the eyes is extremely hazardous. Eyes should immediately be held open and flushed with cool clean water for about 5 minutes; then seek medical treatment at once.

STORAGE BATTERIES

For the most part, trucks use 12-volt automotive-type storage batteries. The storage battery is the energy source for the vehicle's electrical system. It is important to remember that a battery does not store electricity, but rather it stores a series of chemicals, and through a chemical process, electricity is produced. Basically, there are two types of lead in an acid mixture that reacts to produce an electrical pressure called

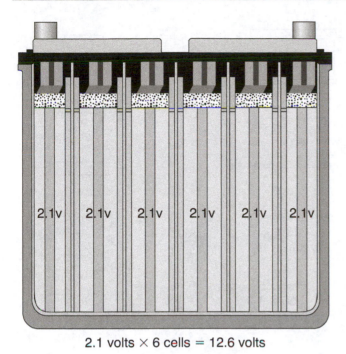

2.1 volts × 6 cells = 12.6 volts

Figure 12-2 A diagram of six battery cells connected together in series to form a 12-volt battery.

voltage. This electrochemical reaction changes chemical energy to electrical energy and is the basis for all automotive-style batteries.

Cell Voltage

Each of the six cell elements in a 12-volt battery produces approximately 2.1 volts, regardless of the size or number of the plates. The cells are connected in series, which produce a total voltage of 12.6 volts.

Figure 12-2 illustrates how cells are connected in series to produce a 12-volt battery.

BATTERY TYPES

Batteries may be further segregated by the amount of maintenance required to keep each at optimum charge level. Batteries can be classified as conventional, low maintenance, and maintenance free. These batteries can be purchased as either wet or dry charged. Low-maintenance and maintenance-free batteries are usually wet charged, because they retain their charge during storage much better than conventional batteries. The major difference between the three types of batteries from a servicing point of view is how often the electrolyte level, terminals, and cables must be checked and maintained. Each of the three types uses different materials for the positive and negative plate construction. This causes different operation and service characteristics.

Conventional Batteries

Conventional batteries require more maintenance than low-maintenance or maintenance-free batteries. This maintenance is due to the chemical composition of the plates, which causes the water in the electrolyte to turn into gas. Therefore, the water must be replenished. When adding water, use only **distilled water,** as minerals and chemical commonly found in drinking water will react with the plate materials and shorten battery life. The water level should be no higher than 1/8 of an inch (3.2 mm) below the bottom of the vent well. To avoid permanent damage, make sure the electrolyte level never drops below the top of the plates. As well, avoid overfilling, because some of electrolyte will be washed away, leaving a weaker solution. Refer to **Figure 12-3** for correct electrolyte level. Gasification can cause the electrolyte to condense on the top of the battery case and cause the battery to discharge by providing a current path between the positive and negative terminals. This also causes corrosion of the battery terminals' cables and cable ends. Conventional batteries are more easily overcharged than low-maintenance and maintenance-free batteries. They also discharge more quickly when stored in a wet charged state. Conventional batteries perform well in **deep cycle** applications. Deep cycling occurs when a battery goes from fully charged down to a low state of charge and is then fully recharged during system operation.

Low-Maintenance Batteries

Low-maintenance batteries are manufactured to require less maintenance than conventional batteries.

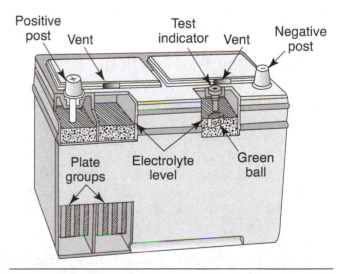

Figure 12-3 Filling the battery to the correct level of electrolyte, making sure the plates are below the surface of the electrolyte.

This is due to the fact that there is less electrolyte **gasification,** so the level of the electrolyte is reduced much more slowly than with conventional batteries. Corrosion of the battery terminals' cables and cable ends is also reduced. Low-maintenance batteries are more resilient to overcharging than conventional batteries but have a shorter life in deep cycle applications.

Maintenance-Free Batteries

Maintenance-free batteries are designed to not require electrolyte replenishment under normal operating conditions. Battery terminals' cables and cable ends require almost no maintenance because of greatly reduced gasification and low water. Maintenance-free batteries retain their charge longer than conventional or low-maintenance batteries when stored in a wet charge state. Maintenance-free batteries generally require a higher voltage regulator setting than low-maintenance or conventional batteries.

BATTERY RATINGS

Storage batteries can be selected for their application by four different ratings: **ampere-hour rating, cold cranking amps rating, reserve capacity rating,** and the **BCI (Battery Council International)** dimensional group number.

Amp-Hour Rating

The most common battery rating is the amp-hour rating. This is a unit measurement for battery capacity obtained by multiplying a current flow in amperes by the time in hours the battery is discharged. (Example: a battery that delivers 5 amperes for 20 hours delivers 5 amperes times 20 hours, or 100 ampere-hours.)

Cold Cranking Amps Rating

Batteries are compared by their cold performance rated cranking amp (CCA) designation. It is the load in amperes that a battery can sustain for 30 seconds at 0°F (−17.8°C) and not fall below 1.2 volts per cell, or 7.2 volts on a 12-volt battery. The CCA rating indicates how much power a battery can deliver in extremely cold conditions. The batteries main function is to start the engine, so it is imperative that the battery have sufficient capacity to accomplish this task.

Figure 12-4 illustrates how a battery's capacity drops as ambient temperature drops. The engine also requires more power to turn and start as the ambient temperature drops.

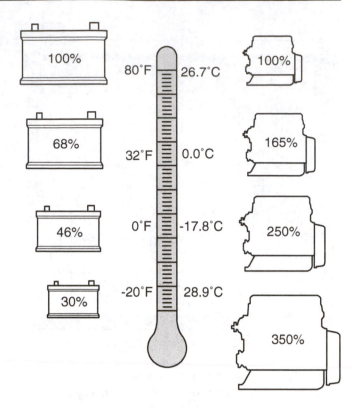

COLD WEATHER AFFECTS THE BATTERY AND ENGINE WHEN STARTING

Figure 12-4 A diagram illustrating how the battery's capacity is severely debilitated as the ambient temperature is reduced and resistance is caused by the engine as it becomes harder to turn in decreasing temperatures.

Reserve Capacity Rating

The reserve capacity rating indicates the number of minutes a new fully charged battery at 80°F (26.6°C) can sustain a load of 25 amperes before the battery voltage drops to 1.75 volts per cell, or 10.5 volts for a 12-volt battery. This is a rating devised by the automotive industry to indicate how long a battery can provide enough power to keep the ignition, head and tail lights, windshield wipers, and heater operating if the charging system were to fail.

BCI (Battery Council International) Dimensional Group Number

The BCI number indicates a battery's physical dimensions. As an example, a group 31 series battery is always 13 inches long (33 cm), 6.8 inches wide (17.3 cm), and 9.4 inches high (23.9 cm).

Figure 12-5 illustrates the BCI dimensional measurements. This rating has nothing to do with the battery's performance capacity, just the physical dimensions.

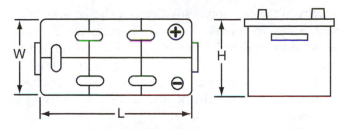

Figure 12-5 A diagram of the battery's dimensions used to configure its BCI number.

A battery's performance characteristic is determined by the internal components, such as the number of plates per cell, not its physical size.

BATTERY MAINTENANCE

As previously stated, conventional batteries require more maintenance than low-maintenance or maintenance-free batteries. However all batteries should be inspected periodically.

Figure 12-6 illustrates a battery requiring maintenance. Listed below are some general battery inspection procedures. If necessary, replace the battery.

- Check for loose or broken terminal posts.
- Check for cracked or broken battery case and/or budges in the case.
- If case is ruptured or terminal is broken, replace battery immediately.
- Inspect case for dirt, moisture, and corrosion.
- The battery case and terminals should be cleaned with a baking soda and water solution. If necessary, use a heavy brush on the battery terminals. Make sure the soda solution is not allowed to mix with the electrolyte in the battery.

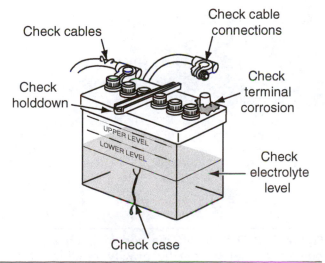

Figure 12-6 The areas a technician should look at when servicing a battery for routine maintenance.

- Battery cables should be checked for cleanliness and tightness. Loose or dirty connections increase resistance to current flow. This condition can cause poor charging and system operation. Replace worn or frayed cables, keeping all connections clean and tight.
- Check the battery electrolyte level. Add water if the electrolyte level is below the top of the plates. Do not overfill, making sure to keep the level below the vent cap openings.

Note: Even maintenance-free batteries should be checked monthly, especially if the battery is in severe service conditions such as extreme heat.

Battery Storage

Wet charged batteries will slowly self-discharge while in storage. The rate at which the battery discharges depends upon the type of battery and the storage conditions it is subjected to. All batteries should be stored in a clean, cool, and dry environment and tested at regular intervals. The best storage temperatures should be between 50°F and 60°F (10°C to 16°C).

True or False

"Never leave batteries on the ground or a concrete floor because all the power will leak out."

This one is not true, as it does not matter what batteries sit on. It probably originated from when batteries were shipped in porous wooden cases. There are some truths behind this myth, though! All batteries do self-discharge over time when they are not being charged. If dust and dirt build up on the battery tops, sulphuric acid will carbonize the grime into an electrical conductor, acting like a short circuit across the terminals and quickly draining the power. Cold temperatures also reduce available power from a battery, and thermal gradients can reduce the life of large battery. This can occur when the air temperature around a battery is much warmer than the surface it is sitting on.

Shop Talk: The batteries state of charge should be checked every 30 to 45 days, and it should be charged whenever the capacity of the battery has been reduced to 75 percent. Battery testing and charging are discussed next.

BATTERY TESTING

There are many ways of testing a batteries state of charge and its ability to perform, but the most common methods are the **open circuit voltage test** and the **hydrometer test** for state of charge and the **load test** for the batteries performance rating (see **Photo Sequence 10**).

Hydrometer Testing

The hydrometer can be used to test a batteries state of charge. This is accomplished by measuring the **specific gravity** (SG) of the electrolyte solution. Water has a SG of 1.000, while sulphuric acid has an SG of 1.835. A fully charged battery contains approximately 65 percent water and 35 percent sulphuric acid. When testing the specific gravity of a fully charged battery, the SG should be 1.265 at 80°F (26.7°C). The SG of the batteries cells will vary with the temperature the battery is subjected to. When performing a SG test on a battery using a hydrometer, you need to compensate for the temperature in order to make an accurate determination on the condition of the battery. In order to do this, 0.004 must be added to the actual reading for every 10°F (5.5°C) above 80°F (26.7°C), and 0.004 must be subtracted from the SG reading for every 10°F (5.5°C) below 80°F (26.7°C).

Figure 12-7 illustrates the proper usage of a hydrometer in order to obtain accurate data. The specific gravity of the electrolyte solution can also be measured with a **refractometer.** With a refractometer, there is no need to correct for temperature. This instrument is much like the one used to test antifreeze strength.

Figure 12-8 illustrates the proper use of a refractometer. Follow the list of instructions when using a refractometer:

1. Wear proper eye protection.
2. Remove battery caps.
3. Starting at one end of the battery, extract a drop of electrolyte and place it on the refractometer lens; close the prism.
4. Holding the refractometer up to the light, take a reading on the chart inside the view finder.
5. Record the reading and do the same on the remaining five cells.

The following table compares the batteries state of charge with the SG reading. A battery should be recharged when its capacity has dropped to 75 percent.

Electrolyte Specific Gravity	Percent of Charge
1.260	100
1.230	75
1.200	50
1.100	0

Note: A difference of more than 0.050 SG between individual cells indicates a defective battery. The unequal consumption of electrolyte is usually due to an internal defect or

1.230 to 1.310 SG

Figure 12-7 The correct method of accurately reading a hydrometer.

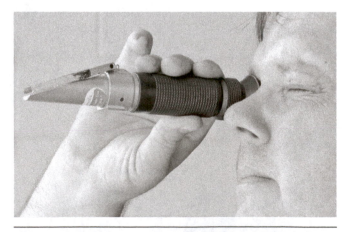

Figure 12-8 The correct method of using a refractometer to get a specific gravity reading of the electrolyte.

deterioration from extended use. A battery found to be in this condition should be replaced.

....

The SG measurement of the battery can also be used to determine at what temperature the battery will freeze. The following table indicates the temperature at which the battery will freeze according to its SG level and state of charge.

Electrolyte Specific Gravity	Freeze Point
1.260	–71.3°F (–57.3°C)
1.250	–62°F (–52.3°C)
1.230	–16°F (–26.6°C)
1.200	0°F (–17.8°C)
1.100	19°F (–7.3°C)

Open Circuit Voltage Test

The open circuit voltage test is another method of determining a battery's state of charge. This test is performed when the battery is at rest and is not being charged or discharged. The battery will contain a surface charge if the vehicle has been running or the battery was just charged. This surface charge must be removed from the battery before testing can be performed. This can be accomplished by applying a 300 amp load for 15 seconds or by disabling the engine and cranking it for a few seconds.

Connect the positive probe of a voltmeter to the battery positive terminal and the negative probe to the battery negative terminal.

Meter Reading	Battery Condition
12.66 volts	100% charged
12.48 volts	75% charged
12.30 volts	50% charged
11.76 volts	0% charge

The battery should be recharged if the open circuit voltage test indicates that the battery voltage is less than 12.4 volts.

Load Test

A load test is a true measurement of a battery's ability to perform. There are two ways of load testing a battery, one with a load tester and the other to use the reefer units engine.

Before a load test is performed, the battery must be fully charged and at room temperature. Always follow the instructions of the instrument manufacturer before performing a load test.

Generally a load test is performed as follows but, as mentioned earlier, follow the manufacturer's instructions.

USING A COMMERCIAL BATTERY LOAD TESTER

After attaching the carbon pile load tester, use the following steps to safely load test the battery:

1. Start by putting on your safety glasses or goggles. If the battery was recently charged or in service, remove the surface charge of the battery by applying a 300 amp load for 15 seconds or by disabling the engine and cranking it for a few seconds.

2. Adjust all mechanical settings on the load tester to zero. Rotate the load control knob fully counterclockwise to the off position.

3. Connect the load tester to the battery. An inductive pickup must surround the wires from the ground terminal if used in the test. Observe correct polarity and be sure the test leads contact the battery posts. Batteries with sealed terminals require adapters to provide a place for attaching the teat leads (see **Figure 12-9**).

4. If the tester is equipped with an adjustment for battery temperature, turn it to the proper setting. For best results, use a thermometer to check

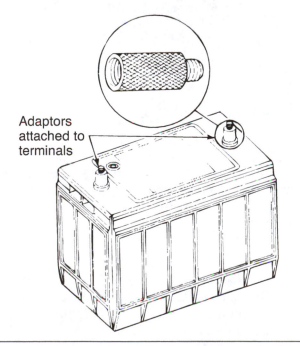

Figure 12-9 Batteries with threaded terminals require the use of adapters for testing or charging.

PHOTO SEQUENCE 10 Testing Truck Batteries

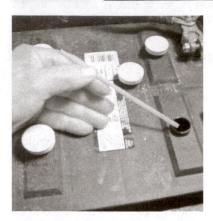

P10-1 Dip the refractometer probe into the battery cell, wetting just the tip of the probe. Throughout this procedure, remember that battery electrolyte is corrosive.

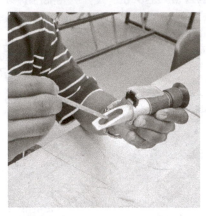

P10-2 Deposit a drop of electrolyte onto the refractometer read-lens as shown. Close the refractive lid.

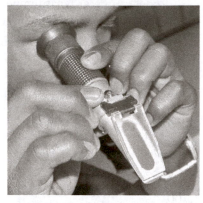

P10-3 Raise the refractometer view scope to your eye and point the refractive window toward a light source, preferably natural. The shaded area in the view finder correlates to a specific gravity reading.

P10-4 Connect a digital AVR to test a battery by connecting the polarized clamps as shown. Then connect amp pick-up lead with its arrow pointing in the direction of current flow.

P10-5 A digital AVR with inductive pick-up connected to a battery bank ready for a load test.

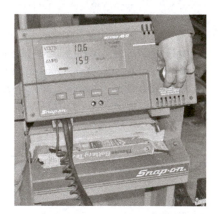

P10-6 Turn the load test knob CW as shown to load the battery. Typically, you will load to 1/2 CCA for 15 seconds and observe the voltmeter reading, which should not drop below 9.6 volts.

the electrolyte temperature in one of the cells. On sealed units, use an infrared thermometer to determine the temperature. Refer to the battery specification to determine its cold cranking amps (CCA) rating.

5. Turn the load control knob on the tester to draw battery current at a rate equal to one half the battery's CCA rating. For example, if a battery's CCA rating is 440, the load should be set at 220 amperes.

6. Maintain specified load for 15 seconds while observing the voltage of the tester. Turn the control knob off immediately after 15 seconds of current draw.

TABLE 12-1	
Battery Temperature	Minimum Test Voltage
70°F (21°C)	9.6 V
60° (15.5°C)	9.5 V
50° (10°C)	9.4 V
40° (4.4°C)	9.3 V
30° (–1.1°C)	9.1 V
20° (–6.6°C)	8.9 V
10° (–12.2°C)	8.7 V
0° (–17.7°C)	9.5 V

7. At 70°F (21°C) or above (or on testers that are temperature corrected), the voltage should not drop below 9.6 volts during the test. If the tester is not temperature corrected, use **Table 12-1** to determine the adjusted minimum voltage reading for the battery temperature. If the voltage reading exceeds the specifications by a volt or more, the battery is supplying sufficient current with a good margin of safety. If the voltage drops below the minimum specification, the battery should be replaced. If the reading is right on the spec, the battery might not have the reserve necessary to handle cranking under demanding low-temperature conditions. Keep in mind that this varies with the state of charge determined before the load test was made. For instance, a battery at 75 percent charge whose voltage dropped to the minimum acceptable load specification is probably in good shape.

BATTERY CHARGING

A battery charger can be used when tests indicate that the battery is in a low state of charge.

When a battery discharges, current flows from the positive terminal through the path of the circuit and back to the negative terminal. Current within the battery flows from the negative to the positive terminal. **Figure 12-10** illustrates the direction of current flow as a battery discharges.

In order to recharge the battery, the direction of current flow must be reversed. **Figure 12-11** illustrates the direction of current flow required to recharge the battery. This action will restore the chemicals to their active state.

Batteries can be charged at one of two different rates: **fast charging** and **slow charging.** When charging

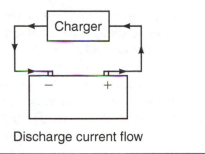

Discharge current flow

Figure 12-10 The direction of current flow as a battery discharges.

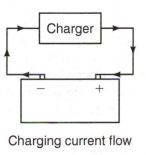

Charging current flow

Figure 12-11 The direction of current flow required to charge a battery.

a battery, it should be recharged at the same rate as it was discharged. A battery should also be at a temperature of between 60°F and 80°F (15.5°C–26.6°C).

Slow Charging

The slow charging method should be used when a battery has been discharged slowly over time. This could possibly be a battery that has been stored for a period of time and not periodically charged, or a small short that slowly drains the battery. The rate at which a slow charge should typically be performed is between 6 to 8 amps over an 8- to 10-hour period. This being said, in a severe case where a battery has stood in a discharged state of a long period of time, the battery may require up to 3 days of charging at a slow rate to bring the battery back up to its full rate of charge.

Fast Charging

The fast rate of charge delivers a high charging rate to the battery for a short period of time. This method of charging should be used when a battery has been discharged quickly, such as a starting failure where the operator of the truck has cranked the engine until the battery can no longer provide enough energy to turn the engine over. In this condition, the battery requires fast charging. A 12-volt battery would typically be fast charged at a rate of 40 amps for up to 2 hours.

When fast charging batteries, it is important to make sure the electrolyte solution does not get too hot. If the temperature of the electrolyte reaches 125°F (51.7°C), or if it **gasses** (bubbles) violently, reduce the charge rate to slow charge, 6 to 10 amps. Overheating causes the plates in the battery to warp or buckle and can cause a short circuit within a cell. If the electrolyte gasses violently, it can strip active material from the surface of the plates, thereby reducing the batteries capacity.

The battery charging guides given in **Table 12-2** and **Table 12-3** show approximately how much recharge a fully discharged battery requires. For partially discharged batteries, the charging current or charging time should be reduced accordingly. For example, if the battery is 25 percent charged (75 percent discharged), reduce charging current or time by one fourth. If the battery is 50 percent charged, reduce charging current or time by one half. If time is available, lowering the charging rates in amperes is recommended.

CAUTION *Always follow safety practices when recharging batteries. Connect the cable from the battery charger to the battery first, before plugging in and turning the charger on.*

TABLE 12-2: BATTERY CHARGING GUIDE—6- AND 12-VOLT LOW-MAINTENANCE BATTERIES

(Recommended rate* and time for fully discharged condition)

Rated Battery Capacity (Reserve Minutes)	Slow Charge	Fast Charge
80 minutes or less	14 hours @ 5 amperes	1³/₄ hours @ 40 amperes
	7 hours @ 10 amperes	1 hour @ 60 amperes
Above 80 to 125 minutes	20 hours @ 5 amperes	2¹/₂ hours @ 40 amperes
	10 hours @ 10 amperes	1³/₄ hours @ 60 amperes
Above 125 to 170 minutes	28 hours @ 5 amperes	3¹/₂ hours @ 40 amperes
	14 hours @ 10 amperes	2¹/₂ hours @ 60 amperes
Above 170 to 250 minutes	42 hours @ 5 amperes	5 hours @ 40 amperes
	21 hours @ 10 amperes	3¹/₂ hours @ 60 amperes
Above 250 minutes	33 hours @ 10 amperes	8 hours @ 40 amperes
		5¹/₂ hours @ 60 amperes

*Initial rate for standard taper charger.

Keep sparks and flames away from the battery and never smoke around a battery. Before using a battery charger, always read the manufacturer's instructions before using the equipment.

Charging Instructions

Follow these steps to charge a battery:

1. Before placing a battery on charge, clean the terminals if needed.
2. Add distilled water until the plates are covered. Fill to the proper level near the end of the charge. If the battery is extremely cold, allow it to warm up before adding distilled water, because the

TABLE 12-3: BATTERY CHARGING GUIDE—12-VOLT MAINTENANCE-FREE BATTERIES

(Recommended rate* and time for fully discharged condition)

Rated Battery Capacity (Reserve Minutes)	Slow Charge	Fast Charge
80 minutes or less	10 hours @ 5 amperes	2¹/₂ hours @ 20 amperes
	5 hours @ 10 amperes	1¹/₂ hours @ 30 amperes
		1 hour @ 45 amperes
Above 80 to 125 minutes	15 hours @ 5 amperes	3³/₄ hours @ 20 amperes
	7¹/₂ hours @ 10 amperes	2¹/₂ hours @ 30 amperes
		1³/₄ hours @ 45 amperes
Above 125 to 170 minutes	20 hours @ 5 amperes	5 hours @ 20 amperes
	10 hours @ 10 amperes	3 hours @ 30 amperes
		2¹/₄ hours @ 45 amperes
Above 170 to 250 minutes	30 hours @ 5 amperes	7¹/₂ hours @ 20 amperes
	15 hours @ 10 amperes	5 hours @ 30 amperes
		2¹/₂ hours @ 45 amperes
Above 250 minutes	20 hours @ 10 amperes	10 hours @ 20 amperes
		6¹/₂ hours @ 30 amperes
		4¹/₂ hours @ 45 amperes

*Initial rate for standard taper charger.

fluid level will rise as the battery warms up. In fact, an extremely cold battery will not accept a normal charge until it warms up.

3. Connect the cable ends of the charger to the battery following the directions of the charger manufacturer. Connect the positive (+) charger lead to the positive battery terminal and the negative (–) lead to the negative terminal. If the battery is in the vehicle, only connect the negative lead to the chassis or engine block ground in older trucks chassis; newer trucks with data bus–driven electronics require that ground clamps be connected to the battery ground posts. Move the charger lead clamps back and forth so the teeth of the clamp bite in and ensure a good connection.

4. Turn the charger on, gradually increasing the ampere output until the desired rate of charge has been reached.

CAUTION *If smoke or dense vapor comes from the battery, shut off the charger and reject the battery. If violent gasses or spewing of electrolyte occurs, reduce or temporarily stop the charging process.*

5. Once the battery has been charged, turn the charger off and disconnect it from the battery.

6. Install the battery back in the truck. If the vehicle's engine does not crank satisfactorily when a recharged battery is installed, capacitance or load test the battery. If the battery passes the load test, the vehicle's fuel, ignition, cranking, and charging systems should be checked to locate and correct the no-start problem. If it does not pass the load test, the battery should be replaced.

Battery Preventive Maintenance

When verifying battery performance during a PM inspection, most forms will require both a refractometer test (if electrolyte can be accessed) and amp/volt/resistance (AVR) test to be performed at minimum.

See **Photo Sequence 10**

JUMP STARTING A TRUCK

Sometimes it may be necessary to get the vehicle started when there is insufficient charge in the battery and no time to properly recharge the battery. This is

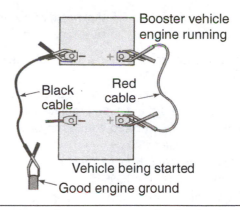

Figure 12-12 The correct method of placing the battery jumper cables when jump-starting or boosting a low battery.

referred to as **jump starting** or boosting a vehicle. In these cases, the vehicle may be successfully started by connecting the stalled battery with a known good battery.

Figure 12-12 illustrates the proper cable placement to jump start a vehicle. To avoid personal injury and damage to the equipment, follow proper safety procedures when jump starting. Be sure the voltage of the jump starting battery is the same as the stalled battery. Make sure the battery used for jump starting is not running another piece of equipment at the time. Cover both batteries with a damp cloth. Follow the proper sequence for connecting the jump starting cables from the stalled to the jump starting batteries:

1. Connect one end of the positive jumper cable to the positive (+) post of the stalled battery.
2. Connect the other end of the jumper cable to the positive (+) post of the jump starting battery.
3. Connect one end of the negative jumper cable to the negative (–) post of the jumper battery.
4. Connect the other end of the negative jumper cable to a good chassis ground connection on the stalled vehicle, away from the discharged battery. Do not connect the jumper cable to the negative (–) post of the stalled battery.
5. Start the stalled unit.
6. Disconnect the jumper cable in exactly the reverse order to that used to connect the jumper cables.

BATTERY REMOVAL AND INSTALLATION

When removing or installing a battery, always follow battery safety procedures.

Check the orientation of the positive battery and install the new battery in the same manner:

1. Remove the ground (negative) cable first before the positive cable.
2. Inspect the battery mounting tray and hold-down assembly; clean and replace as necessary. This can be performed with a mixture of baking soda and water.
3. Check the condition of the battery cables and terminal ends. Replace if insulation is worn or frayed or if corrosion is a problem.
4. Make sure that the new battery is fully charged before it is installed into the unit.
5. Use a battery lifting strap to lower the battery into the battery tray.
6. Install the battery hold-down clamps evenly and securely. Do not overtighten. A battery must be mounted securely; vibration can loosen connections, crack the case, and loosen internal components.
7. Clean battery terminals and cable clamps before installation. Install the positive battery cable first before the negative terminal. Note: this is the reverse of the sequence for removal. This will prevent shorting the battery while installing the positive cable if the wrench comes into contact with a chassis ground. Once the cable ends are installed, the terminals should be coated with a dielectric grease to prevent corrosion.

CHARGING SYSTEMS

The battery is responsible for providing the energy to start the engine of the truck. **Figure 12-13** illustrates the circuit of a battery delivering power to a load. Once the battery has performed this task, the energy it has expended must be replenished to ensure that the battery will have enough power to again restart the unit.

Figure 12-14 illustrates the circuit for a charging system to replenish the battery as well as supply

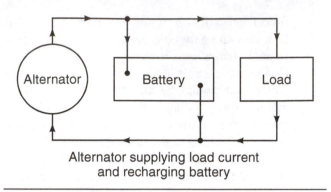

Alternator supplying load current and recharging battery

Figure 12-14 The current flow when the alternator is charging the battery as well as supplying the power for any electrical load on the unit.

current to the load. The charging system converts mechanical energy to electrical energy when the engine is running. It is the job of the alternator to supply the current to recharge the battery and provide the power supply to operate electrical components used to control the refrigeration, heating, and defrosting functions. During peak operation, the battery may be required in addition to the alternator to provide current for system loads. **Figure 12-15** illustrates the circuit when the battery and alternator supply load current.

The rate at which the alternator charges the battery is controlled by a voltage regulator. The voltage regulator controls the alternator output by monitoring the charge condition of the battery.

Alternator Output Test

An alternator output test checks an alternator's ability to deliver its rated output of voltage and current. This test should be performed any time an overcharging or undercharging condition is suspected. The output current and voltage should meet the specifications of the alternator, or there may be a problem with the alternator or regulator.

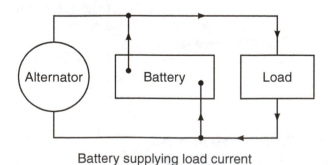

Battery supplying load current

Figure 12-13 The current flow of a charging system when the alternator is not producing voltage.

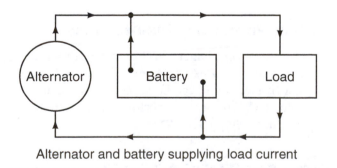

Alternator and battery supplying load current

Figure 12-15 A charging system when the unit is running at peak electrical capacity; in this case, both the alternator and the battery supply current to the load.

Figure 12-16 A tester used to load test batteries, starters, and alternators. *(Courtesy of John Dixon)*

Figure 12-16 shows a typical tester used for testing the capacity of batteries and alternators.

CAUTION *Always follow the manufacturer's instructions for testing alternator output.*

The following is a typical procedure for performance testing the alternator's output:

1. Make sure the knob controlling the load is backed off all the way.
2. Ensure the volt and amp meters are zeroed, and adjust if necessary.
3. Connect the tester load leads to the battery terminals: red to positive, black to negative.
4. Connect the clamp-on amps pickup around the battery ground (−) cable.
5. Start the vehicle up and run fast enough to obtain maximum output from the alternator (typically 1,500 rpm).
6. With the engine running, adjust the load on the battery to obtain the highest ammeter reading possible without causing the battery voltage to drop below 12.7 volts.
7. Read the ammeter.
8. Compare the amperage reading with the manufacturer's specifications. If the output is within 10 percent of the alternator's rated output capacity, the alternator is good. If the output is more than 10 percent below specs, **full field test** the alternator.

Full Field Testing

A full field test will allow the technician to determine whether the problem with the charging system is the fault of the alternator or the voltage regulator that is supposed to control the alternator output. This is accomplished by applying full battery voltage directly to the field windings in the rotor. When this is performed, the alternator should have an output within 10 percent of its rated capacity. If the alternator passes this test, the problem is more likely the voltage regulator. See the service manual for proper procedures for the alternator you are trying to diagnose.

Alternator Removal and Installation

When removing and reinstalling an alternator, technicians should follow a few safety measures:

1. The battery must be disconnected when installing or removing an alternator. The negative battery cable should be removed first.
2. Alternator polarity must be identified before connecting the alternator to the battery. If polarity is incorrect, the rectifier diodes will be ruined.
3. Make sure that the alternator pulley and the drive pulley are in alignment to ensure maximum alternator belt life. See **Figure 12-17** for proper alternator belt alignment.
4. The alternator belt should wrap around the pulley by a minimum of 100 degrees to prevent belt slippage belt and pulley ware and damage to the front alternator bearing due to overheating. See **Figure 12-18** for correct belt installation.

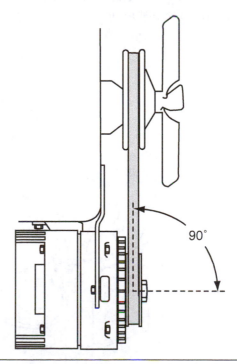

Figure 12-17 Proper alternator pulley to drive pulley alignment, critical to ensuring long belt life.

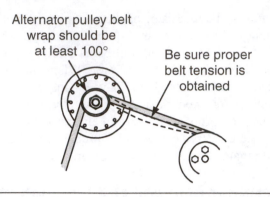

Figure 12-18 The amount of belt contact required to prevent the belt from slipping; as the alternator output increases, so does the alternator's resistance to turn. Slippage is prevented by a balance between proper belt pulley contract, along with recommended belt tension.

5. Check belt tension and verify by consulting the maintenance manual for the correct belt tension value.

STARTERS

Getting the vehicle started is possibly the most important function of the electrical system. The starting system performs this function by changing electrical energy from the battery into mechanical energy in the starter motor. The starter motor can then transfer this mechanical energy through gears to the engine flywheel, thus turning the engine's crankshaft. As the crankshaft turns, the piston draws the air fuel mixture into the cylinders and is then compressed and ignited to start the engine. Most engines require a cranking speed of around 200 revolutions per minute (rpm) to start. Starter motors are rated by power output in kilowatts; the greater the output the greater the cranking power.

Starter Motor Types

There are types of starters used by vehicle manufacturers. The first type (used on older model vehicles) uses a conventional starting motor, and the newer vehicles use a gear reduction starter motor.

STARTER TESTING

A starter amp draw test is a common way of ensuring a starter is in good condition. This test provides a quick check of the entire starting system. This test also tests battery cranking voltage. The following is an example of a starter draw test, but always follow the recommendations of the equipment manufacturer. Always refer to the service manual for the actual amp draw range that the starter should be expected to perform at. A typical starter on a reefer application would be in the range of 150 to 300 amps at 12.5 volts.

Shop Talk: Testing starters should always be performed with the engine at operating temperature.

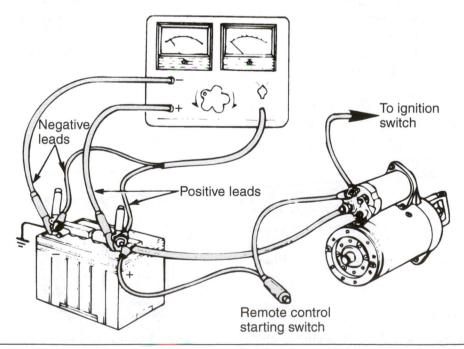

Figure 12-19 Test connection for a current-draw test. (*Courtesy of International Truck and Engine Corporation*)

1. Verify that the battery used for the draw test is in serviceable condition. Charge the battery if necessary (see **Figure 12-19**).

2. The next step is to prepare the load tester by ensuring the load knob is rotated all the way off, counterclockwise. Check the meters and adjust to zero if necessary. Connect the leads of the load tester to the battery: red on positive and black on negative. Note that the battery voltage should be at least 12.2 volts (50 percent charged). If not, recharge the battery. Adjust ammeter to zero using the adjustment control knob.

3. Connect the amp inductive pickup clamp around the negative battery cable or cables.

4. Disable the engine to prevent it from starting.

5. Crank the engine over while watching the voltmeter.

6. Stop cranking. Adjust the carbon pile until the voltmeter reading matches the reading taken in step 5.

7. Note the ammeter reading.

8. Remove the leads from the tester and prepare the engine to run.

Tech Tip: Digital multimeters can be used with an amp clamp to make this same measurement. The amp clamp is installed on the battery ground cable so that the amperage can be read off the meter as the engine is cranking.

Test Results

If the starter draw current is higher than specified and cranking speed is low, this usually indicates a problem with the starter itself. However, slow cranking speed can also be an indication of engine problems. A low cranking speed with low current draw, but high cranking voltage, usually indicates high resistance in the starter circuit. Keep in mind that the battery must be fully charged and all terminal connections tight to ensure accurate results.

LIFTGATES

A **liftgate** is designed to move cargo from the vehicle's floor height to ground level or vice versa. This is an extremely useful option when no loading dock is available. As with any piece of equipment, there are inspections and periodic maintenance that must be performed to ensure the safe working operation of the liftgate (see **Figure 12-20**).

Figure 12-20 Typical liftgate. *(Courtesy of John Dixon)*

There are many manufacturers of truck and trailer liftgates available in today's marketplace. Always follow the manufacturer's service manuals for specific scheduled maintenance and inspection intervals. The following maintenance procedures are from the Maxon Lift Corporation and Holland Hitch.

General Liftgate Safety Precautions

Before attempting to operate a truck/trailer liftgate, you must be fully trained in how the equipment is to be used and know the capabilities and the limitations of the equipment. When making repairs to the equipment, always use the proper tool for the job, and always replace worn or damaged tools before work begins.

Operating Safety Precautions

- Before operating a lift gate, be aware of your surroundings by making sure the area is clear of obstacles and other personnel.
- When operating the liftgate from the ground level, always stand to one side, clear of the platform. Make sure hands, arms, legs, and feet are clear of the moving liftgate (see **Figure 12-21**).
- Never exceed the maximum rated capacity of the liftgate (see **Figure 12-22**).
- If an emergency occurs, immediately release the control switch to stop the operation of the liftgate. **Figure 12-23** and **Figure 12-24** show an operation switch and an operation lever.
- Always make sure the liftgate is in its stowed position when not in use.
- Never jump on or off a moving liftgate or a raised platform.
- Ensure proper footing when dismounting the liftgate platform.

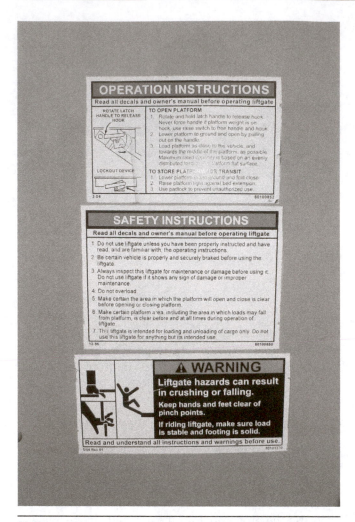

Figure 12-21 Safety precaution decal. *(Courtesy of John Dixon)*

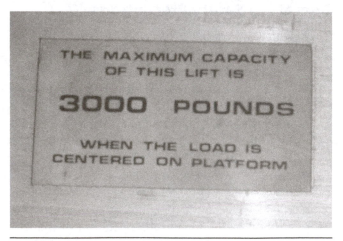

Figure 12-22 Maximum capacity decal. *(Courtesy of John Dixon)*

Servicing Safety Precautions

- Never work underneath a raised platform without making sure the platform is properly supported so that it cannot possibly be lowered.

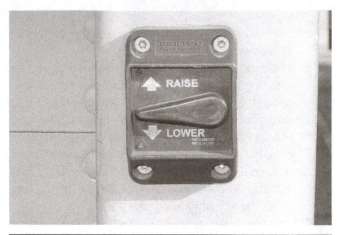

Figure 12-23 Liftgate switch. *(Courtesy of John Dixon)*

Figure 12-24 Control lever. *(Courtesy of John Dixon)*

- Use the proper tool for the job.
- Always wear proper safety attire to protect yourself against pressurized fluid spray or any airborne debris.
- When servicing the liftgate, disconnect the electrical power to the motor(s) to prevent accidental movement of the lift or possibly violent hydraulic fluid leaks.
- Whenever welding repairs are required on the liftgate, disconnect the battery ground cable(s) so that all electronic equipment is completely isolated and the work area is free of any combustible items before welding begins.
- Always have a fire extinguisher on hand before welding.

CAUTION *When hydraulic fluid is under pressure, it has the ability to penetrate the skin, causing serious personal injury, blindness, or death. If hydraulic fluid is injected beneath the* *skin, it must be surgically removed by a qualified doctor familiar with the treatment of this type of injury. Failure to adhere to this warning may result in serous injury or death.*

CAUTION *Never use your bare hand to check for fluid leaks!*

- Hydraulic fluid leaks under pressure may not be visible, so it is important to protect yourself whenever looking for a possible fluid leak. Always wear work gloves and a face shield, and use a piece of wood or cardboard to detect a fluid leak.
- Hydraulic pressure may remain present in the system, even though the hydraulic pump motor is shut off or disabled. Make sure there is no pressure in any of the hydraulic cylinders or hoses before working on hydraulic components or disconnecting hoses.
- Always observe local environmental protection regulations when disposing of waste hydraulic fluid.

General Maintenance

- Check the hydraulic fluid level in the pump reservoir. See the maintenance manual for correct level.
- If hydraulic fluid appears contaminated, change the hydraulic fluid. Refer to the maintenance manual for correct procedure.
- Keep track of the type of hydraulic fluid used in the system, and never mix two different grades of hydraulic fluid.
- Check all hoses and fittings for any signs of hydraulic fluid leakage or chaffing of the hoses due to movement or vibration. Repair as necessary.
- Check electrical wiring for any chaffing, ensuring all connections are tight and corrosion free. Repair as necessary.
- Check and make sure all warning and instructional decals are in place and legible.
- Check that all roll pins are in place and protrude evenly from both sides of the hinge pin collar. Replace roll pins if required.

CAUTION *When checking the hydraulic fluid level, keep dirt, water, and other contaminants from entering the hydraulic system. Before opening the reservoir filler cap, drain plug or*

Figure 12-25 Liftgate components. *(Courtesy of John Dixon)*

hydraulic lines and clean up any contaminants that could possibly enter the system.

Daily Maintenance

- Before use, inspect the liftgate for any signs of damage to the mounting assemblies, mechanical parts, or hydraulic components or hoses. Repair any defect found before attempting to operate the liftgate to avoid damage and personal injury. **Figure 12-25** illustrates some of the liftgate components.

Monthly Maintenance

- Every month, lubricate all pivot joints using EP2 chassis grease. In severe winter conditions, it may be necessary to lubricate more frequently. **Figure 12-26** shows a liftgate pivot point.

Figure 12-26 Pivot point. *(Courtesy of John Dixon)*

- Lubricate all linkage pivot points with penetrating oil or light engine oil.
- Check the oil level in the reservoir.
- Inspect entire liftgate for any worn parts or broken welds.

Yearly Maintenance

- Once per year, drain and replace the hydraulic fluid in the reservoir and refill with the manufacturer's recommended fluid. Always use a funnel with a fine mesh screen to prevent contamination from entering the hydraulic system.

- Check and make sure all warning and instructional decals are in place and legible.

Bi yearly Maintenance

- Every two years, the liftgate should be disassembled and all pivot point inspected for wear. Replace any pins or part that shows signs of wear.
- Replace the seal kit in the hydraulic cylinder and any hydraulic hoses that appear worn.

LIFTGATE TROUBLESHOOTING

Fault	Cause	Remedy
Platform moving down very slowly or may not lower	1. Not enough lubrication	Lubricate all pivot points.
	2. Excessive wear of mechanical components causing binding	Check freedom of all moving parts.
	3. Incorrect or contaminated oil in system	Oil should be clean; drain and replace as necessary.
	4. Restriction in hydraulic hose	Check hoses for external damage or pinching.
	5. Blocked hydraulic hose	Check for blockage in hoses.
	6. Battery is drained, insufficient voltage to solenoid.	Re-charge or replace battery.
	7. Electrical connections corroded or disconnected	Check connections.
	8. Mechanical components seized or damaged	Lubricate all pivot points and replace all damaged items.
Platform appears tilted	1. Incorrect platform adjustment, installation, or broken mounting components	Adjust levelling bolts; repair any broken components.
Platform will not lift its rated capacity	1. Hydraulic relief valve setting too low	Adjust relief valve setting.
	2. Hydraulic pump worn	Change worn parts or entire pump.
	3. Hydraulic cylinder leaking internally	Replace seals or complete cylinder.
	4. Hydraulic system not plumbed correctly	Check to see that the hose connected to the high pressure port is connected to the rod end of the cylinder.
Platform moves down too slowly	1. Flow control valve blocked	Replace valve.
	2. Flow control incorrectly installed	Check location and direction of arrow on valve.

Fault	Cause	Remedy
Oil expels from reservoir when the lift is lowered	1. Too much oil in reservoir	Ensure that level of oil allows displacement of cylinders.
	2. Wrong flow control fitting	Check that the number stamped on flow control fitting is opposite to the number of cylinders on your tailgate.
	3. Motor is not running when lowering platform	Check for faulty down button on push button control box.
Pump will not operate	1. Battery flat	Recharge or replace battery.
	2. Electrical coupling to trailer not complete (tractor/trailer only)	Connect coupling.
	3. Electrical connection to pump poor or broken	Check wiring to pump.
	4. Remote control switch broken or corroded	Check wiring.
	5. Solenoid switch on pump faulty	Check solenoid switch.
Control handle won't operate to lower gate from folded position	1. Latches binding on lock pins	Move handle to raise position to lift gate to clear latches, then lower.
	2. Linkage broken or disconnected	Repair linkage.
	3. Linkage won't move freely	Lubricate linkage, repair broken or seized parts or check for interference between linkage and rear sill of tailgate.
Platform will not raise	1. Battery flat	Recharge or replace battery.
	2. Voltage too low	Check voltage.
	3. Insufficient oil in reservoir	Top up hydraulic fluid as required.
	4. Pump solenoid switch burned out	Replace pump solenoid switch.
Motor runs but will not raise platform	1. Liftgate is loaded past rated capacity	Reduce the load on the platform.
	2. Hose or fitting leak	Check and retorque fittings.
	3. Leaking check valve or relief valve	Clean or replace valve.
	4. Broken pump shaft	Replace shaft or coupling as necessary.
	5. Pivot points seized	Lubricate.
	6. Worn or scored piston	Replace cylinder.
	7. Worn piston seals	Replace seals.
	8. Pump is worn out	Replace pump.
Platform will not raise smoothly	1. Insufficient oil in reservoir	Fill reservoir.
	2. Air lock in hydraulic system	Operate raise control for a few seconds at top of stroke. Repeat twice, pausing between operations.
	3. Undue mechanical wear or lack of lubrication	Lubricate pivots; replace any worn parts.
Platform creeps down when stationary	1. Cylinder seal leaking	Replace seals or cylinder if scored.
	2. Dirt under check valve	Clean valve.

(continued)

Fault	Cause	Remedy
Platform raises slowly	1. Battery flat	Recharge battery.
	2. Liftgate is loaded past rated capacity	Decrease load on platform.
	3. Poor cable connections	Check connections.
	4. Pump motor had poor ground connection	Check chassis ground connection.
	5. Leaking hydraulic hose	Repair hose connection.
	6. Not lubricated or excessive wear	Lubricate all pivot points; replace worn parts.
	7. Internal leak at cylinder	Replace seals or cylinder.
	8. Incorrect relief valve setting	Check relief valve setting.
	9. Worn pump or clogged filter	Check pump and filter.
Platform tips up when lowered to the ground	1. The lifting mechanisms are worn from severe wear or overload	Requires a rebuild. Replace parallel arms, shackles, pins, and possibly the lift frame.

Summary

- Battery electrolyte contains sulphuric acid, which can cause severe personal injury (burns) and damage to clothing and equipment.

- Storage batteries are the energy source for the vehicle's electrical system.

- Each of the six cell elements in a 12-volt battery produces approximately 2.1 volts.

- Batteries can be classified as conventional, low maintenance, and maintenance free.

- Use only distilled water when adding water to the battery.

- Maintenance-free batteries are designed to not require electrolyte replenishment.

- Storage batteries can be selected for their application by four different ratings. They are ampere-hour rating, cold cranking amps rating, reserve capacity rating, and the BCI (Battery Council International) dimensional group number.

- All batteries should be stored in a clean, cool, and dry environment and tested at regular intervals.

- For testing a batteries state of charge and its ability to perform, the most common methods are the open circuit voltage test and the hydrometer test for state of charge, and the load test for the batteries performance rating.

- Batteries can be charged at one of two different rates: fast charging or slow charging.

- It may be necessary to get the vehicle started when there is insufficient charge in the battery. This is referred to as jump starting or boosting a vehicle.

- The charging system converts mechanical energy to electrical energy when the engine is running.

- The voltage regulator controls the alternator output by monitoring the charge condition of the battery.

- An alternator output test checks an alternator's ability to deliver its rated output of voltage and current.

- A full field test will allow the technician to determine whether the problem with the charging system is the fault of the alternator or of the voltage regulator.

- Getting the vehicle started is possibly the most important function of the electrical system. The starter changes electrical energy from the battery into mechanical energy in the starter motor.

- A starter amp draw test is a common way of ensuring a starter is in good condition.

- A liftgate is designed to move cargo from the vehicle's floor height to ground level or vice versa.

- Before attempting to operate a truck/trailer liftgate, you must be fully trained.

- Hydraulic fluid under pressure has the ability to penetrate the skin, causing serious personal injury, blindness, or death.

- Never use your bare hand to check for fluid leaks.

- When checking the hydraulic fluid level, keep dirt, water, and other contaminants from entering the hydraulic system.

Review Questions

1. What are the dimensions of a group 31 storage battery?

   A. 13 inches long, 6.8 inches wide, and 9.4 inches high

   B. 12 inches long, 10 inches wide, and 9 inches high

   C. 11 inches long, 11 inches wide, and 11 inches high

   D. 10 inches long, 10 inches wide, and 9 inches high

2. What type of battery will need its water level topped up more often during normal use?

   A. Maintenance-free batteries

   B. Low-maintenance batteries

   C. Conventional batteries

   D. Dry charged batteries

3. When jump starting a vehicle with a known good battery, what is the last connection to be made before starting the stalled vehicle?

   A. Connect one end of the positive jumper cable to the positive (+) post of the stalled battery.

   B. Connect one end of the negative jumper cable to the negative (–) post of the jumper battery.

   C. Connect the other end of the jumper cable to the positive (+) post of the jump starting battery.

   D. Connect the other end of the negative jumper cable to a good chassis ground connection on the stalled unit, away from the discharged battery.

4. Which characteristic about a battery can the cold cranking amp rating tell you?

   A. It indicates the number of minutes a fully charged battery at 80°F (26.6°C) can sustain a load of 25 amps before the battery voltage drops to 1.75 volts per cell, or 10.5 volts for a 12-volt battery.

   B. It indicates how much power a fully charged battery can deliver when ambient temperatures become extremely cold.

   C. It indicates the load in amperes that a battery can sustain for 30 seconds at 0°F (–17.8°C) and not fall below 1.2 volts per cell, or 7.2 volts on a 12-volt battery.

   D. Both statements A and B are correct.

   E. Both statements B and C are correct.

5. How should you charge a battery that has lost its charge over a long period of time?

   A. The battery should be fast charged at a rate of 40 amps for up to 2 hours.

   B. The battery should be slow charged at 6 to 8 amps for 8 to 10 hours.

6. How many volts are produced in each cell of a battery?

   A. 2.1

   B. 6.0

   C. 9.6

   D. 12.0

7. A batteries reserve capacity in measured in:

   A. amperes.   C. watts.

   B. amp-hours.   D. minutes.

8. What is the state of charge of a battery that has a specific gravity of 1.190 at 80°F (26.7°C)?

   A. Fully charged   C. About 1/2 charged

   B. About 3/4 charged   D. Completely discharged

9. A battery being load tested with a commercial load tester is discharged at half its CCA rating for _____ seconds.

   A. 5   C. 15

   B. 10   D. 20

10. Which component provides the power to operate electrical components used to power all accessories on the vehicle when the engine is running?

    A. Voltage regulator   C. Stator

    B. Battery   D. Alternator

11. When performing a starter draw test, if low current draw is detected, the most likely cause would be:

    A. high resistance.   C. a discharged battery.

    B. a bad starter.   D. a short in the starter.

12. When performing a starter draw test, if higher than specified current draw is detected, the most likely cause would be:

    A. a discharged battery.   C. battery terminal corrosion.

    B. high resistance.   D. engine problems or a defective starter.

13. Before performing any service work on a hydraulic liftgate, the technician should:

    A. select the proper tool for the job.   C. disconnect the electrical power to the motor(s).

                                              D. check the level of the hydraulic fluid.

    B. know the rated capacity of the lift.

14. Which of the following safety policies should be adhered to when working on a hydraulic liftgate?

    A. Never work underneath a raised platform that is not supported.   C. Always wear protective safety equipment when working on hydraulics.

                                                                         D. Disconnect battery ground whenever welding must be performed on the liftgate.

    B. Always use the proper tool for the job.

                                                                         E. All of the above

CHAPTER

13 Coupling Systems

Objectives

Upon completion and review of this chapter, the student should be able to:

- Describe some of the different styles and types of fifth wheels available in the trucking industry.
- Outline the operation of the Holland, Fontaine, and ConMet fifth wheels.
- Explain the importance and list the consequences of incorrectly locating the fifth wheel on the tractor.
- Describe the locking principles of each type of fifth wheel.
- Explain how to couple and uncouple a tractor trailer.
- Perform general service procedures to common fifth wheels.
- Describe the procedure required to overhaul a fifth wheel.
- Explain the term high hitch and how it can be avoided.
- Describe the operating principle of pintle hooks/couplers and draw bars.
- Describe the benefits of cushioning to a coupling system.
- Outline prescribed maintenance for pintle hooks/couplers and drawbars.
- Outline the function of the kingpin and upper coupler assembly.

Key Terms

articulation

bolster plate

compensating fifth wheel

couplers

drawbars

elevating fifth wheel

fifth wheels

fully oscillating fifth wheel

gross axle weight rating (GAWR)

high hitch

kingpin

kingpin setting

oscillate

pintle hooks

sand shoes

semi-oscillating fifth wheel

semi-trailer

throat

towed vehicle weight (TVW)

upper coupler

INTRODUCTION

The coupling device, along with brakes and steering, is an area of maintenance that can never be overlooked. Regardless of the type or style of the coupling device, a technician must know how to perform maintenance and inspections on this extremely important item. A coupling device that fails can allow a trailer to become disconnected, causing untold damage to property and human life.

There are many different coupling devices, from fifth wheels to pintle hooks and drawbars.

SYSTEM OVERVIEW

In this chapter, we start by examining some of the more common styles of **fifth wheel** by some of the larger manufacturers of these components, as well as the kingpin that ties the tractor and trailer together. We will also look at other common coupling devices such as pintle hooks, couplers, and drawbars, that are also used extensively in the trucking industry. We will also cover adjustment, maintenance, and inspection procedures that all technicians must be able to perform correctly.

FIFTH WHEELS

By far the most common coupling device in North America would be the fifth wheel. The purpose of the fifth wheel is to connect a power unit to a trailer, or in the case of tractor/trailer train configurations, would connect a trailer to another trailer. The fifth wheel permits **articulation** (pivoting movement) between the tow and towed vehicle. This movement is essential to steer a tractor and **semi-trailer** or tractor and multiple trailer and multiple trailer combination properly on of off road. A fifth wheel also must transfer a portion of the semi-trailer's weight. A full trailer is one in which the entire weight of the trailer is supported on its own axles, much like a wagon. The majority of trailers are classified as semi-trailers, because the weight of the front portion of the trailer is supported by the tractor. In the case of some tractor/trailer train configurations, the load of the second semi-trailer is transferred to the rear portion of the first semi-trailer. A full description of tractor/trailer combinations and truck trains is provided in another chapter.

A fifth wheel is a wheel-shaped deck plate usually designed to **oscillate** or tilt on pivoting mounts. The entire assembly is bolted to the frame of the towing vehicle. A section at the back portion of the wheel is

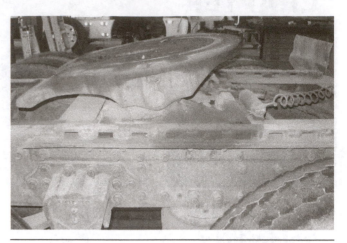

Figure 13-1 Fifth wheel assembly. *(Courtesy of John Dixon)*

cut away (referred to as the **throat**) to allow a trailer **kingpin** to connect with the locking jaws in the center of the fifth wheel. The trailer kingpin is mounted to the trailer's **upper coupler** assembly. The upper coupler is highly reinforced frame assembly that is either weld-bolted or riveted into the front section of the trailer. The kingpin protrudes through a steel plate called a **bolster plate.** This bolster plate transfers the weight of the semi-trailer through to the top of the fifth wheel when connected (see **Figure 13-1**).

The kingpin itself is a hardened steel cylindrical flanged stub that extends through the bolster plate. The condition of the kingpin is vital, as it holds the towing unit to the towed unit. As a tractor is backed into a trailer, the trailer's bolster plate is the first point of contact with the fifth wheel plate. The kingpin then enters the throat, guiding it into the locking jaws of the fifth wheel. The fifth wheel jaw assembly is spring-loaded to surround the kingpin. Most fifth wheels have both primary and secondary locking systems to maximize coupling safety. The flange at the bottom of the kingpin stub prevents vertical separation from occurring.

Fifth Wheel Types

The majority of fifth wheels manufactured in North America are produced by two companies: Holland Hitch and Fontaine. Holland Hitch bought out ConMet Simplex (formerly American Steel Foundries [ASF]) in 2005. Each type of fifth wheel uses its own distinctive locking mechanism. A truck technician must understand exactly how each type of fifth wheel lock operates before attempting service and overhaul procedures. There are many different styles of specialty fifth wheel, but by far the **semi-oscillating fifth wheel** is the most common.

Semi-Oscillating Fifth Wheels

The semi-oscillating fifth wheel (refer to **Figure 13-1**) articulates (or rocks) about an axis perpendicular to the vehicle centerline. What this means is that the fifth wheel will tip forward and backward. The top plate of the fifth wheel is either a cast steel or a pressed steel plate. The fifth wheel plate has a pair of mounting bosses on either side. The plate rocks back and forth on a pair of saddle brackets fitted on a base plate or frame brackets. Pins are inserted through the bosses on the fifth wheel plate and the saddle brackets. These pins are bushed in plastic sleeves encased in rubber to provide a cushioning effect during an up shock load. The pins are what permit the fifth wheel to articulate. The base plate is either rigidly fitted to brackets that in turn are bolted to the frame rails, or it can be mounted to a slider assembly that is bolted to the frame rails. A slide assembly allows the fifth wheel to slide forward and backward to change weight distribution on the tractor.

Many fifth wheels are mounted to the tractor in a stationary position. A stationary fifth wheel is used when the axle loading, trailer **kingpin setting** (location), and vehicle combinations are consistent throughout the fleet for which the tractor is specified.

Sliding Fifth Wheels

A sliding fifth wheel in most cases uses a semi-oscillating fifth wheel that can slide fore and aft on the tractor frame. This will change the point where the weight of the trailer loads the tractor. The sliding of the fifth wheel allows the tractor to be more flexible. Because different jurisdictions have different weight-over-axle regulations, highway trucks can meet a wider range of operational requirements using sliding fifth wheels. The versatility of a tractor with a sliding fifth wheel also makes the vehicle more attractive at the time of resale.

Sliding fifth wheels allow the operator to redistribute the trailer weight at different positions and also permit the same tractor to accommodate trailers with a multitude of kingpin settings. As well, the tractor/trailer combination length can be altered (see **Figure 13-2**).

Before a fifth wheel on a sliding mechanism can be used to pull a trailer, the slider mechanism must be locked into position. The fifth wheel base plate mechanically locks to toothed rails on the slide plate. Two methods are used to release the slide locks. The

Figure 13-2 A manual fifth wheel sidling mechanism. *(Courtesy of John Dixon)*

manual release mechanism is used for applications that do not require the fifth wheel to be moved often. The second method is an air release mechanism. This system allows the driver to release and lock the slider mechanism from the cab and is used when the position of the fifth wheel must be adjusted often. The driver should still visually inspect that the slider mechanism is locked before hauling a trailer.

Drivers also benefit when using a sliding fifth wheel. When axle loading or vehicle length is not an issue, the driver can slide the fifth wheel back close to the centerline of the bogie (or center of the axle on a single). Most tractors will ride more smoothly in this position, rewarding the driver with a much more comfortable ride.

Fully Oscillating and Compensating Fifth Wheels

A **fully oscillating fifth wheel** is used only in applications where the center of gravity of the trailer is located below the fifth wheel top plate, such as gooseneck or low-bed trailers. A fully oscillating fifth wheel permits articulation from front to rear like a semi-oscillating, but also side-to-side articulation. This side-to-side movement of the wheel can reduce much of the damaging torsional stress that can be transferred through tractor and trailer frames when a unit must go into uneven terrain like an off-road situation construction site (see **Figure 13-3**).

A **compensating fifth wheel** is designed to rock front to back and also allow movement from side to side. It is intended for applications where the center of gravity of the loaded semi-trailer does not exceed

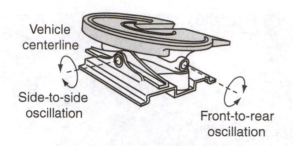

Figure 13-3 Fully oscillating fifth wheel.

Figure 13-5 Compensating fifth wheel. *(Courtesy of John Dixon)*

44 inches (1.1 m) above the top plate (see **Figure 13-4**). The compensating fifth wheel was developed for trailers that allow no tensional twist, such as a tank trailer. Tank trailers will not flex, so any flexing will be transmitted to the tractor frame, causing poor handling and eventually damaging the tractor frame or trailer itself (see **Figure 13-5**). Compensating fifth wheels have made it possible to manufacture B-train trailer frames with aluminum alloy by minimizing the trailer torque transfer effect. As previously mentioned, the compensating fifth wheel pivots forward and backward just like a semi-oscillating style and uses shoes that slide side to side in a concave track. In this way, the trailer is cradled as it slides up the track. A heavy spring re-centers the top plate after the turn (see **Figure 13-6**).

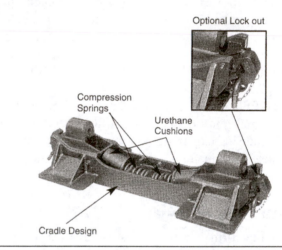

Figure 13-6 Holland Kampensator cradle view and lockout mechanism. *(Courtesy of SAF-HOLLAND, Inc.)*

Specialty Fifth Wheels

A number of special applications exist for fifth wheels. In most cases, the locking mechanism of specialty fifth wheels is no different from any other type of fifth wheel. A popular specialty fifth wheel is the **elevating fifth wheel.** These fifth wheels are used by many yard shunt tractors. The elevating fifth wheel allows a trailer to be pined to the tractor quickly without the need for the operator to climb down from the cab to raise the trailer landing gear. They can also be used for unloading operations that require the trailer to be tilted. The fifth wheel is first engaged to the upper coupler and then elevated so that the landing gear sand shoes clear the ground by a safe margin. Elevating is achieved either hydraulically or pneumatically (see **Figure 13-7** and **Figure 13-8**).

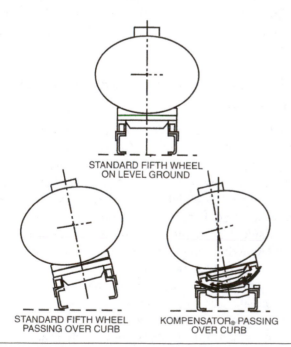

Figure 13-4 Holland Kampensator operating principle. *(Courtesy of SAF-HOLLAND, Inc.)*

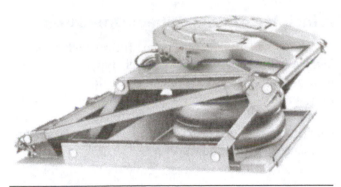

Figure 13-7 Air-actuated elevating fifth wheel. *(Courtesy of SAF-HOLLAND, Inc.)*

Figure 13-8 Hydraulically actuated elevating fifth wheel. *(Courtesy of SAF-HOLLAND, Inc.)*

Rigid fifth wheels are used in some vocational applications. A rigid fifth wheel is used in some frameless dump trailers equipped with oscillating bolster plates. This results in the articulation taking place at the trailer rather than the fifth wheel (see **Figure 13-9**).

Turntable or stabilized fifth wheels are used in some converter dolly assemblies. A turntable fifth wheel allows the top plate to rotate with the trailer bolster plate and allows use of a four-point support on the fifth wheel deck.

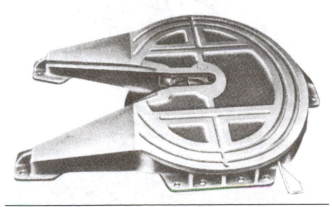

Figure 13-9 Rigid-type fifth wheel. *(Courtesy of SAF-HOLLAND, Inc.)*

FIFTH WHEEL CAPACITIES, HEIGHT, AND LOCATION

Before selecting a fifth wheel, it is important to look at the application type of vehicle and usage for which the unit will have to perform.

Fifth Wheel Capacity

The loading that a fifth wheel will encounter over its service life is a critical decision that must be part of the selecting process. A fifth wheel that is underrated for the application may result in an unsafe operating condition as well as damage to the fifth wheel. The consumer should spec out a fifth wheel with a higher capacity rating than they would normally require, taking into consideration the **towed vehicle weight (TVW)** to be pulled, maximum drawbar load expected, vertical load to be carried by the fifth wheel, and type of operation. If the vehicle is expected to be operated in an off-road application, it will normally require a higher capacity rating.

Generally, the higher the capacity of the fifth wheel and length of sliding mechanism, the more it will weigh. The choice of fifth wheel should be a balance between operational cost, strength, and durability. Any additional weight (over your requirements for operation) will add additional expense. This expense will be in the initial cost of the unit, the lost carrying capacity, and added fuel costs for the extra weight.

Fifth Wheel Height

Most trailers are designed with an intended 47-inch (1.19 m) bolster plate height. All trailers operating on North American roads cannot exceed a height of 13 feet 6 inches (4.1 m). This makes the positioning of the fifth wheel on a tractor, trailer, or bogie frame a critical dimension. The fifth wheel is designed to operate with the top plate level, so every attempt should be made to match the fifth wheel height with the trailer bolster plate height. Keep in mind that a final overall trailer height check must be performed with the trailer unloaded so that spring deflection will not be a factor.

One more consideration in dealing with fifth wheel height is developing a standard fleet specification. The advantage of having all tractors with the same fifth wheel height (distance from the ground to the top of the fifth wheel) is that the driver will not have to adjust the trailer landing gear before pining up to a fleet trailer, as it will have be dropped at the correct height.

Fifth Wheel Location

The placement of the fifth wheel on the tractor frame is extremely important, especially when using a stationary fifth wheel. There are three reasons why this is critical:

1. To allow the tractor to be loaded as close to the **gross axle weight rating (GAWR)** as possible without overloading the steering axle.

2. To ensure proper overall vehicle combination length, as well for proper vehicle steering stability. The fifth wheel centerline must be located forward of the rear axle on single axle tractors and forward of bogie centerline on tandem axle trailers. If the centerline of the fifth wheel is moved behind the axle or bogie centerline, weight will be removed from the steering axle affecting steering, stability, and braking.

3. To ensure clearance for articulation between the tractor and the trailer. It is not only important to have clearance between the tractor and trailer in a straight line, but there also must be enough clearance to turn the tractor 90 degrees to the trailer without contact. There are two points of contact that must be taken into consideration. First, there must be clearance between the rear of the tractor (cab or exhaust, whichever protrudes rearward the most) and the front wall of the trailer (or reefer). There must also be enough clearance between the rear of the tractor frame (fender or mud flap hangers) and the trailer landing gear. Holland Hitch Company has extremely good literature that can be purchased or downloaded free that will walk you through the procedures for placement for fifth wheels.

http://www.hollandhitch.com/Portal/#fastlatch

Select literature, then fifth wheels, then Fifth Wheel Catalog & Selection Guide.

Principles of Fifth Wheel Operation

Because the manufacturers of fifth wheels used on highway trucks each use distinct types of locking mechanisms, the truck technician should take nothing for granted when working on them. It is beneficial for a technician to have a basic understanding of how each mechanism operates. A few of the more popular locking mechanisms are introduced here.

HOLLAND HITCH A-TYPE MECHANISM

One of most commonly use fifth wheel locking mechanism in the industry is the Holland A-type. The A-type locking mechanism uses a single swinging lock jaw and plunger assembly. The A-type locking mechanism is often referred to as the FleetMaster in the industry (see **Figure 13-10**).

TYPE A LOCK

Bolster plate flush with fifth wheel

No gap

Plunger

Figure 13-10 Type A lock.

Operation

When the kingpin enters the throat of the fifth wheel and exerts pressure on the swinging lock, it pivots, enclosing the kingpin. This action causes a spring-loaded plunger to jam into a concave recess in the swing lock. Slack caused by mechanical wear of the kingpin or jaw mating surface can be adjusted with the use of an adjusting bolt located in the throat of the fifth wheel top plate.

HOLLAND HITCH B-TYPE MECHANISM

The Holland B-type locking mechanism, often referred to as a 3500 fifth wheel, can easily be identified by the adjusting rod and nut assembly protruding from the front of the top plate. This type uses a pair of opposed jaws and a yoke-locking mechanism (see **Figure 13-11** and **Figure 13-12**).

TYPE B LOCK

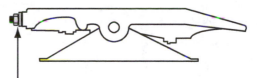

When locked with locks properly adjusted, the yoke will be seated as shown and the nut and washer will be snug against the fifth wheel.

Figure 13-12 Type B lock.

B-Type Coupling

In the open position, the yoke spring applies pressure to the yoke, which applies pressure to cam notches on the lock jaws to hold them in the open position (see **Figure 13-13**). This will prevent the

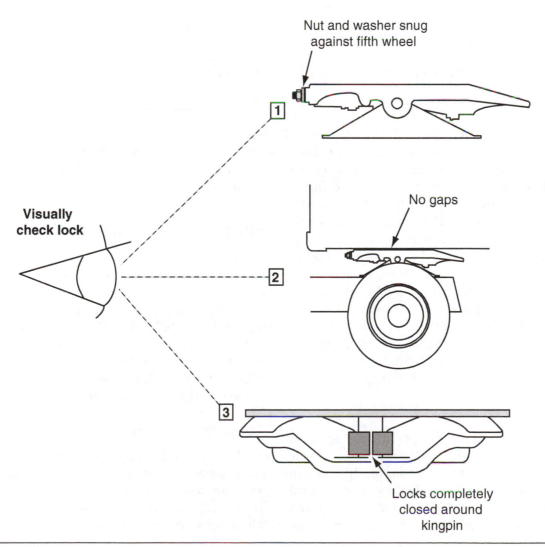

Figure 13-11 Visually check lock.

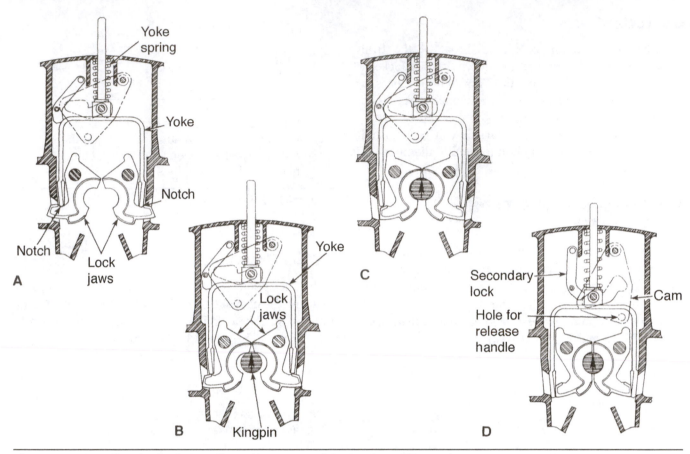

Figure 13-13 Holland B-lock coupling. *(Courtesy of SAF-HOLLAND, Inc.)*

jaws from accidentally locking as the vehicle travels over uneven terrain. When the fifth wheel is backed under a trailer, the kingpin enters the throat of the fifth wheel. When the kingpin contacts the inside bore of the lock jaws, it forces them to pivot and close around the kingpin. The tips of the jaws will pivot in and away from the tips of the yoke assembly. The yoke spring forces the yoke across the holes in the casting of the top plate, preventing the jaws from opening. Movement of the yoke assembly also causes the cam to rotate. The spring-loaded secondary lock follows the outer profile of the cam and permits the secondary lock to rotate in behind the yoke, thereby locking the yoke into a closed position. The jaws are now unable to open to release the kingpin.

B-Type Release

To release the locking mechanism, the operator pulls the fifth wheel's release handle outwards (see **Figure 13-14**). This causes the cam to pivot, which simultaneously swings the secondary lock outward away from the yoke and forces the yoke up and away from its wedged position behind the cam

notches on the locking jaws compressing the yoke spring. As the kingpin is further removed, it spreads the lock jaws, forcing them into their fully retracted position.

CONMET FIFTH WHEELS

The old standard ConMet Simplex 400 (the old ASF) locking mechanism is described here. The ConMet 400 consists of sliding front and rear jaws linked by a hinge pin assembly. The front jaw is backed by a rubber block. This is a major advantage and has made this fifth wheel popular among tanker operators because of its ability to dampen the surge effect of the load. See **Figure 13-15** for an x-ray view of the ConMet assembly in the locked position.

ConMet Locking

When the fifth wheel is unlocked, the front jaw is on a horizontal plane with its inner bore exposed rearward to the throat. The lower jaw is swung downward, pivoted by the hinge pins. When the fifth wheel is backed under the trailer kingpin, the kingpin

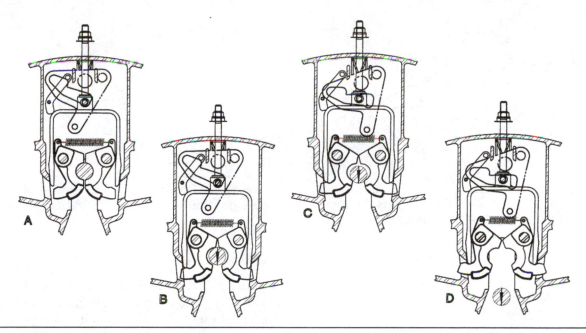

Figure 13-14 Holland B-lock uncoupling. *(Courtesy of SAF-HOLLAND, Inc.)*

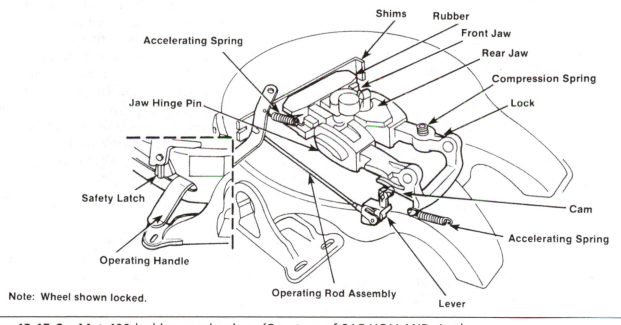

Figure 13-15 ConMet 400 locking mechanism. *(Courtesy of SAF-HOLLAND, Inc.)*

contacts the front jaw and forces it to slide toward the rubber block. As the front jaw is driven backward, the rear jaw is pulled up onto a horizontal plane, enclosing the kingpin. In the locked position, the rubber block is loaded under some compression (crush) by the lock arm assembly, which acts on the rear jaw (see **Figure 13-16**). When the fifth wheel is engaged, the operating/release handle slides into its locked position (see **Figure 13-17**). A safety latch on the operating handle prevents accidental release.

ConMet Release

The safety latch located on the side of the fifth wheel is raised, allowing the operating/release lever to be forced forward (see **Figure 13-18**). This action rotates a cam, permitting the lock arm assembly to swing away and unjamming the rear jaw; a detent on the cam prevents reengagement of the lock arm assembly (see **Figure 13-19**). When the tractor separates from the trailer, both sets of jaws slide rearward until the rear jaw swings down, clearing the kingpin, as shown in **Figure 13-20.**

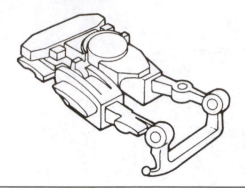

Figure 13-16 ConMet 400 sliding jaws in locked position. *(Courtesy of SAF-HOLLAND, Inc.)*

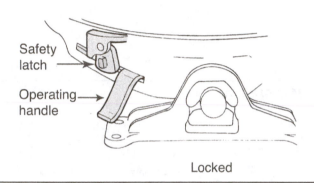

Safety latch

Operating handle

Locked

Figure 13-17 Side view of a ConMet 400 fifth wheel when fully locked. *(Courtesy of SAF-HOLLAND, Inc.)*

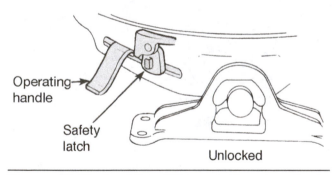

Operating handle

Safety latch

Unlocked

Figure 13-18 ConMet 400 operating handle in release position. *(Courtesy of SAF-HOLLAND, Inc.)*

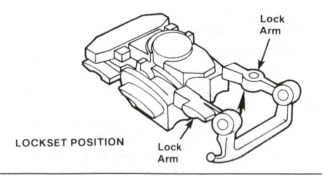

Lock Arm

LOCKSET POSITION

Lock Arm

Figure 13-19 ConMet 400 lock arms in release position. *(Courtesy of SAF-HOLLAND, Inc.)*

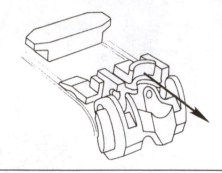

Figure 13-20 ConMet 400 rear jaw dropped into release position. *(Courtesy of SAF-HOLLAND, Inc.)*

SIMPLEX

The ConMet ASF Simplex fifth wheels are now known as Holland Simplex. They now incorporate a J-shaped rod handle, which makes it easier for drivers to release the unit. The locking jaw also has been redesigned, and the new jaw assembly can be retrofit to Simplex and Simplex II fifth wheels. A major advantage of the Simplex fifth wheel is the small number of parts used in the locking assembly. The locking mechanism used on Simplex II fifth wheels is known as Touchloc® and is illustrated in **Figure 13-21.**

The Simplex ll fifth wheel locking mechanism operates very much like the Holland A-type. Fifth wheel adjustment is done by removing and turning a cam-shaped jaw pin to the next of three positions. This action will remove the play between the jaw assembly and the kingpin. This wheel, when properly adjusted, should have 1/16 inch of clearance.

CAUTION *Never force the jaw pin to the next position; it must be able to be rotated by hand. Always install a new cotter pin.*

FONTAIN NO-SLACK® II FIFTH WHEEL

The Fontain fifth wheel is lighter in comparison to the Holland products that use a cast top plate. Fontain uses a pressed steel top plate that tends to be lighter than their equivalently rated competitor versions, an important consideration when weight is a predominant factor in choosing a fifth wheel. **Figure 13-22** shows an exploded view of a Fontaine No-Slack® II fifth wheel. There are three models manufactured: standard duty, moderate duty, and severe heavy duty. The standard duty has an improved locking mechanism.

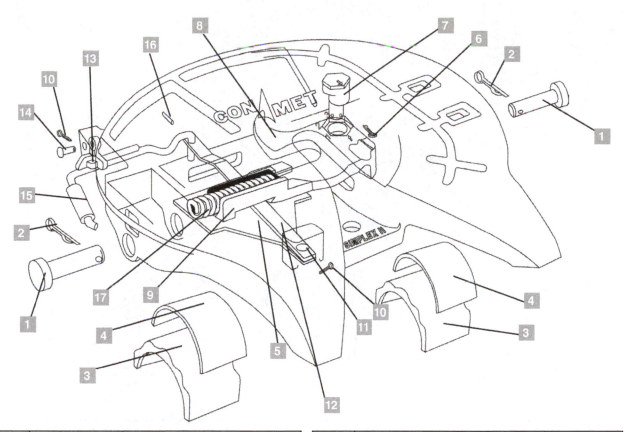

Item	Description	Item	Description
1	Bracket Pins (2 required)	10	Cotter Pin (2 required)
2	Clinch Pins for Bracket Pins (2 required)	11	Lever Bar Pin
3	Bracket Shoes (2 required)	12	Lever Bar
4	Bracket Pads (2 required)	13	Safety Indicator
5	Cover Plate	14	Safety Indicator Pin
6	Clinch Pin for Jaw Pin	15	Operating Rod
7	Jaw Pin (standard)	16	Slotted Spring Pin
7a	Jaw Pin (oversize—not shown)	17	Lock Spring
8	Jaw	18	Grease Fittings (2 required)
9	Lock	N/A	SIMPLEX 11 Top Plate Assembly

Figure 13-21 Holland Simplex II fifth wheel components. *(Courtesy of SAF-HOLLAND, Inc.)*

Operation Locking

The kingpin enters the throat and moves up until it contacts the front jaw and simultaneously the bumper and timer assembly. This action trips the spring-loaded rear jaw, which moves into position, blocking the rear of the kingpin. As the rear jaw moves into a position, a serrated wedge set is forced by spring pressure against the serrated back side of the rear jaw, holding it in position. This action takes out any play between the kingpin and the jaw assembly. Unlike other locking mechanisms, it takes very little force to lock the assembly.

Operation Release

To release the kingpin, the operator pulls the release handle, which in turn pulls the wedge and rear jaw assembly out of the way, allowing the kingpin to be removed through the throat.

For more information and video demonstrations on Fontaine fifth wheels, check out http://www.fifth-wheel. com/.

FIFTH WHEEL MAINTENANCE

Although maintenance procedures for fifth wheels differ only slightly between manufacturers, you should always follow the manufacturer's specific

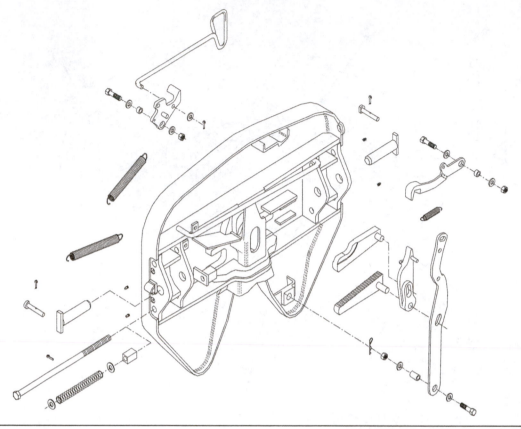

Figure 13-22 Exploded view of Fontaine No-Slack® II fifth wheel. *(Courtesy of Fontaine International)*

recommendations. This section will deal with some of Holland Hitch's recommended procedures.

Maintenance Interval

To obtain the best performance and service life, the fifth wheel should be serviced within the first 1,000 to 1,500 miles (1,600–2,400 kms). After this initial service, periodic maintenance is recommended at intervals of 30,000 miles (50,000 kms) or 3 months, whichever comes first, to ensure trouble-free operation.

See **Figure 13-23. Figure 13-24. Figure 13-25. Figure 13-26,** and **Figure 13-27** for some of Holland Hitch's recommended maintenance procedures.

Maintenance Procedures

The two most important factors to obtain trouble-free operation of a fifth wheel are:

- Proper PM procedures for the fifth wheel (as needed, lubricate)
- Proper coupling procedures

AS-NEEDED LUBRICATION

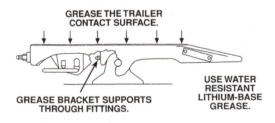

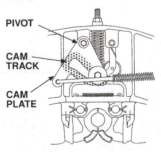

Lubricate the cam track and pivot with a light oil or diesel oil.

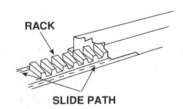

On sliding fifth wheels, spray a light oil or diesel oil on the rack and slide path.

Figure 13-23 Holland's as-needed lubrication guide. *(Courtesy of SAF-HOLLAND, Inc.)*

REQUIRED INSPECTIONS AND ADJUSTMENTS

Perform the following every six months or 60,000 miles, whichever comes first. Thoroughly steam clean all components before inspecting or adjusting.

GENERAL FIFTH WHEEL INSPECTION:

1. Inspect fifth wheel mounting and fifth wheel assembly.

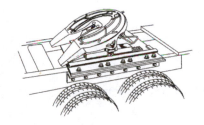

- Check fastener torque
- Replace missing or damaged bolts
- Replace bent, worn or broken parts (Use only genuine Holland parts.)

Figure 13-24 Holland general fifth wheel inspection. *(Courtesy of SAF-HOLLAND, Inc.)*

INSPECTION – LOCKING MECHANISM:

1. Verify operation by opening and closing locks with Holland Kingpin Lock Tester model no. TF-TLN-1500 or TF-TLN-5001.

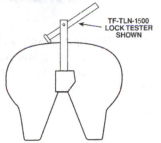

Properly closed fifth wheel

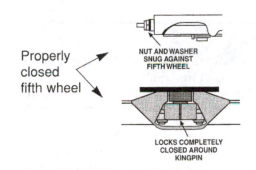

Figure 13-25 Inspection of locking mechanism. *(Courtesy of SAF-HOLLAND, Inc.)*

ADJUSTMENT – LOCKING MECHANISM:

1. Close locks using Holland Lock Tester.

2. Rotate rubber bushing located between the adjustment nut and casting.

3. If the bushing is tight, rotate nut on yoke shank <u>counterclockwise</u> until bushing is snug, but still can be rotated.

4. Verify proper adjustment by locking and unlocking with the lock tester.

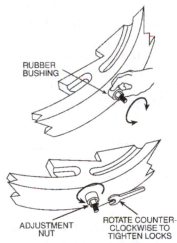

The bushing should be snug, but you should be able to rotate it.

Figure 13-26 Holland adjustment of locking mechanism. *(Courtesy of SAF-HOLLAND, Inc.)*

Holland recommends the following fifth wheel maintenance procedures:

1. Observe fifth wheel operation when coupled to a trailer, checking for slack. This will require the attention of two maintenance personnel, one to make the observation and the other to rock the tractor back and forth. This can also be performed with the use of the appropriate lock tester.

Note: If the amount of slack is too much, do not automatically assume that the fifth wheel is the problem. Always check the kingpin for wear using a kingpin gauge.

2. Uncouple the fifth wheel from the trailer.
3. Pressure wash or steam clean the top plate assembly.

ADJUSTMENT – FIFTH WHEEL SLIDE MECHANISM:

1. Loosen locknut and turn adjustment bolt out (counterclockwise).

2. Disengage and engage the locking plungers. Verify that plungers have seated properly as shown.

3. Now tighten adjustment bolt until it contacts the rack.

4. Turn the adjustment bolt clockwise an additional 1/2 turn then tighten the locknut securely.

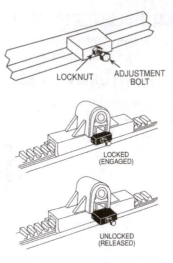

⚠ **CAUTION:** Proper adjustment of the sliding bracket locking plungers must be performed at installation and maintained at regular intervals using the adjustment bolts for both plungers. Proper adjustment is required for proper operation, load transfer, and load distribution.

5. If plungers do not release fully to allow fifth wheel to slide:
   A. Check the air cylinder for proper operation. Replace if necessary.
   B. Check plunger adjustment as explained above.
   C. If a plunger is binding on the plunger pocket, remove the plunger using a Holland TF-TLN-2500 spring compressor. Grind the top edges of the plunger 1/16" as shown. Re-install and adjust the plungers as explained above.

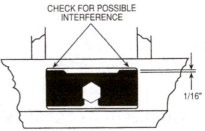

6. If the locking plungers are too loose:
   A. Check plunger adjustment as explained above.
   B. Check plunger springs for proper compression. Replace if necessary.
   C. Check for plunger wear. If necessary, replace as described above.

After inspection and adjustment, relubricate all moving parts with a light, rust resistant oil.

Figure 13-27 Adjustment of fifth wheel slide mechanism. (*Courtesy of SAF-HOLLAND, Inc.*)

4. Inspect the top plate and brackets for cracks, and replace all damaged, loose, or missing parts.
5. Check all mounting bolts for proper torque.
6. Visually inspect the fifth wheel and mounting plate welds.
7. Check the fifth wheel locking mechanism with a lock tester and adjust as necessary.
8. Relubricate the fifth wheel.

Never attempt to inspect a fifth wheel without removing the grease. The best way to do this is to first scrape off as much as possible, and then apply some solvent before hot pressure washing. Check the top plate, throat, and saddle/pivot assembly for cracks. Check the dynamic action of the locking mechanism and the secondary lock integrity. Check all the mounting fasteners and welds.

HOLLAND FIFTH WHEEL LOCKING ADJUSTMENT

Type-A Lock Adjustment

The lock adjustment bolt is found on the right side of the throat:

1. Using the appropriate lock tester, install the tester, locking the fifth wheel.
2. Tighten the adjuster bolt (clockwise) with a 1/2-inch Allan wrench or 1/2-inch Allan socket extension.

Figure 13-28 2-inch diameter plug. *(Courtesy of John Dixon)*

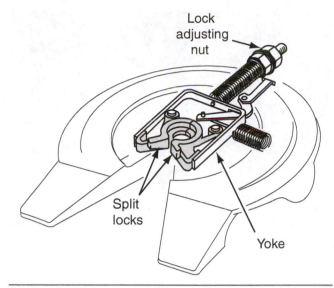

Figure 13-29 B lock mechanism.

3. Loosen the adjustment bolt counterclockwise a half turn. The locking mechanism is now properly adjusted.
4. Verify the operation of the fifth wheel by locking and unlocking the tester several times.
5. After inspection and adjustment, lubricate all moving parts with a light, rust-resistant oil.
6. Apply water-resistant, lithium-based grease to the trailer contact surface of the fifth wheel top plate.

Type-B Lock Adjustment

1. Carefully close the locking jaws and insert a 2-inch diameter plug (Holland P/N TF-0237) in the locks (see **Figure 13-28**).
2. The plug should fit snugly, but you should be able to rotate it by hand if you apply some force.
3. If the plug fits loosely, tighten jaws by turning the lock nut protruding from the front of the top plate counterclockwise. It may be necessary to tap the end of the yoke shank lightly to allow the nut to seat against the top plate.
4. Check the operation of fifth wheel by installing and releasing the lock tester several times (see **Figure 13-29**).
5. After inspection and adjustment, lubricate all moving parts with a light, rust-resistant oil.
6. Apply water-resistant, lithium-based grease to the trailer contact surface of the fifth wheel top plate.

ROUTINE SERVICE LUBRICATION

1. Apply water-resistant, lithium-based grease to the trailer contact surface of the fifth wheel top plate.

2. Apply grease to the bearing surface of the support bracket through the grease fittings on the side of the fifth wheel plate. Do not overgrease, as this will only make a mess on the deck plate. The top plate should be tilted back slightly to relieve weight on the brackets when applying the grease.
3. Spray an oil and diesel fuel mixture (80 percent engine or gear oil and 20 percent diesel fuel) on the rack and bracket slide path of sliding fifth wheels.

Lubrication after Cleaning

1. Apply a light grease to all moving parts, or spray the with diesel/oil mixture.
2. Perform the routine service lubrication procedure.

Sliding Fifth Wheels

A sliding fifth wheel allows greater versatility of a tractor by allowing weight to be shifted fore and aft to maximize axle capacities.

Sliding Procedures

The fifth wheel should be fully engaged to the trailer for this procedure:

1. The tractor and trailer should be parked in a straight line on level ground. Engage the trailer parking bakes.

CAUTION *The trailer must be stationary with its parking brakes applied to prevent*

Figure 13-30 Air-actuated sliding fifth wheel.

damage to the tractor or trailer by uncontrolled sliding of the fifth wheel.

2. Release the sliding locking plungers. For an air slide release (see **Figure 13-30**), put the cab control valve into the unlocked position. For a manual slide release, pull the release lever (see **Figure 13-31**).
3. Visually check that both plungers have disengaged from the rail teeth. If the locking plungers are jammed in the rack teeth, try lowering the landing gear to relieve pressure on the plungers. This should allow the fifth wheel to slide more easily.
4. Release the tractor brakes and drive the tractor forward or backward slowly to position the fifth wheel.

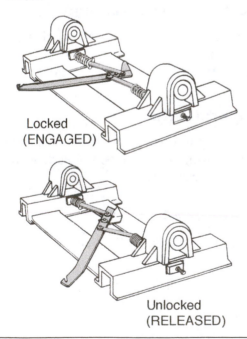

Figure 13-31 Manual slide release plunger position. *(Courtesy of the Holland Group, Inc.)*

5. After sliding the fifth wheel to the desired setting, engage the slide locking plungers. For an air slide release, put the control valve in the lock position to engage the plungers to the rack teeth. For a manual slide release, trip the release arm to allow the plungers to engage with the rack teeth.
6. Visually check to see that both plungers are fully engaged to the rail teeth so that no forward or rearward movement is possible. Leaving the trailer brakes locked and nudging the tractor slightly may be required to engage the plungers in the rack teeth. Raise the landing gear to the fully retracted position.

CAUTION   *Do not operate the vehicle if the plungers are not fully engaged and the landing gear fully retracted, because damage to the tractor, trailer, and landing gear can occur.*

KINGPINS

The kingpin stub is mounted on the trailer upper coupler assembly. It is designed to securely fasten the trailer to the fifth wheel while still allowing rotary motion between the two units. This motion allows the tractor trailer to turn. There are two SAE standard kingpin sizes, 2 inches and 3.5 inches. The kingpin is required to have a SAE-specified Brinell surface hardness that will produce a yield strength of 115,000 psi and a tensile strength of 150,000 psi.

The majority of kingpins are welded to the upper coupler assembly, but both bolt-on and removable fasteners are also used (see **Figure 13-32**).

Figure 13-32 Kingpin and bolster plate. *(Courtesy of John Dixon)*

Checking Kingpin Dimensions

During the service on a trailer, the kingpin should be cleaned and inspected for wear, integrity to the upper coupler, and cracks. Most shops use a template or gauge to check the kingpin for wear. Holland's TF-0110 gauge can also be used to check kingpin length and squareness to the bolster plate and for checking straightness and flatness of upper coupler or bolster plate.

Inspection of Kingpin Diameter

- Install the gauge over the kingpin using the open center access hole.
- Try to slip gauge slots over indicated portions of the kingpin. Rotate the gauge a full 360 degrees around the kingpin to check the diameter in all directions.
- The gauge is a "No Go" gauge and indicates 1/8 inch (3.2 mm) of wear if the appropriate 2 inches (or 3-1/2 inches) kingpin enters the slot of the gauge.
- If the gauge does slide on, recheck the measurement using a more accurate device. Replace if the diameter is worn 1/8 inch (3.2 mm) (see **Figure 13-33**).

Inspecting Kingpin for Squareness and Height

- Place the correctly corresponding edge of the gauge against the kingpin and the trailer bolster plate.

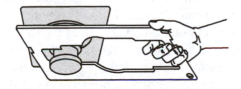

Inspection of the kingpin diameter

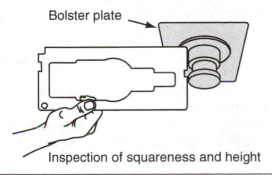

Bolster plate

Inspection of squareness and height

Figure 13-33 Inspection of the kingpin diameter using a gauge.

- If the gauge doesn't fit into the kingpin, the height of the kingpin is incorrect or the kingpin is bent.
- Rotate the gauge 360 degrees around the kingpin to check the squareness with the plate.
- A bent kingpin or kingpin of incorrect length must be replaced.

KINGPIN SERVICE

There is really no maintenance that can be performed on a kingpin. Manufacturers of kingpins recommend that if the kingpin wear exceeds the limits, which is easily verified with the kingpin gauge, the component should be replaced. A vehicle with an undersized kingpin should never be put back into service without being properly repaired. The replacement of kingpins in most trailer applications is a large task because the upper coupler may also have to be rebuilt to access the kingpin. Mobile repair of kingpins has become an acceptable alternative to replacement of the kingpin. This type of repair is performed on site and claims to meet SAE J-133 and J-140 standards. Although mobile repair of kingpins may be controversial, because of the cost and time savings, it has become a fact of life in the trucking industry.

Kingpin Maintenance

1. **Upper Coupler Plate Flatness**
   Using a 48-inch straightedge, check the flatness in all directions. Any bumps, valleys, or warping will cause uneven loading of the fifth wheel, which could result in damage to the top plate and poor lock life. Replace the trailer upper coupler plate if flatness exceeds the specification shown in **Figure 13-34.**
2. **Inspect the Kingpin for Straightness**
   Using a square or Holland kingpin gauge (TF-0110), check to see if the kingpin is bent. A bent kingpin accelerates lock wear and may interfere with proper fifth wheel locking. This also may indicate damage. The kingpin should be replaced if it exceeds 1 degree from square in any direction.
3. **Inspect the Kingpin for Proper Length**
   Using a Holland kingpin gauge, check the length of the kingpin. Caution: if a lube plate is used in your operation, make sure to check the kingpin length. The kingpin must be sized to compensate for the thickness of the lube plate. Otherwise, the kingpin will be too short. If the kingpin length is improper, the kingpin should be replaced.

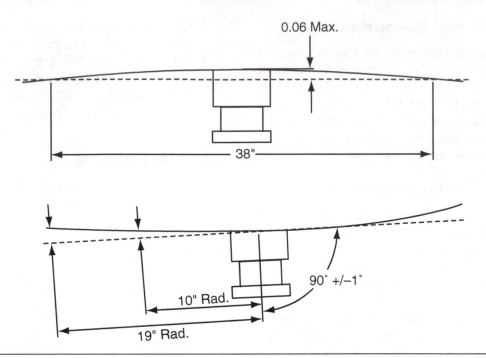

Figure 13-34 Checking upper coupler flatness.

4. **Inspect the Kingpin for Wear**

Using a Holland kingpin gauge, check for wear on both the 2-inch and 2.88-inch diameters. Wear of 1/8 inch (3.2 mm) is indicated if the appropriate diameter enters the gauge slot. Replace the kingpin if the gauge slides into the appropriate gauge slot.

5. **Check the Kingpin Mounting**

In addition to being a safety hazard, a loose mount will cause excessive chucking and rapid lock wear. Reinstall or replace any kingpin that is not securely mounted.

6. **Check Kingpin for Damage**

Inspect the kingpin for any nicks, gouges, deformation, or cracks, which may interfere with or affect the safe use of the kingpin. Replace the kingpin if any damage is noted.

FIFTH WHEEL SERVICING TOOLS

Refer to **Figure 13-35** for some of the basic tools required to test and adjust fifth wheels and kingpins. The fifth wheel lock tester is available in the SAE standard kingpin dimensions and is used to test the locking action of the fifth wheel assembly. The C-clamp fixture is used for removing and installing springs, retainers, and pins of the slide release mechanisms of Holland sliding fifth wheel assemblies. The kingpin gauge is used to verify that kingpin dimensions are within specifications. It is extremely easy to use and will indicate an undersized condition and

the need of replacement for 2-inch and 3-1/2-inch SAE kingpins. This gauge can also be used to check for proper kingpin length and straightness as well as flatness of the upper coupler of bolster plate. The 2-inch plug is used in checking and adjusting the kingpin locks of the fifth wheel, especially type-B locks during periodic maintenance. It is also used in lock installation and adjustment during rebuilding procedures.

COUPLING PROCEDURES

The locking procedures for fifth wheels are more or less the same, but the locking mechanisms differ between manufacturers. The inspection procedures for proper coupling will vary. Because the fifth wheel is such a critical safety component of a tractor/trailer combination, be sure you are informed on what to look for to ensure the unit is properly coupled; this means that the jaws are fully closed, the locks are secured, and a high hitch condition does not exist.

1. First make sure the jaws are in the open positions, that the fifth wheel locks are not engaged, and that the fifth wheel is tilted down at the back. Check the condition of the fifth wheel's mounts and also check tightness. Make sure the top plate is properly lubricated. A fifth wheel that is not adequately lubricated can cause a hard steering complaint.

2. Line up the truck and trailer, making sure the throat of the fifth wheel is in perfect alignment with the trailer's kingpin and stop.

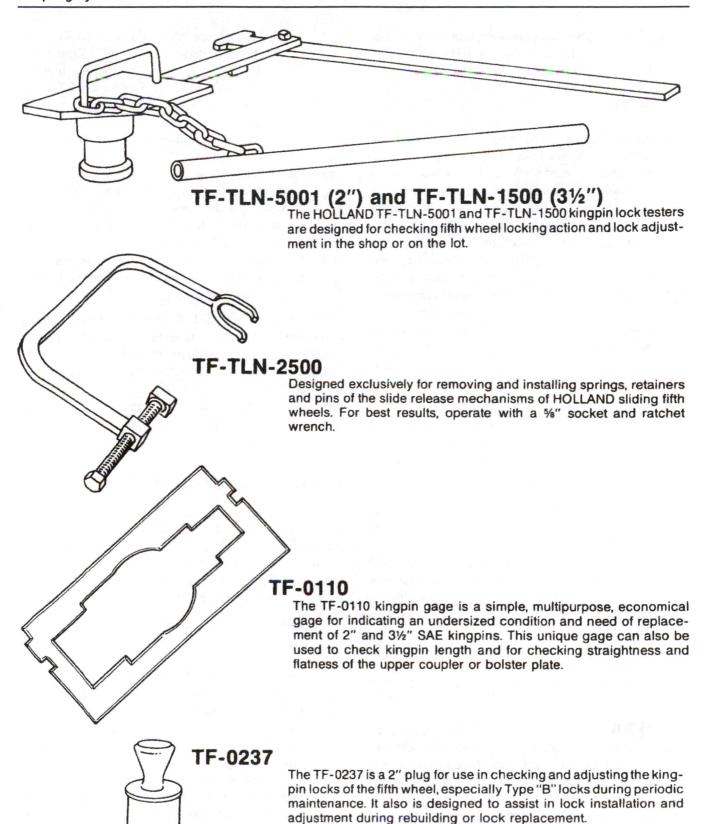

TF-TLN-5001 (2") and TF-TLN-1500 (3½")

The HOLLAND TF-TLN-5001 and TF-TLN-1500 kingpin lock testers are designed for checking fifth wheel locking action and lock adjustment in the shop or on the lot.

TF-TLN-2500

Designed exclusively for removing and installing springs, retainers and pins of the slide release mechanisms of HOLLAND sliding fifth wheels. For best results, operate with a ⅝" socket and ratchet wrench.

TF-0110

The TF-0110 kingpin gage is a simple, multipurpose, economical gage for indicating an undersized condition and need of replacement of 2" and 3½" SAE kingpins. This unique gage can also be used to check kingpin length and for checking straightness and flatness of the upper coupler or bolster plate.

TF-0237

The TF-0237 is a 2" plug for use in checking and adjusting the kingpin locks of the fifth wheel, especially Type "B" locks during periodic maintenance. It also is designed to assist in lock installation and adjustment during rebuilding or lock replacement.

Figure 13-35 Holland fifth wheel and kingpin service tools. *(Courtesy of SAF-HOLLAND, Inc.)*

3. Check that the trailer parking brakes are applied and block the trailer wheels. Hook up brake lines and electrical cord.

4. Check the height of the trailer's upper coupler to the height of the fifth wheel. The trailer should be in a position so that the fifth wheel top plate contacts the trailers bolster plate 4 to 6 inches behind the fifth wheel pivot point. Using low gear, raise or lower the trailer landing gear as required to obtain the proper coupler height. This will prevent the possibility of a **high hitch.** A high hitch is a situation where the locks of the fifth wheel close and lock only around the lower portion of the kingpin. In this condition, there will be space between the top plate and the bolster plate. All the trailer weight is transferred through the lock jaws. The trailer is extremely dangerous to operate, as there is nothing to prevent the kingpin from moving upward and releasing the trailer.

CAUTION *Attempting to couple with the trailer at an improper height could result in a false or improper coupling such as a high hitch and cause damage to the tractor, fifth wheel, or trailer.*

5. Back the tractor slowly into the trailer, maintaining alignment of the fifth wheel throat to the kingpin.

6. After picking up the trailer and leveling the fifth wheel, stop the truck and then continue backing until the fifth wheel locks firmly to the kingpin. Stopping will help prevent hitting the kingpin too hard.

7. Next, nudge the tractor forward with the trailer brakes still applied to test the completeness of the coupling as an **INITIAL** check.

CAUTION *To ensure that a proper coupling has occurred, a direct visual inspection must be performed. Even an improperly coupled fifth wheel can pass a pull test, and you should never rely on the sound of the jaws snapping closed; this is also unreliable. You must get out of the cab and use a flashlight to verify that the jaws are closed properly and there is no gap between the top plate and the bolster plate.*

8. Visually inspect that the fifth wheel jaws are closed properly around the kingpin and that the kingpin is not overhanging the fifth wheel or

caught in a grease groove. There should be no gap between the top plate and the bolster plate. Check for proper coupling by looking into the throat of the fifth wheel. Ensure the locks are closed like those in **Figure 13-10, Figure 13-11,** and **Figure 13-12.**

9. If your fifth wheel uses a secondary lock, engage it.

10. Raise the landing gear in low speed until the **sand shoes** clear the ground, then shift landing gear to high and raise till fully retracted. Fold down or remove the crank handle and place it in the crank handle holder.

11. Recheck air line and electrical connections.

12. Remove the wheel chocks and continue with your pre-trip inspection.

Photo Sequence 11-1 illustrates a typical coupling procedure. Once again, students are reminded that the coupling procedure required by driver license certification testing may differ because this does not account for default to park brake systems. Driver testing may require that the trailer be pneumatically coupled before mechanically engaging the tractor fifth wheel with the trailer upper coupler.

UNCOUPLING PROCEDURES

The release lever for the fifth wheel may be found on the left or right side of the wheel. The release levers may pull out linearly or slide axially. If you are unfamiliar with the release mechanism of the wheel, ask someone who knows. The following is a general procedure for uncoupling a fifth wheel from a trailer:

1. Park the tractor/trailer combination and apply the trailer brakes. If the trailer is loaded, make sure the parking area is level and the surface can support the landing gear sand shoes. Back the tractor into the trailer and set the tractor parking brakes. This procedure will relieve any pressure on the lock jaws.

2. Lower the landing gear in high speed until the sand shoes just contact the ground. Switch to low speed and crank **a few extra** few turns to take some of the weight off the tractor. Fold down or remove the crank handle and place it in the crank handle holder.

3. Disconnect the air lines and electrical cord and attach to dummy air line couplers to keep foreign material out of the air lines.

4. Activate the release lever. Some fifth wheels have a secondary lock that must be released before the main release lever can be moved. Others have a two-lever release mechanism.

Trailer Coupling

P11-1 Inspect the fifth wheel for damaged, worn, or loose components and mountings. Be sure that the fifth wheel jaws are open and the handle is in the unlocked position. The fifth wheel must be properly lubricated and tilted down at the rear.

P11-2 Check that the trailer parking brakes are applied and block the trailer wheels as a precaution.

P11-3 Adjust the trailer landing gear so that the trailer bolster plate is just below the fifth wheel.

P11-4 Slowly back up the tractor, maintaining the alignment between the fifth wheel throat and the trailer kingpin.

P11-5 When the trailer kingpin enters the fifth wheel throat and the trailer bolster plate rides on the fifth wheel, back up the tractor until there is full trailer resistance.

P11-6 Connect the air hoses and electrical connectors between the tractor and trailer.

P11-7 Inspect the trailer bolster plate to be sure that it is supported evenly on top of the fifth wheel. No gap should be visible between the bolster plate and the fifth wheel surface.

P11-8 Inspect the safety latch, making sure that it swings freely and that the operating handle is behind the safety latch positioned to the rear of its operating slot.

P11-9 Crawl under the tractor and use a flashlight to verify that the kingpin is properly engaged to the fifth wheel jaws.

(Continued)

Trailer Coupling (Continued)

P11-10 With the trailer service brakes applied, place the transmission in the lowest gear and partially engage the clutch to create a pulling force between the fifth wheel and the kingpin. The fifth wheel jaws should remain securely locked on the kingpin.

P11-11 When coupling is completed, apply the tractor and trailer brakes and crank the trailer landing gear to the fully upward position. Remove the blocks from the trailer wheels.

5. Release the tractor brakes and pull ahead slowly, allowing the trailer to slide down the fifth wheel and pick-up ramps, being careful that the tractor landing gear is not shock loaded. If the tractor is equipped with an air suspension, pull tractor ahead just until the kingpin is released from the lock, stop and dump the air in the suspension before pulling the top plate out from under the bolster plate. This will prevent damage to the tractor suspension.

PINTLE HOOKS, COUPLERS, AND DRAWBARS

Pintle hooks, couplers, and **drawbars** are another means of connecting a truck to trailer or trailer to trailer. These components are designed to be used primarily for the towing of vehicles and where backing up of vehicles is limited. Damage to these coupling components, fasteners, mounting structure, and safety chains can occur from the drawbar binding in the pintle horn or from jackknifing if the vehicle is turned while backing up.

Pintle Hook

A pintle hook is a coupling device that uses a towing horn that is fixed. The towing horn is attached to a drawbar, which is connected to the towed vehicle. To make a coupling with the setup, the drawbar eye must be lifted up over the pintle horn and then secured with a pivoting swing-down latch. A pintle hook may be one of two mounting types: ridged or swiveling (see **Figure 13-36**).

Couplers

A coupler is slightly different from a pintle hook in that the towing horn is not fixed but pivots. To make a coupling with the setup, the coupler with jaw open is

Figure 13-36 A pintle hook may be one of two mounting types. *(Courtesy of John Dixon)*

Figure 13-37 A coupler assembly. *(Courtesy of John Dixon)*

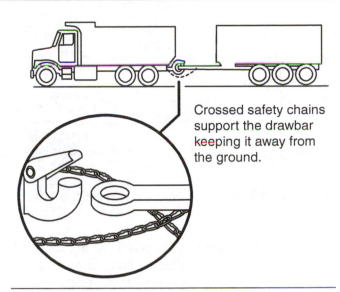

Crossed safety chains support the drawbar keeping it away from the ground.

Figure 13-39 Crossed safety chains support the drawbar, keeping it away from the ground.

driven into the drawbar. The drawbar activates the pivoting horn (or jaw), causing it to rotate up through the center of the drawbar eye, securing the drawbar to the coupler. Couplers are generally a ridged mount design (see **Figure 13-37**).

Drawbars

The drawbar is the mating part to the pintle hook or coupler in the coupling system. The drawbar (also called a Lunette) is a steel doughnut-shaped component that is attached to the towed vehicle by various bolting or welding configurations. Drawbars are available as a ridged or swivel mount design (see **Figure 13-38**).

Over the Road Applications

Pintle hooks and couplers that are used for on-road applications require a secondary locking system. In addition, all pintle hooks and couplers require the use of safety chains. Safety chains are intended to keep the towing and towed vehicles together and in directional control in the event of an improper coupling or a coupling device failure (see **Figure 13-39**).

Safety chains are required by law on all vehicles equipped with pintle hook/coupler and drawbar devices to pull a trailer. These requirements are outlined in the Federal Motor Carrier Safety Regulations 393.70. The Society of Automotive Engineers (SAE) J697 and the Truck Trailer Manufactures Association (TTMA) RP-6 have similar recommendations. These regulations require the following items regarding safety chains or cables:

1. The safety device must not be attached to the pintle hook/coupler device.
2. The safety device must have no more slack than is necessary to permit the vehicle to be turned properly.
3. The safety device, and the means of attaching it to the vehicles, must have an ultimate strength of not less than the gross vehicle weight of the vehicle or vehicles being towed.
4. The safety device must be connected to the towed and towing vehicle and the tow bar in such a way that they will prevent the tow bar from contacting the ground in the event that the device fails or becomes disconnected.

Figure 13-38 A drawbar assembly. *(Courtesy of John Dixon)*

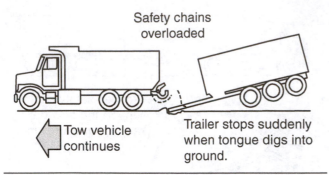

Figure 13-40 Trailer draw bar catastrophic failure.

Crossing the safety chains beneath the drawbar will prevent it from possibly digging into the ground, resulting in the sudden overload of the safety chains (failure) (see **Figure 13-40**).

Another benefit of crossing the safety chains or cable is directional control in the event of a coupling failure. The safety chain on the right side of the towing vehicle will pull the trailer to the trailer to the right, while the safety chain on the left side of the towing vehicle will pull the trailer to the left. These two forces will balance each other out, resulting in proper tracking, so that the equipment can be controlled while the vehicle is stopping. Failure to cross the safety chains can result in the towed vehicle swaying back and forth into oncoming traffic (see **Figure 13-41**).

Off-Road Applications

When a coupling device is used for an off-road application, the rated capacity of the device must be reduced by 25 percent to provide for the increased dynamic shock loads generated by operating over tough terrain. In addition to this, either the pintle hook or the drawbar must be of a swivel mount design. The swivel provides additional side-to-side rotation (oscillation) over uneven terrain.

Note: Utilizing safety chains of the proper size and condition is a very important step toward safe vehicle operation.

Cushioning

There are two types of cushioning available for pintle hooks and drawbars:

- One provides shock absorption upon acceleration.
- The other provides a snubbing of the clearance between the drawbar eye and the pintle or coupler horn as well as shock absorption upon deceleration (braking). Because the snubbing action eliminates most of the movement between the

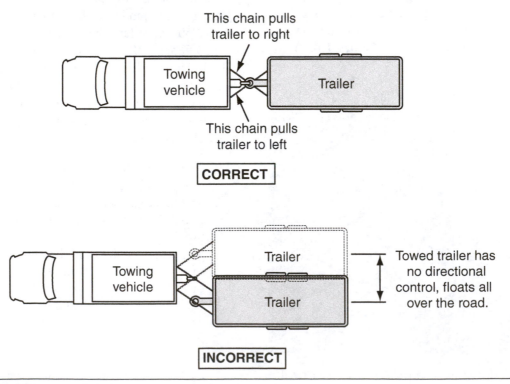

Figure 13-41 Safety control chains.

drawbar and the pintle horn, service life is greatly increased.

Both are available on numerous pintle hooks and drawbars but not on couplers. If a coupler is used and cushioning is desired, a cushioned drawbar must be used.

CUSHIONING MAINTENANCE

Check plunger adjustment during routine maintenance. The plunger should be adjusted to hold the drawbar firmly against the coupling device when it is energized, to provide and to provide adequate clearance for coupling and uncoupling when de-energized.

CAUTION *Never use both a swiveling drawbar and swiveling coupler component at the same time.*

Pintle Hook/Coupler Inspection

All coupling devices and safety chains should be regularly inspected as outlined in the maintenance section of the appropriate component specification sheet.

The devices should be inspected immediately if there is:

- Overloading
- Binding of the drawbar upon backing
- Jackknifing
- Exceeded limit of articulation
- Exceeded limit of oscillation
- Any suspected abuse

Replacement of the device is required if any cracks are found or if wear exceeds the limits found in the component's specification sheet.

Pintle Hook/Coupler Maintenance

For proper maintenance, Holland Hitch recommends the following steps that should be performed every 30,000 miles (50,000 kms) or 3 months, whichever comes first:

1. Clean and check for proper operation. Inspect for missing parts. Replace as required with only OEM components.
2. Inspect the coupling contact area and periodically disassemble to inspect for wear on the shank mounting flanges using the appropriate specifications for your component. For example, the specifications for a Holland PH-760 (swiveling pintle hook) states that you must replace any component when wear exceeds 1/8 inch (.0125 inch) from the original surface profile.

3. Regularly lubricate the pivotal latch with a light oil lubricant.
4. Check mounting fasteners for proper torque. All fasteners are grade 8 and torqued to specifications.

Weld Repairs

Holland Hitch states that their pintle hooks, drawbars, and couplers are made from quenched and tempered alloy steels. **Weld repairs to either repair a broken part or to build up a worn surface are strictly forbidden.** Such weld repairs can locally alter the chemistry, heat treatment, and strength of the coupling device, leading to its failure, or create a stress concentration, which could initiate a fatigue failure.

OUT-OF-SERVICE CRITERIA FOR COUPLING DEVICES

Fifth Wheels

1. **Mounting of Frame**
   A. More than 20 percent of fasteners on either side missing or ineffective
   B. Any movement between mounting components
   C. Any mounting angle iron cracked or broken

Special Note: Any repair weld cracking, well-defined (especially open) cracks in stress or load-bearing areas, cracks through 20 percent or more original wield or parent metal

2. **Mounting Plate and Pivot Brackets**
   A. More than 20 percent of fasteners on either side missing or ineffective
   B. Any welds or parent metal cracked

Special Note: Any repair weld cracking, well-defined (especially open) cracks in stress or load-bearing areas, cracks through 20 percent or more original wield or parent metal.

   C. More than 3/8 inch (9.5 mm) horizontal movement between pivot bracket pin and bracket
   D. Pivot bracket pin missing or not secured

3. **Sliders**

A. More than 25 percent of latching fastener per side ineffective

B. Any fore or aft stop missing or not securely attached

....

Note: A moveable fifth wheel that is secured with vertical pin does not need fore or aft stops.

....

C. Movement of more than 3/8 inch (9.5 mm) between slider bracket and slider base

4. **Operating Handle**

A. Operating handle not in closed or locked position

5. **Fifth Wheel Plate**

A. Cracks in fifth wheel plate

....

Special Note: Any repair weld cracking, well-defined (especially open) cracks in stress or load-bearing areas, cracks through 20 percent or more original wield or parent metal.

....

Exceptions: (1) Crack in fifth wheel approach ramps, and (2) casting shrinkage cracks in the ribs of the body of a cast fifth wheel.

....

6. **Locking Mechanism**

A. Locking mechanism parts missing, broken, or deformed to the extent that the kingpin is not securely held

Upper Coupler Assembly (Including Kingpin)

1. Horizontal movement between the upper and lower fifth wheel halves exceeds 1/2 inch (13 mm)
2. Kingpin can be moved by hand in any direction.

....

Note: This item is to be used when uncoupled semi-trailers are encountered, such as a terminal inspection, and it is impossible to check item 1 above. Kingpins in coupled vehicles are to be inspected using item 1 above and items 3 and 4 below. Vehicles are not to be uncoupled.

....

3. Kingpin not properly engaged.
4. Any semi-trailer with a bolted upper coupler having fewer effective bolts than shown in the following table.

Minimum Total Quantity of Bolts

(Total minimum quantity of bolts must be equally divided with half on each side of the coupler.)

Bolt Size	Bolt Size
½ inch (13 mm)	5/8 inch (16 mm) or larger
10 (5 bolts on each side)	8 (4 bolts on each side)

....

Note: This bolt size table applies to trailers having 68,000 lb maximum gross vehicle weight rating (GVWR). Such trailers are typically used in tractor–semi-trailer combinations with a maximum gross combination weight rating (GCWR) of 80,000 lb.

....

5. Any welds or parent metal cracked

....

Special Note: Any repair weld cracking, well-defined (especially open) cracks in stress or load-bearing areas, cracks through 20 percent or more original wield or parent metal.

Pintle Hooks

Mounting and Integrity

1. Loose mounting, missing or ineffective fasteners, or insecure latch

....

Note: A fastener is not considered missing if there is an empty hole in the device but no corresponding hole in the frame and vice versa.

....

2. Cracks anywhere in the pintle hook assembly, including mounting surface and frame cross member
3. Any weld repairs to the pintle hook assembly
4. Section reduction visible when coupled

....

Note: No part of the horn should have any section reduced by more than 20 percent.

Drawbar Eye

Mounting and Integrity

1. Any cracks in attachment welds or drawbar eye
2. Any missing for ineffective fasteners
3. Any weld repairs to the drawbar eye
4. Section reduction visible when coupled

....

Note: No part of the eye should have any section reduced by more than 20 percent.

....

Drawbar/Tongue

1. Slider (Power/Manual)

A. Ineffective latching mechanism
B. Missing or ineffective stop

C. Movement of more that 1/4 inch (6 mm) between the slide and housing
D. Any leaking air or hydraulic cylinders, hoses, or chambers (other than slight oil weeping normal with hydraulic seals)

2. Integrity

A. Any cracks
B. Movement of 1/4 inch (6 mm) between sub frame and drawbar at point of attachment

Safety Devices

1. Missing
2. Unattached or incapable of secure attachment
3. Improper repairs to chains and hooks including welding, wire, small bolts, rope, and tape

Summary

- The most common coupling device in North America is the fifth wheel.

- The fifth wheel permits articulation between the tow and towed vehicles.

- The kingpin protrudes through a steel plate called a bolster plate.

- The kingpin is vital, as it holds the towing unit to the towed unit.

- Each type of fifth wheel uses its own distinctive locking mechanism.

- The semi-oscillating fifth wheel is the most common style.

- A sliding fifth wheel offers the advantage of re-distributing trailer weight on the tractor axles as well as the ability to change vehicle combination length.

- Before a fifth wheel on a sliding mechanism can be used to pull a trailer, the slider mechanism must be locked into position.

- Because the manufacturers of fifth wheels used on highway trucks each use distinct types of locking mechanisms, the truck technician should take nothing for granted when working on them.

- The kingpin stub is mounted on the trailer upper coupler assembly. It is designed to securely fasten the trailer to the fifth wheel while still allowing rotary motion between the two units.

- During the service on a trailer, the kingpin should be cleaned and inspected for wear, integrity to the upper coupler, and cracks.

- Pintle hooks, couplers, and drawbars are another means of connecting a truck to trailer or trailer to trailer.

- A pintle hook is a coupling device that uses a towing horn that is fixed.

- A coupler uses a towing horn that is not fixed but pivots.

- The drawbar is the mating part to the pintle hook or coupler in the coupling system.

- All pintle hooks and couplers require the use of safety chains.

- Weld repairs to pintle hooks, couplers, and drawbars to repair a broken part or to build up a worn surface are strictly forbidden.

Review Questions

1. The fifth wheel height is defined as:

   A. the distance of the trailer from base of the wheel to roof of the trailer when the trailer is coupled.

   B. the distance from ground level to the top of the fifth wheel.

   C. the distance from ground level to the base of the fifth wheel.

   D. the distance from the bottom of the fifth wheel to the top plate.

2. Which type of fifth wheel is most commonly used on highway tractors?

   A. Fully oscillating

   B. Semi-oscillating

   C. Stabilized

   D. Elevating

3. The air release slide mechanism is preferred over the manual release when:

   A. slide length is short.

   B. slide length is long.

   C. frequent adjustment is necessary.

   D. driver comfort is important.

4. Which of the following is an advantage of a sliding fifth wheel?

   A. Weight-over-axle on the tractor can be altered.

   B. Weight-over-axle on the trailer can be altered.

   C. Trailer bridge formula can be altered.

   D. Both A and C are correct.

   E. All of the above

5. Which type of fifth wheel permits the articulation of any angle for use with gooseneck trailers operating in off-road applications?

   A. Fully oscillating

   B. Semi-oscillating

   C. Elevating

   D. Compensating

6. Which type of fifth wheel uses a locking mechanism with a pair of transversely opposed lock jaws?

   A. Holland A lock

   B. Holland B lock

   C. Fontaine

   D. All fifth wheels

7. Holland recommends that the fifth wheel should be inspected every:

   A. 50,000 miles or 80,000 km.

   B. 40,000 miles or 65,000 km.

   C. 30,000 miles or 50,000 km.

   D. 20,000 miles or 32,000 km.

8. When lubricating the fifth wheel, technician A lubricates all moving parts with a water-resistant lithium-based grease. Technician B lubricates the rails on a sliding fifth wheel with a mixture of oil and diesel fuel. Who is correct?

   A. Technician A is correct.

   B. Technician B is correct.

   C. Both technicians A and B are correct.

   D. Neither technician A nor B is correct.

9. The jaw pair of the fifth wheel lock mechanism should be replaced when wear exceeds:

    A. 3/8 inch.

    B. 1/4 inch.

    C. 1/8 inch.

    D. 1/16 inch.

10. To ensure that the tractor and trailer are properly coupled, the best test is to:

    A. try to pull forward with the trailer brakes set.

    B. listen for the jaws to snap closed when the tractor and trailer are coupled.

    C. look for a gap between the fifth wheel and the trailer bolster plate.

    D. visually inspect to make sure the fifth wheel jaws have clamped around the kingpin.

11. In what application would you most likely find an elevating fifth wheel?

    A. Highway tractor

    B. Construction vehicle

    C. Shunt tractor or yard spotter

    D. All of the above

12. Which is the most common method for attaching a kingpin to a trailer upper coupler assembly?

    A. SAE Grade 8 fasteners

    B. SAE Grade 5 fasteners

    C. Welding

    D. Lockpins

13. If you have trouble uncoupling a fifth wheel, what should you do first?

    A. Crank the landing gear down to relieve the pressure on the plungers.

    B. Lubricate all moving parts of the fifth wheel with penetrating fluid.

    C. Release tractor brakes and back into the trailer and reapply the truck's parking brakes.

    D. Make sure the fifth wheel slide mechanism works properly.

14. What is the first thing a technician does before performing a fifth wheel inspection?

    A. Check all cotter pins for proper installation.

    B. Perform a thorough steam cleaning or hot power washing.

    C. Check the torque values of all mounting bolts.

    D. Ensure the safety catch operates properly.

15. For proper steering stability, the fifth wheel centerline must be located:

    A. forward of the rear axle centerline on a single axle tractor.

    B. behind the rear axle centerline on a tandem axle tractor.

    C. forward of the bogie centerline on a tandem axle tractor.

    D. Both A and C are correct.

    E. Both A and B are correct.

16. Trailer weight is transferred to the fifth wheel plate by which of the following?

    A. Bolster plate   C. Pintle hook

    B. Kingpin   D. Drawbar

17. Which type of fifth wheel is designed to minimize the twist transmitted between trailer and tractor frame?

    A. Semi-oscillating   C. Elevating

    B. Rigid   D. Compensating

18. All off-road applications using a coupling device require both a swiveling drawbar and a swiveling pintle hook.

    A. True   B. False

19. In the event of a coupler failure, what prevents the drawbar from hitting the ground?

    A. Safety chains   C. Duct tape

    B. Bungee cords   D. Crossed safety chains

20. If a coupling device is to be used in an off-road application, the rated capacity must be decreased by:

    A. 5 percent.   D. 20 percent.

    B. 10 percent.   E. 25 percent.

    C. 15 percent.

21. Upon inspection of a coupling device, if cracks are noted or if wear exceeds the limits outlined in the detail sheet, they should be welded to repair or build up the surface back to its original specifications.

    A. True   B. False

22. What is the normal maximum vehicle height on a tractor/trailer combination used on our highways?

    A. 144 inches (366 cm)   C. 13 feet (4.0 m)

    B. 12 feet 6 inches (3.8 m)   D. 13 feet 6 inches (4.1 m)

23. Which of the following dimensions is equivalent to a standard SAE kingpin size?

    A. 1.5 inches (38 mm)   C. 3.5 inches (89 mm)

    B. 2.5 inches (64 mm)   D. 4.5 inches (114 mm)

24. Technician A says that when a sliding fifth wheel is moved forward on a tractor, the percentage of total vehicle weight supported by the trailer is reduced. Technician B says that when a tractor sliding fifth wheel is moved rearward, weight is removed from the tractor steer axle. Who is correct?

    A. Technician A   C. Both technicians A and B are correct.

    B. Technician B   D. Neither technician A nor B is correct.

14 Truck and Trailer Refrigeration Maintenance

Objectives

Upon completion and review of this chapter, the student should be able to:

- List the various components that must be checked and maintained on a routine PM service on a refrigeration unit.
- Accurately check and adjust engine coolant strength.
- Describe refrigeration maintenance.
- Explain evacuation techniques.
- Explain a pre-trip inspection.
- Describe leak testing procedures.
- List structural items that should be checked on a PM.

Key Terms

bleeding
pre-trip
preventive maintenance service (PM)

INTRODUCTION

Proper maintenance of the refrigeration unit is extremely important in order to ensure the reliability and performance of the unit. Loads carried in the truck or trailer can be very expensive, possibly worth $100,000 and more. Making sure the reefer unit is reliable is a big concern for the shipper and the receiver. Often, shipping receivers have a temperature-tracking device hidden in with the cargo so that they are able to establish if the product was shipped at the correct temperature. If the receiver is not satisfied with the results, they may even refuse to accept the load, resulting in lost profit margins for shippers and possibly insurance negotiations for the shipping company.

SYSTEM OVERVIEW

In this chapter the technician will learn the procedures that are necessary to perform a complete **preventive maintenance service (PM)** on a reefer unit. Included are engine service, fuel system service, compressor service, drive belt service, refrigeration system checks, glow plug system checks, and structural maintenance checks.

Only a unit passing an extensive PM should be put in service and loaded with product. In addition to proper preventive maintenance, a technician should be able to diagnose problems with the refrigeration system, including refrigerant flow problems, leak testing, compressor oil change, and filter drier change.

ENGINE LUBRICATION SYSTEM

Engine oil and filters should be changed at regular intervals according to the manufacturer's recommended intervals. Due to the fact that mileage is not an issue with refrigeration units, the intervals will be determined by hours of engine operation recorded on the engine hour meter.

Engine Oil Change

Before the engine lubrication is changed, the engine should first be warmed up to operating temperature. This allows the oil to drain faster from the oil pan and more completely than if the engine is cold. Once the engine is warm, shut the unit off and remove the drain plug from the oil pan or uncap the oil pan extension hose. Refer to **Figure 14-1** for drain plug and oil filter location.

Have a large enough container to capture all the oil in the engine, approximately 17 quarts (16 liters) for a trailer unit. The unit and trailer must be kept in a level position to ensure all the oil is allowed to drain from the oil pan. It is important to get as much of the oil out as possible, as most of the dirt particles are in the last few quarts (liters) of oil to drain from the oil pan. When

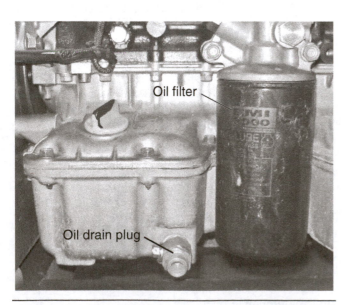

Figure 14-1 The location of the drain plug and oil filter. *(Courtesy of John Dixon)*

drainage is complete, reinstall the oil pan plug or cap and refill the oil pan with 16 to 17 quarts (16 liters) of oil and check the engine oil dipstick. Run the unit and again check the oil level and adjust as necessary.

Shop Talk: Always follow the manufacturer's recommendations for oil viscosity ratings for the ambient conditions the piece of equipment will be subjected to.

Oil Filter Replacement

When changing engine oil filter(s), have a catch basin ready, because there will be some oil loss. Use an oil filter wrench or large water pump pliers to remove the filter. Before installing the new oil filter(s), lightly lubricate the filter gasket with clean engine oil. Fill oil filter(s) with oil from a clean container, making sure no foreign material enters the filter. Failure to prime the filter(s) with oil may allow the engine to operate for a period with no oil supplied to the bearings. Tighten the new filter until the gasket comes in contact with the filter base, and then tighten an additional half turn. Start the engine and check for leaks.

Fuel Filter Replacement

The fuel filters are another item that should be changed while servicing the engine. **Figure 14-2** shows a typical spin-on fuel filter.

■ Start by removing the primary and secondary filters and discard. Using clean diesel fuel, lubricate the rubber gasket on the filter.

Shop Talk: Do not attempt to prime the fuel filters before installation, as unfiltered fuel can damage the injection pump.

■ Tighten the secondary fuel filter until the rubber gasket makes contact with the filter base, and tighten an additional half turn.
■ Install the primary fuel filter and spin it on, leaving the filter loose (rubber gasket not contacting the filter base).
■ Operate the manual hand pump located on the injection pump until fuel bubbles are around the top of the primary filter.

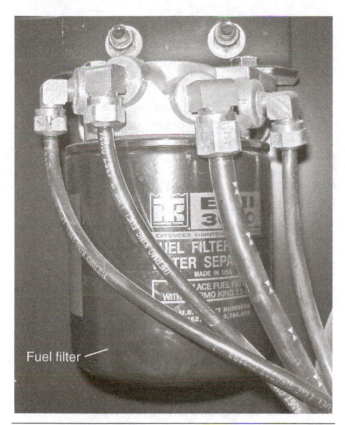

Figure 14-2 A typical spin-on fuel filter. *(Courtesy of John Dixon)*

Figure 14-3 A bleeder screw as used on a Thermo King refrigeration unit. *(Courtesy of John Dixon)*

Figure 14-4 A manual hand pump used to prime the fuel injection pump. *(Courtesy of John Dixon)*

- Tighten the primary fuel filter until the rubber gasket makes contact with the filter base and tighten an additional half turn.
- Start engine and inspect for potential leaks.

BLEEDING THE FUEL SYSTEM

Bleeding of the fuel system will be necessary if the unit was allowed to operate until all the fuel in the tank is depleted. (The bleeding process removes unwanted air from the fuel system.) To bleed the fuel system:

- Start by filling up the fuel tank for the refrigeration unit.
- Next crack the bleeder screw located on top of the injection pump. **Figure 14-3** shows a bleeder screw as used on a Thermo King refrigeration unit.
- Unscrew the pump plunger (counterclockwise) from the manual hand valve. **Figure 14-4** shows the manual hand pump used to prime the fuel injection pump.
- Operate the manual hand valve by depressing the plunger; spring force will allow the plunger to return for another stroke. This may take a few minutes of operation.

- Operate the hand pump quickly until air bubbles escaping from the bleeder screw are replaced by clean running fuel with no bubbles.
- Tighten bleeder screw while continuing to operate hand pump.
- Screw hand pump plunger back into the pump assembly (clockwise).
- Operate glow plugs if necessary and start unit.
- Allow the unit to operate until the engine runs clean (engine not missing). This will ensure all the air is worked out of the fuel system. Also make sure the alternator is not charging heavily. If so, allow it to recharge the battery or install a battery charger to bring the battery back up to full charge state. **Figure 14-5** shows a typical battery charger with boost capabilities.

Figure 14-5 A typical battery charger with boost capabilities. *(Courtesy of John Dixon)*

....

Note: The battery is usually discharged almost completely from operators' trying to unsuccessfully start the unit, so the battery should be checked and charged as necessary, especially in cold ambient conditions.

....

For Carrier refrigeration units using a mechanical fuel pump, a bleeder valve is located on the top of the fuel injection pump. The bleeding process for these units is very similar to that of the Thermo King unit.:

- Start by turning the bleed valve (red) counter-clockwise until it is completely open.
- Unscrew the pump plunger (counterclockwise) from the manual hand valve.
- Operate the hand pump quickly until a positive pressure (resistance) is felt on the plunger, indicating that fuel is flowing.

- Screw hand pump plunger back into the pump assembly (clockwise).
- Operate glow plugs if necessary and start unit.
- Allow the unit to operate until the engine runs clean (engine not missing). This will ensure all the air is worked out of the fuel system.
- Close the bleed valve by turning it (clockwise) until it is completely closed.

Bleeding Fuel System with Electric Fuel Pump

If the refrigeration unit is equipped with an electric fuel pump (usually mounted on the fuel tank mounting bracket), proceed as follows:

- Turn the bleed valve (red) counterclockwise until it is completely open.
- Operate glow plugs if necessary and start unit.
- Allow the unit to operate until the engine runs clean (engine not missing). This will ensure all the air is worked out of the fuel system.
- Close the bleed valve by turning it (clockwise) until it is completely closed.

AIR FILTER SERVICE/ REPLACEMENT

Air cleaners filter all air entering the engine for the combustion process. In time, the air cleaner will become dirty, causing the air to be restricted from entering the intake of the engine and resulting in lost horsepower, increased fuel consumption, and shortened engine life. The speed with which the air cleaner becomes fouled is proportionate to the conditions the unit is operating in. For example, the air cleaner will become dirty faster if the unit is operated in dusty areas like secondary or gravel roads. Air cleaners should be serviced or replaced at every PM service. There are two styles of air cleaners used in the transport refrigeration industry: oil bath type and dry type. Older model refrigeration units used air cleaners of the oil bath type while current units use a much more efficient positive dry type air filter.

Oil Bath Type

The oil bath air cleaner is a serviceable unit. Refer to **Figure 14-6** to see the components of an oil bath air cleaner.

To service an oil bath cleaner:

- Remove the lower cup of the assembly and discard the old dirty oil.

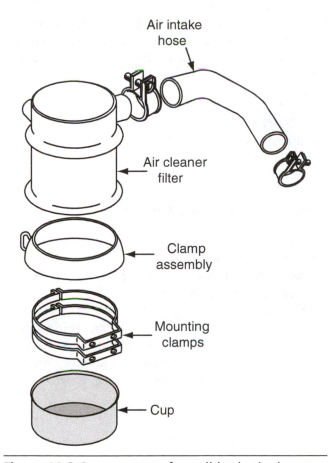

Figure 14-6 Components of an oil bath air cleaner.

Figure 14-7 The air restriction indicator shows when the air cleaner is restricting the flow of fresh air to the engine's intake, and warns when it is time to change the air filter. *(Courtesy of John Dixon)*

Figure 14-8 A typical dry type air cleaner and intake system. *(Courtesy of John Dixon)*

- Wash the cup in solvent to remove sludge that has adhered to the cup assembly.
- Dry the cup of any solvent residue.
- Fill the cup to the oil level mark with clean oil of the same weight as that in the engine's crankcase.
- DO NOT OVERFILL THE CUP.

Dry Type Air Cleaner

Many manufacturers using dry type air cleaners will install an air restriction indicator in the air intake elbow. Refer to **Figure 14-7** to see the air restriction indicator. This indicator should be inspected periodically to confirm that the air cleaner is not restricted. Replace the dry air filter cartridge when the red signal remains in view with the engine shut off.

To replace the dry type air filter:

- Stop the engine and remove necessary clamps.
- Discard old air cleaner and install new unit.
- Be certain all clamps are tight and fitment is correct. Any leaks in the air intake will allow foreign material directly into the engine, causing damage and premature wear. **Figure 14-8**

shows a typical dry type air cleaner and intake system.

- Depress the reset button on top of the restriction indicator after replacement has been performed.

DRIVE BELTS

During every service, the drive belts should be inspected, adjusted, or replaced as necessary. It is important for V-belts to be in good condition with proper tension to provide adequate air movement across the evaporator and condenser coils. Look for signs of cracking, scuffing, or wear. Belt tension should always be checked during these routine services. A belt that is too loose can whip (long belts) as well as slip, and belts that are too tight create unnecessary strain on bearings and the belt itself. The use of a belt tensioning tool is recommended in order to achieve the recommended belt tension. Follow manufacturer's recommendations for belt tension specifications.

Tech Tip: Always loosen belt adjustments before installing or removing a belt. Trying to pry a belt over a pulley will result in shortened belt life due to internal belt cord damage.

CAUTION   *Never try to make any belt adjustments with the unit running. Due to the fact that many units today have an automatic stop-start feature, the best way to service belts on a refrigeration unit is to first disconnect the negative battery terminal. This will prevent accidental starting of the unit and serious personal injury or even death.*

GLOW PLUGS

The glow plugs should be checked whenever the unit comes in for service. The glow plugs preheat the combustion chamber to aid in quick starting, especially in cold ambient conditions. The glow plug circuit is energized whenever the preheat switch/start switch is toggled (older units) and is shut off when the operator releases the switch. Some new units eliminate the preheat switch, and the glow plug circuit is energized by the processor during the start-up procedure.

If glow plugs become defective, the unit will start to have trouble starting, especially in cold ambient conditions.

Glow Plug Test

The glow plugs can be tested with the unit ammeter, or each plug can be individually tested with an ohmmeter. To test for a burnt-out glow plug, energized the glow plug circuit. There should be a draw of approximately 28 to 30 amps. If the amperage rating is below 28 amps, there could be a defective glow plug (burnt out) or a poor electrical connection at one or more of the glow plugs. To identify the defective plug or connection, remove the power terminal bar from each of the glow plugs and clean any corrosion as necessary. With an ohmmeter, using the engine block as the ground, test the resistance of each of the glow plugs individually. Each glow plug should indicate a resistance of around

1.55 ohms or a current draw of about 7 amps. A shorted glow plug will be indicated by a full current discharge on the ammeter or a blown current limiter (fuse) at the ammeter.

Note: Refer to the service manual for exact current and resistance values for the unit you are working on.

ENGINE COOLING SYSTEM

Engine cooling systems for transport refrigeration units are similar to those used by trucks. Refer to Chapter 5 for complete antifreeze strength tests and tests for antifreeze condition.

All equipment that contains antifreeze requires periodic maintenance and inspections, refrigeration units being no different. This maintenance is performed to ensure that the antifreeze provides the following benefits for the reefer's cooling system:

- Prevent the coolant from freezing down to –30°F (–34°C).
- Slow down the formation of rust and mineral scale than can cause the engine to run hot or overheat.
- Slow down the corrosion (acid) that attacks the internal components of the cooling system.
- Provide the necessary lubrication for the water pump seal.

A good practice used by many manufacturers is to drain, flush, and replace the antifreeze mixture every 2 years (unless extended-life coolant is used) to maintain total cooling system protection. Failure to maintain the condition of the coolant can result in scaling and higher acidity readings. If antifreeze is to be replaced with an ethylene glycol–based coolant, it is recommended to use a 50/50 mixture of antifreeze and demineralized water. This is true even of refrigeration units that will not be exposed to freezing temperatures. In the summer, the accumulator can get cold enough to freeze water as it circulates around the perimeter of the tank. Also, a 50/50 mixture provides the necessary corrosion protection and water pump lubrication required. During a PM, always check the coolant level and top up or test for coolant leaks as necessary. Refer to Chapter 5 for more information on pressure testing cooling systems.

Coolant Replacement

As previously stated, many manufacturers recommend replacement of the coolant every 2 years. The following is a basic guideline to correctly remove and replace coolant in a refrigeration unit:

Note: Always comply with environmental regulations in your area for properly disposing of coolant.

- Warm up the engine until it reaches operating temperature.
- Open the drain cock located on the engine block and allow coolant to drain, observing the color of the coolant. If the coolant is dirty, the cooling system should be flushed before you replenish with new coolant. See "Flushing the Cooling System" next in this chapter.
- Run clean water into the top of the radiator or coolant expansion tank and allow it to flow through the system until the water draining out of the block is also clear. **Figure 14-9** illustrates the coolant expansion tank with cooling system pressure cap. For radiator cap testing, refer to Chapter 5 of this text.
- Visually inspect all the coolant hoses for deterioration and check the hose clamps for tightness.
- Remove the tension from the water pump belt and inspect the bearing for looseness and signs of coolant leakage from the seal.

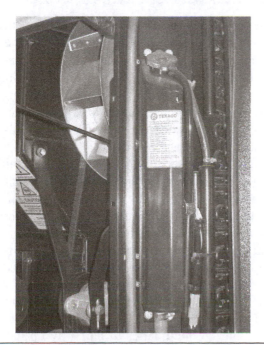

Figure 14-9 The coolant expansion tank with cooling system pressure cap. *(Courtesy of John Dixon)*

- Test the radiator cap. Please refer to Chapter 5 for complete radiator cap test procedures.
- Premix a 50/50 solution of antifreeze and demineralized water in a container before adding to the cooling system. (Extended life coolants are already premixed.)
- Close drain cock on the engine block.
- Refill the cooling system and warm the engine up with the radiator cap on loosely. This allows trapped air to be removed from the cooling system. Top up system as necessary and secure radiator cap.

CAUTION *Avoid direct contact with hot engine coolant.*

Flushing the Cooling System

When the coolant in the system is very dirty, flush using the following sequence:

- Run clean water into the top of the radiator and allow it to flow through the system until the water draining out of the block is also clear.
- Close the drain cock on the engine block.
- Install a commercially available radiator and block flushing agent. Run unit following the flush agent manufacturer's instruction for correct procedures.
- Open the engine drain cock and drain the flushing agent.

Shop Talk: If the engine has already developed an overheating problem caused by scaling, the flushing agent will have little effect on the cooling system. In this case, the engine would have to be disassembled and the cylinder block and heads boiled in a soak tank.

DEFROST SYSTEM

The defrost system should be checked on every service to check components. Start by running the unit in high speed cool until the unit box temperature indicates it is at or below 38°F (3.3°C). Once this has been achieved, depress the manual defrost switch. The unit should now shift from the cool mode to the defrost mode. If the unit fails to go into the defrost mode, consult the service manual for defrost cycle checkout procedures.

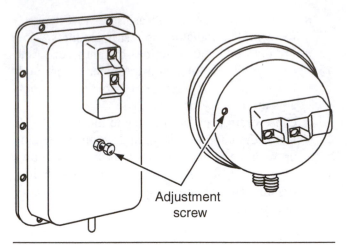

Figure 14-10 Two different styles of air switches responsible for initiating a defrost cycle.

Many units use a defrost air switch to initiate the defrost cycle. The defrost air switch senses the air pressure on the inlet side and the outlet side of the evaporator coil. **Figure 14-10** illustrates two different styles of air switches.

When ice builds up on the coil, the pressure created by the evaporator fan will increase in comparison to the outlet side of the evaporator coil. When this happens, the bellow section within the air switch moves, tripping a micro switch initiating a defrost cycle. The defrost cycle terminates when the defrost termination switch mounted on the evaporator reaches 56°F (13.3°C). Air switches are preset at the manufacturing level and usually do not require recalibration unless the switch does not function properly or you want to test a new switch before installation.

REFRIGERATION UNIT PRE-TRIP

Modern microprocessor-controlled refrigeration units incorporate a feature called a **pre-trip.** When a pre-trip is selected on the control panel by the unit operator, the unit will cycle itself through all different modes of operation. In each mode the processor evaluates the unit's performance and will signal a fault if parameters are not met. This test should be performed by the operator before the unit gets loaded with product.

TESTING REFRIGERANT LEVEL

If a refrigeration unit is run with an inadequate supply of refrigerant, the suction pressure will be lower than normal, causing the refrigerant in the evaporator to boil well before it reaches the end of the coil. This will cause the box temperature to rise and a complaint of

insufficient cooling by the operator. The suction pressure will continue to drop with refrigerant loss.

Testing of the refrigerant level should be performed every the time the unit comes in for routine maintenance. It should also be checked when recharging completely or topping up after major or minor service work of the refrigeration system. If there is doubt that there might not be a sufficient charge of refrigerant in a unit, the following test may be performed.

The following pressures are for units operating on R-12. Check specific unit manual for actual specifications:

1. Operating the unit in cool mode, there should be at least 150 psi of discharge pressure. (It might be necessary to cover the condenser to achieve this head pressure if ambient temperature is low.)
2. Suction pressure and box temperature are about equal.
3. The ball should just be floating in the receiver tank.

The preferred method would be to lower the box temperature to 0°F (–18°C). For trailers loaded with products that must be carried at a temperature above 32°F (0°C), it is recommended that an insulated test bag be placed over the evaporator so that a temperature of 0°F (–18°C) can be achieved.

If it is impossible to meet these conditions, then the unit could possibly be low on refrigerant.

COMPRESSOR OIL LEVEL CHECK

The compressor oil level should be checked whenever a refrigerant leak is suspected or when the unit has had refrigeration components serviced or replaced. To check the oil level of the compressor, follow the manufacturer's recommendations for compressor oil level checks. Most manufacturers have very similar procedures, but the procedures may vary due to the use of different refrigerants and pressures. Usually the unit is run for 15 to 20 minutes in high speed cool. This will allow any oil that may have been trapped in the low side of the system to be drawn back into the compressor. (Sometimes oil can get trapped in the evaporator if the unit had last been running in the heat or defrost mode.) The oil level sight glass can then be checked. The compressor oil level sight glass is positioned in the body of the compressor close to where the oil pan is bolted to the compressor body. Generally the oil level in the sight glass should be one-quarter to half way up the window of the sight glass. **Figure 14-11** shows the location of the sight glass on a compressor.

Figure 14-11 The location of the sight glass on a compressor. *(Courtesy of John Dixon)*

STRUCTURAL MAINTENANCE

During all PM services, check the following items and service as required.

Mounting Bolts

The mounting bolts fixing the unit to the trailer and engine mount bolts should be checked and tightened if necessary. Check unit service manual for mounting bolt specifications.

Unit Visual Inspection

Visually inspect the unit for signs of oil leaks on the engine and compressor. Check condition of wires and hardware; repair as necessary. Check for any physical damage that could possibly influence the performance of the unit.

Condenser

The condition of the condenser should be checked on every PM to ensure that it is clean and that there is no buildup of road dirt or insects that can restrict air flow and in turn hinder the condenser's heat transferability. Condensers should be washed frequently with a garden hose, detergent, and a soft nylon bristle brush. Inspect the coil and fins for damage and repair as necessary.

Note: High-pressure washers should never be used to clean a condenser, as the high-pressure water can damage the delicate fins and can impede air flow through the condenser.

Defrost Drain Hoses

During the PM, make sure the drain hoses are free of obstructions. This can be accomplished by running water into the evaporator drain pan and making sure water flows from both the drain hoses (ensure unit is level). If there is an obstruction, it can usually be cleared by blowing compressed air up through the drain tubes.

Evaporator

The condition of the evaporator should be checked on every PM to ensure that it is clean and that there is no debris that can restrict air flow and in turn hinder the evaporator's heat transferability. The evaporator fan can draw shrinkwrap as well as particles from cardboard boxes and deposit them on the evaporator service. Evaporators should frequently be washed with a garden hose, detergent, and a soft nylon bristle brush. Inspect the coil and fins for damage and repair as necessary.

CAUTION *High-pressure washers should never be used to clean an evaporator, as the high-pressure water can damage the delicate fins and can impede air flow through the evaporator.*

Defrost Damper Door

If the unit is equipped with a defrost damper door, check the damper door bushing for wear as well as wear to the shaft. Ensure the blade of the shaft seals the flow of air and that the damper door solenoid is bottomed out when energized. Make any adjustments as necessary. Some manufacturers do not use a defrost damper door in their defrost systems. Instead, they may choose to use an electric fan that can be switched off during the defrost cycle, and still others use an electromagnetic clutch to effectively turn or turn off the fan shaft to the evaporator section of the unit (Carrier).

Summary

- A PM on a reefer unit includes engine service, fuel system service, compressor service, drive belt service, refrigeration system checks, glow plug system checks, and structural maintenance checks.

- Engine oil and filters should be changed at regular intervals according to the manufacturer's recommended intervals.

- Fuel filters are another item that should be changed while servicing the engine.

- The bleeding process removes unwanted air from the fuel system.

- Air cleaners filter all air entering the engine for the combustion process.

- Belt tension and condition should always be checked during every PM.

- Glow plugs preheat the combustion chamber to aid in quick engine starting.

- The condition of engine coolant should be checked periodically. Coolant not capable of performing can damage the engine as well as the cooling system.

- The cooling system should be flushed when it appears to be very dirty.

- The purpose of the defrost system is to remove accumulated ice from the evaporator coil.

- The defrost air switch is responsible for initiating a defrost cycle. (Electronic timers as well as a microprocessor can also initiate a defrost cycle.)

- The defrost termination switch is responsible for ending the defrost cycle.

- Refrigerant level should be checked during every PM. Follow instructions in the unit service manual for proper procedures for checking the level for that particular unit.

- The compressor oil level should be checked whenever a refrigerant leak is suspected or when the unit has had refrigeration components serviced or replaced.

- The evacuation process removes air and moisture that is able to enter the refrigeration system when it has been opened for service.

- The structural integrity of the unit should be examined on a PM. This would include unit engine and fuel tank mounting, a visual inspection of the unit, cleanliness of the coils, evaporator drainage, and the defrost damper door system.

Review Questions

1. When servicing a diesel engine, it is important to drain the engine oil after the engine has warmed up so that the oil will flow out faster and more completely than if it were cold.

   A. True   B. False

2. Bleeding the fuel system is necessary to remove unwanted air from the fuel system when the unit has:

   A. been run out of fuel.   C. had maintenance performed to the fuel system.

   B. had fuel or fuel filters   D. All of the statements are correct.
      replaced.
                                              E. None of the statements is correct.

3. The receiver drier can be replaced while the unit is in which of the following states?

   A. A low side pump down   D. Both statements A and B are correct.

   B. A compressor pump down   E. Both statements A and C are correct.

   C. A state where all the
      refrigerant has been removed
      from the system

4. The defrost termination switch senses the temperature of the:

   A. condenser coil.   C. compressor discharge.

   B. evaporator.   D. hot gas line.

5. If during a refrigeration unit pre-trip the unit found a detected a fault, the unit would:

   A. shut down.   C. restart the pre-trip.

   B. signal a fault.   D. None of the above

6. Manufactures of refrigeration units recommend that fuel filters should be primed with fuel just like the oil filter before installation.

   A. True   B. False

7. The majority of new refrigeration units use oil-bath style air filters because of their superior filtering efficiency over dry positive type air filters.

   A. True   B. False

8. In order to initiate a defrost cycle the box temperature of the compartment must be below:

   A. 78°F (25.5°C).   C. 48°F (8.9°C).

   B. 58°F (14.4°C).   D. 38°F (3.3°C).

Trailer Service and Inspection

Objectives

Upon completion and review of this chapter, the student should be able to:

- List the benefits of a well-planned trailer preventive maintenance program.
- Explain the construction of aluminum sheet and post vans.
- List and explain four methods of leak testing a van trailer.
- Describe FRP trailer construction.
- Explain maintenance and inspection procedures for FRP trailers.
- Describe the pros and cons of swinger style doors and roll-up doors, and describe maintenance procedures for each.
- Describe trailer wood floor maintenance procedures.
- List safety procedures that must be adhered to while performing service work on dump trailers.
- Perform a trailer axle alignment inspection.
- Perform an inspection on a rear impact guard to ensure compliance.
- Perform a trailer service on a van, flatbed, or dump-style trailer.

Key Terms

barn style	hinged	swinger doors
conspicuity tape	interior cube	unibody
dog track	post	
FRP	roll-up	

INTRODUCTION

In this chapter we look at some of the maintenance items that are specific to the trailer. Many truck mechanics will at some point in their career will work on or service a trailer. Trailers outnumber power units by a ratio of approximately three to one, and higher in many areas, where warehousing goods in a trailer is cheaper than warehouse space.

Most van and flatbed trailers can be fairly straightforward to service, but many specialty trailers, along with subsystems, can become very complex to maintain. In this chapter we cover some of the more common areas of trailer maintenance.

SYSTEM OVERVIEW

This chapter starts by reviewing the benefits of establishing a good trailer PM program. This is followed by a look at the service requirements of an aluminum sheet and post style van trailer as well as FRP van trailers.

Trailer loading door maintenance is covered because this area of the trailer usually receives much wear and tear, along with the maintenance of the trailer flooring. Trailer underride guard maintenance requirements as well as trailer alignment are also covered. The chapter concludes by running through all the steps to perform a complete trailer maintenance inspection.

TRAILER PREVENTIVE PROGRAM

A preventive maintenance program involves the servicing and maintenance of the vehicle as a whole.

Maintenance programs are necessary to ensure the designed life expectancy of the vehicle and its components.

The results of underservicing are the premature wear of parts and components, frequent road failures, and higher than normal maintenance costs.

An effective preventive maintenance program minimizes maintenance costs while ensuring that all required maintenance is performed.

A well-planned and executed PM program offers the following results:

A. The lowest possible maintenance costs
B. Maximum vehicle (trailer) availability
C. Improved fuel economy
D. Reduced road failures (breakdowns), resulting in improved dependability
E. A reduction in the possibility of accidents due to faulty equipment
F. A reduction in driver complaints

Although there are many factors that make up a successful PM program, the program will not be successful without careful planning:

- It is the responsibility of the maintenance manager and all personnel to perform their tasks.
- Inspections must be performed as scheduled and performed properly, with accurate records being kept.
- Pre-trip inspections must be performed by the driver to detect impending problems or failures.
- Inspections made by the technician are performed at specified intervals.

- Drivers and technicians must be trained in the process of inspection, with the inspection being performed properly.
- Drivers must fill out driver inspection forms.
- Technicians work with PM forms, repair orders, vehicle files, and historic files.
- In some operations, maintenance records are required by the DOT.
- Records can be used to track the performance or failure of components for the purpose of corrective action.
- Records can be helpful in the planning of or adjustment to a PM program.
- A vehicle file containing maintenance records can be beneficial in the event of a serious accident.
- The age of the fleet and the nature of the business will have a bearing on the type of PM program for any particular operation.
- Following the service recommendations supplied by the component or trailer manufacturer is a sound basis for a PM program.

ALUMINIUM SHEET AND POST VANS

Aluminium sheet and **post** vans are constructed so the posts act as the major sidewall structural members connecting the top and bottom side rails together (see **Figure 15-1**).

Post spacing is typically on 24-inch centers with some manufacturers opting for 16-inch spacing depending on the type of loads the trailer will be carrying. The aluminum sheeting does help to tie the side rails together, although its main function is to prevent the walls from "racking" as well as to protect the

Figure 15-1 Van trailer. *(Courtesy of John Dixon)*

Figure 15-2 Trailer interior with scuff liner. *(Courtesy of John Dixon)*

contents from the elements. The inside of these trailers is typically lined with plywood with the addition of an optional scuff-liner that can help prevent damage to the post and to the lower side rail connection caused by fork-lifted pallets (see **Figure 15-2**).

Floor cross-members are "I" beam style, with high-strength steel being used in some applications. Cross-members are usually spaced on 12-inch centers with 6-inch spacing being popular near the rear of the trailer (see **Figure 15-3**).

Flooring materials include hardwood, steel, and aluminium, with hardwood being the most popular. The two types of hardwood flooring construction are ship-lap and tongue and groove.

Aluminium Sheet and Post Van Maintenance

Due to the construction methods (**unibody** no main frame rails) used for these types of trailers, it is imperative that all rivets or fasteners connecting the top and bottom rails as well as all cross-member members are present and secure.

As part of the PM program, the entire trailer body should be inspected for loose or damaged rivets, and those found to be defective should be replaced. If looseness is suspected, a sharp rap with a hammer will produce a dull sound in a loose rivet. A further

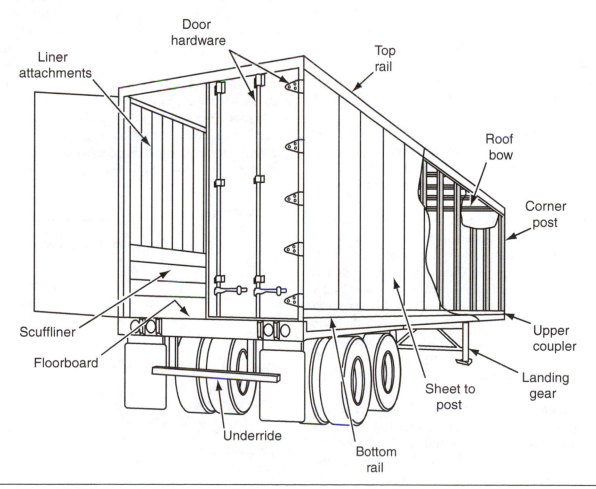

Figure 15-3 Van trailer components.

check is to coat the rivet head with penetrating oil and wait a few seconds for the oil to penetrate. Wipe the rivet clean and then rap it again with a hammer. If penetrating fluid appears around the edges again, the rivet is loose and should be replaced. Also the appearance of movement or shifting between riveted parts, such as bare or shiny areas, is a good indication of the need for replacement. Rust or corrosion around a rivet does not necessarily indicate looseness; however, it could be a good reason to investigate further.

Van Trailer Leakage Testing

During a PM, the trailer should be inspected to ensure that it is watertight, so that no water is able to come in contact with cargo. In most instances, when water is found inside a van trailer, it is not at the actual site of the leak. It may even be necessary to remove liner panels to pinpoint the actual leakage site. There are four methods that can be use by technicians to pinpoint a water leak:

1. **Light Test:** The light test is the easiest and most common test that can be performed to locate water leaks. Move the trailer outside on a bright day (works best) and enter the trailer. Have a coworker close the door and carefully visually inspect the interior of the trailer, looking for any areas where light leaks that are potential water leaks.

2. **Water Test:** This test is performed by running the trailer through a truck/trailer wash under relatively high pressure. Directly afterward, inspect the complete interior of the trailer for the presence of water.

3. **Smoke Test:** This test is performed by lighting a smoke bomb placed in a pan on the trailer floor and closing all trailer doors and vents. Allow the smoke to build for several minutes and then inspect the exterior of the trailer looking for any leaking smoke. This test must be performed indoors unless there is little or no wind outdoors. These smoke bombs are usually available at theatrical supply stores and are one of the most effective methods of identifying leaks.

4. **Ultrasonic Leak Detection:** A sonic generator is placed inside the trailer and the doors are closed. An electronic testing device is scanned over the entire van exterior. The detection device has a signal strength meter, which is used to precisely locate the leak site.

FRP PLYWOOD TRAILERS

FRP is an acronym taken from the words "Fibreglass Reinforced Plastics." FRP plywood is multiple layers of matt fibreglass cloth bonded to both sides of a plywood core, providing excellent strength and corrosion resistance. The use of FRP trailers and van bodies has been recognized as a viable alternative to traditional sheet and post van bodies. FRP offers advantages over other materials, including more interior width and a smooth interior and exterior finish, and composite construction that resists punctures. However, it has characteristics, such as vulnerability to water damage, that must be considered in the design and construction of trailers and van bodies utilizing FRP.

Bottom Side Rails

There are two basic designs of bottom rail construction used in FRP trailers and van bodies.

Figure 15-4 shows one design of single shear construction. The bolts are in a single shear because the rail is in two pieces with a rub rail separate from the trailer frame. The bolt-to-frame load is carried through the joint between the frame and panel only. Single shear rails offer adequate strength and allow easy access to the panel for repair. All single shear bottom side rails should have continuous fastener bearing strips on both the inner and outer sides to prevent fasteners from pulling through the panel and to provide adequate clamping force.

Figure 15-5 shows a double shear bottom side rail design. The double shear design offers significantly greater strength than the single shear design, but it does not allow easy access to the panel for repair.

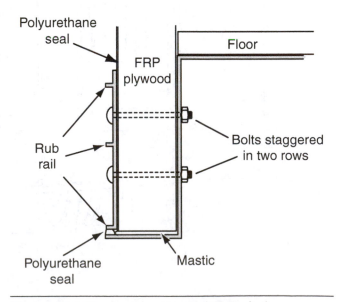

Figure 15-4 Single shear construction design.

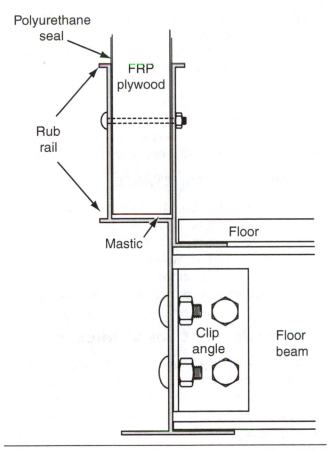

Figure 15-5 Double shear bottom side rail design.

FRP Periodic Maintenance Schedule

Every 6 months, calking should be inspected for leaks in the seal along the lower and upper rails and front and rear post assemblies.

Once per year, all fasteners should be checked for tightness along the lower and upper rails and front and rear post assemblies.

Every 3 months, sidewalls should be visually inspected on both the interior and exterior for cuts in the fibreglass skin. All cuts should be filled with polyester-based reinforced putty or covered with PVF repair tape on the exterior until repairs can be performed. For more information, contact the FRP supplier for information on repair.

TRAILER BRAKES

Trailer brakes generally used on highway trailers are S-cam 16-1/2-inch × 7-inch. Options may include wedge brakes or disc brakes. Both tractor and trailer brakes are essentially the same, within size, brand, and gross axle weight rating (GVWR). Brake shoes and linings, springs, anchor pins and bushings, rollers

and retainers, camshaft hardware, and slack adjusters are similar on both tractors and trailers. The service requirements are also much the same for all brakes. For more information on braking systems, refer to Chapter 9.

VAN DOORS

The two types of doors used on van type trailers are **hinged** and **roll-up** (see **Figure 15-6**).

Hinged doors, sometimes referred to as **barn style** or **swinger doors,** are available in single width or double width and are typically as wide as the body,

A

B

Figure 15-6 Van doors: (A) swing doors; and (B) roll-up or overhead door. *(Courtesy of American Trailer Industries, Inc.)*

providing a full rear door opening. Hinged doors also provide greater headroom versus a roll-up door.

The core of these doors is plywood, with a thin sheet of steel or aluminium bonded to both sides. A weather strip that incorporates a "U" shaped channel that fits over the core seals the edge of the core from moisture and creates a seal between the door and the frame. A latching mechanism runs vertically and secures the door to the door header and sill.

Roll-Up Doors

Although not as popular as hinged doors, roll-up doors are used in some applications where the door has to be opened constantly.

- Roll-up doors have a reduced opening size as well as reducing the **interior cube** (space) of the trailer.
- These doors are usually made up of seven panels stacked one on top of another and joined together with hinges.
- A series of hinges are used to join the panels together, and a pair of tracks run up each side of the door opening and then curve and run along the underside of the roof (see **Figure 15-7**).
- Door rollers are used to guide the door in the track assembly.
- A special opening mechanism consisting of a cross-shaft, drums, and cables that run down the outside of the door and attach to the bottom panel are used in conjunction with heavy coil spring(s) to aid in lifting the door.
- Closing the door loads the coil spring (s), which in turn makes it possible to lift the door with minimal effort, as this type of door is very heavy.
- A latching mechanism, which is attached to the bottom panel of the door, locks the door to the lower sill.

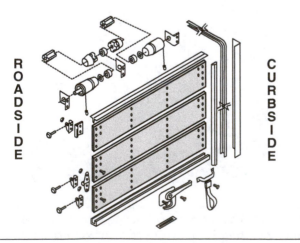

Figure 15-7 Components of a roll-up door.

Proper counterbalance adjustment will allow the door to remain in any given position. If when in the closed position, the door opens, then the counterbalancing spring(s) are adjusted too tightly and should be wound back off, relieving some of the tension. If, on the other hand, the door lowers after it is opened, the counterbalancing spring(s) do not have enough tension and should be wound slightly tighter until the door will stay in any position without movement.

ROLL-UP DOOR MAINTENANCE

- Start by lubricating the door rollers, counterbalance, hinges, and latches with a dry lubricant.
- Inspect the condition of the cables and pull strap. Replace if frayed.
- Inspect the condition of the door panels. If gouged, repaint as necessary to seal the panel.

Swing Style Van Door Maintenance

During a PMI on a van trailer with swinger doors, inspect the condition of the door panel itself. Check hinges and lock rods for damage or distortion resulting from impacts. Inspect for doors with cams that do not properly fit into place or have loose bearing plates that will not permit the rear doors to seal. Inspect the door seals for wear or damage. The door seals should be considered a consumable item and must be repaired or replaced regularly to keep the trailer sealed and dry.

TRAILER FLOOR CARE AND MAINTENANCE

The trailer's floor system should be inspected on every PMI to ensure it can safely handle the loads it is intended to support and ensure its service life. Floors should be checked for loose or missing fasteners, separated lamination, or failing floorboards deformed by rear impact damage or other forklift traffic. Undercoating should be checked regularly and reapplied if necessary.

The wood should be protected by mopping with an application of linseed oil on the top surface. A urethane finish is used in some applications.

1. Wood floors should not be steam cleaned.
2. Do not exceed load capacity or concentrate load beyond established limits.
3. Do not nail into floor.
4. Use dock boards for entrance of forklifts into trailers to prevent shock loads.
5. Do not expose floors to corrosive materials.

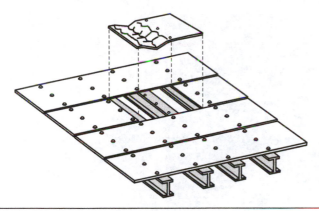

Figure 15-8 Illustration showing the removal of a damaged trailer floorboard.

6. Do not wash inside a van and close the doors. Allow time to dry.
7. Trailer doors should be kept closed except when loading and unloading to prevent exposure to elements.

Wood Floor Repairs

When a floorboard is damaged and must be replaced, the board should be removed two cross-members ahead and two cross-members behind the damage (see **Figure 15-8**). The new floorboard must be supported by four cross-members. The new floorboard must not end at the same point as either of the two boards on either side of it. If required, extend the replacement board to the next cross-member. The new floorboard must be the same material and thickness as the existing flooring material. Butt joints must fall on the cross-member centrelines (see **Figure 15-9**). Butt joints must be tight with a maximum gap of 1/16 inch, with the gap being caulked.

TRAILER CERTIFICATION PLATE

Every trailer is required to have a certification plate mounted near the front, and it should contain the following information (see **Figure 15-10**):

1. Verification or conformity to all applicable United States/Canadian or other motor vehicle safety standards as of the date of manufacture
2. Gross vehicle weight rating (GVWR), the maximum combined weight of a vehicle and its payload based on structural capacity alone
3. Gross axle weight rating (GVWR), the rated load-carrying capacity of an individual axle and wheel end assembly
4. Trailer serial number
5. Date of manufacture
6. Manufacturer's name
7. Tire size
8. Air pressure

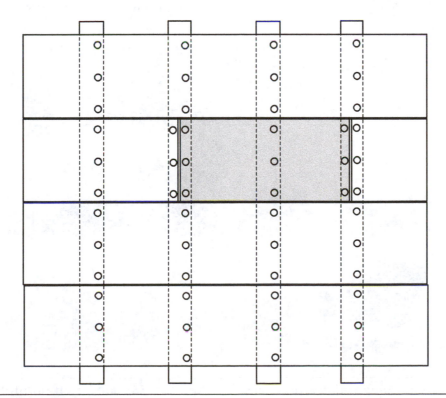

Figure 15-9 Butt joints must fall on the cross-member centerlines.

Figure 15-10 Trailer certification plate. *(Courtesy of John Dixon)*

DUMP TRAILERS

Dump trailers are used primarily to haul aggregates such as sand, gravel, and stones, but they can also be used to transport a variety of solid commodities that range from soil to slag from steel mills. Some examples of dump trailers are shown in **Figure 15-11.** They may be manufactured from steel or aluminium, and many use replaceable liners, especially when subjected to the kind of abrasive wear that occurs in aggregate haulage.

Dump Trailer Maintenance

While performing maintenance on dump trailers, always follow safety precautions:

- Always block the truck or trailer dump bed with an approved device before placing your body between the bed and the frame.
- Check tire pressures daily.
- Ensure that repairs to the dump bed leave bottom and sides clear of obstructions to the smooth flow of materials.

Figure 15-11 Selection of dump trailers: (A) frame-type bathtub; (B) slant front bathtub; (C) rock dump; (D) bottom hopper; and (E) bottom hopper doubles. *(Courtesy of American Trailer Industries Inc.)*

Figure 15-12 Dump trailer hydraulic cylinder. *(Courtesy of John Dixon)*

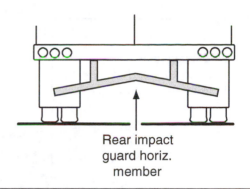

Rear impact
guard horiz.
member

Figure 15-14 Underride bent upward in the middle.

- Inspect and repair suspension systems.
- Lubricate pins and bushings for smooth operation.
- Inspect and repair hydraulic lift cylinders regularly (see **Figure 15-12**).

TRAILER REAR IMPACT GUARD

The rear impact guard, or underride guard, should be inspected regularly for cracked welds, cracked or fractured vertical members, cuts and tears in any member, and for dimensional conformity (see **Figure 15-13**). These dimensional requirements became effective on October 1, 1999, requiring equipment users to maintain the underride guard in a close-to-new condition.

A common type of damage is a vertical bend upward as illustrated in **Figure 15-14**. Studies indicate that this type of damage does not affect a particular guard's ability to meet the strength or energy absorption requirement of the new laws if the deflection is 3 inches (76 mm) or less. If the bend does exceed 3 inches (76 mm), it should be

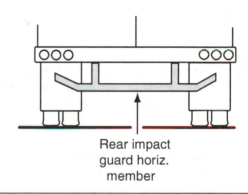

Rear impact
guard horiz.
member

Figure 15-15 Horizontal members are bent upward or downward.

repaired to conform to the dimensions illustrated in **Figure 15-12.**

In some cases the ends of the horizontal member are bent upward or downward (see **Figure 15-15**) or even toward the front of the trailer (see **Figure 15-16**). If it is bent upward, it may be bent downward to its original position. If it is bent downward, it can be left alone or bent upward to its original position at the equipment owner's option.

NOTE: All dimensions are for unloaded trailers on a flat surface.

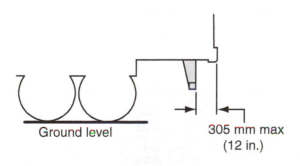

Ground level

305 mm max
(12 in.)

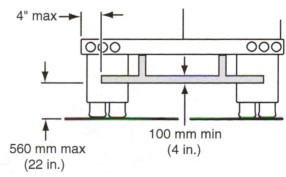

4" max

560 mm max
(22 in.)

100 mm min
(4 in.)

Figure 15-13 Underride dimensions.

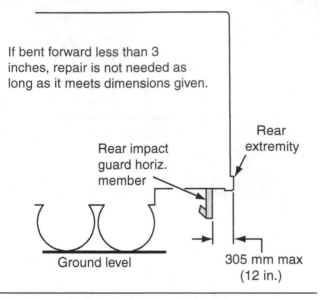

Figure 15-16 In some cases, the ends of the horizontal member are bent toward the front of the trailer.

Shop Talk: When bending underride guards, only cold bending methods should be used. Heating the guard may affect its strength. Also, repetitive bending should be avoided since it will have a detrimental effect on strength.

If the middle of the horizontal member is bent toward the front of the trailer, the trailer manufacturer should be consulted for repair guidelines (see **Figure 15-17**). Additionally, in some instances there may be accompanying damage to the cross-members at the rear of the trailer. The rear cross-members, rear sill, vertical members, and last 6 feet of the floor must also be inspected for damage, and repaired as needed. If replacement of the underride guard is recommended, it should be replaced with an OEM-approved replacement member having the same metal alloy and dimensions as the original or better. Installation must follow manufacturer's instructions.

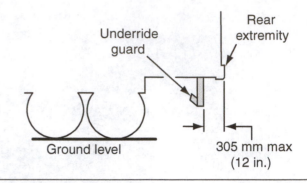

Figure 15-17 Underride guard bent forward in the middle.

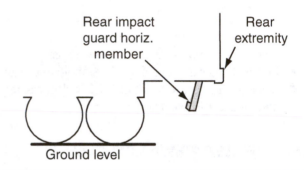

Figure 15-18 Entire guard is bent inward to the front of the trailer.

When there is no damage to the member but the entire guard is bent inward to the front of the trailer (see **Figure 15-18**), the trailer manufacturer should be consulted for repair guidelines.

TRAILER ALIGNMENT

Poorly aligned trailer axles will cause the trailer to **dog track** (a term used to describe a vehicle not tracking properly), resulting in unnecessary tire wear (see **Figure 15-19**). With longer-wearing radial tires, alignment is a significant maintenance issue to increase tire life. Tires on a trailer have the potential of lasting longer than tractor tires and are therefore more

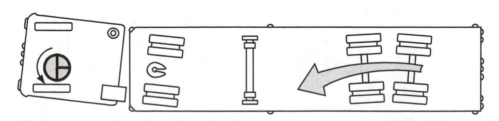

Figure 15-19 Dog tracking trailer.

Figure 15-20 Kingpin extension. *(Courtesy of John Dixon)*

susceptible to irregular wear and misalignment than any other tires on the vehicle.

A misalignment of 1/4 degree (15 minutes) produces 218 miles of side scrub per 100,000 miles (350 km/160,000 km).

A recent study found almost one-half of all trailers in service were out of spec for alignment. Misalignment increases fuel consumption from tire scrub and increased wind resistance. From a safety point of view, a trailer that is out of alignment has poor directional stability, especially when braking and turning, and a greater potential for jackknifing.

Misalignment Causes

Some of the possible causes that can put a trailer suspension out of alignment are damage from accidents, worn or broken suspension parts, and replacing suspension components without rechecking axle alignment afterward.

Pre-alignment Checks

Before axle alignment is checked, the suspension should be in a natural relaxed state. This can be accomplished by moving the trailer forward and backward over a level floor two to three times, with the last movement forward. This movement causes the suspension to shift the axle into its running position.

Make sure the trailer is empty, on even ground, level from side to side, and with tires correctly inflated. Make sure all suspension components are in satisfactory condition.

Trailer Axle Alignment

The axle alignment check starts at the kingpin. This will be considered the referencing center point for the vehicle. Remove the hubcaps from the wheel and screw on axle extenders. These will allow you to measure to the outer wheel centers without interference with the tires.

Install and level the bazooka gauge under the fifth wheel (see **Figure 15-20**). Using a 50-foot steel tape, measure the distances A and B, referring to **Figure 15-21**. These distances must be equal to within 1/16 inch. If required, adjust front axle of the tandem until you are within this specification.

With the front axle of the tandem within specification, install screw on wheel extenders (see **Figure 15-22**) to the rear axle and measure C and D referring to **Figure 15-20.** These measurements must also be within 1/16 inch of each other. Adjust rear axle if required.

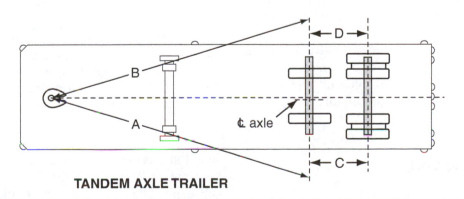

TANDEM AXLE TRAILER

Figure 15-21 Axle measurements.

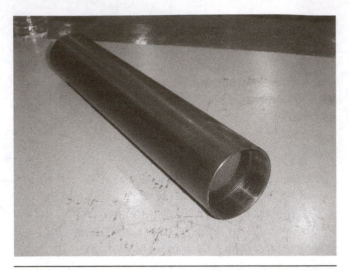

Figure 15-22 Axle extender. *(Courtesy of John Dixon)*

Note:

- Always measure from the axle ends for greatest accuracy.
- Avoid measuring to rims, suspension brackets, brake drum, and so on. These measuring locations will result in improper alignment.
- If difficulty is encountered in achieving a true alignment, check and repair or replace worn or bent suspension components.

REFLECTIVE TAPE

The National Highway Safety Administration Regulations require that red and white alternating striped reflective tape, also known as **conspicuity tape,** be applied to the full trailer width at the rear of the trailer, near the van base. The tape height should be as close to 4 feet as possible. But trailer builders have some leeway, especially when it comes to specialty vehicles like auto transporters. At the rear, white reflective tape should outline the vehicle extremities, that is, corners or curves (see **Figure 15-23**).

Reflective tape needs only to be applied to a clean surface to adhere. Once installed, it requires no more maintenance than occasional washing or wiping. The tape functions by reflecting headlight glare, so it consumes no electrical power.

VAN TRAILER PMI

In this section, we will describe the PMI procedure for a van trailer in detail. This section may be referred to when going through a PM inspection sheet

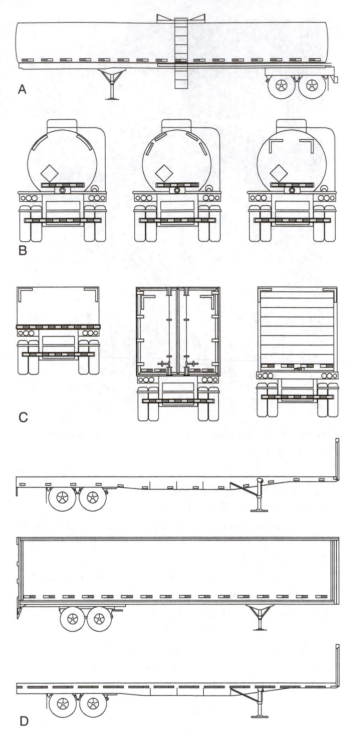

Figure 15-23 Conspicuity marking: location of reflective tape on various trailers: (A) tanker curbside; (B) three alternate rear views; (C) rearview (left to right), platform, swing doors, and roll-up door; (D) sideview layout.

found in this chapter. In some cases, the items listed both on the PMI record sheet and in this section are self-explanatory. In other cases, the items listed on the PMI record sheet are explained in detail in this section.

The following are based upon the recommendations of The Maintenance Council (TMC) American Trucking Associations. The TMC is a great resource for any technician to access because all information is kept current, as rules and regulations within the industry do change regularly to improve the safety of commercial vehicles.

SECTION I

Start Inspection at the Front of the Trailer

1. **Check the registration holder and the annual inspection sticker.**
   - Ensure the registration is current and the federal annual inspection sticker and all appropriate permits are in order.

2. **Inspect the SAE J560 seven-pin receptacle.**
   - Inspect the condition of the receptacle for proper operation and signs of damage and/or corrosion.

3. **Inspect gladhands.**
   - Inspect the gladhands and seals for damage or wear. Replace as necessary.

4. **Check the operation of all trailer lights.**
   - Using an appropriate light tester or known good tractor as a power source, check the illumination of all trailer lights.

5. **Check all light lenses and reflectors.**
   - Inspect for missing or damaged lenses or reflectors.

6. **Inspect the trailer's wiring harness.**
   - Inspect the electrical wiring for signs of damage and security to the trailer.

7. **Inspect rear underride guard.**
   - Check for damage, missing hardware, and security. See Figure 15-13 for detailed dimensions.

8. **Inspect rear door(s), handle(s), hinges, lock(s), tie backs, and cables.**
   - Check for damage, missing hardware, and security. For more information see the "Van Doors" section in this chapter.

9. **Inspect operation of rear door(s).**
   - Check rear door(s) for ease of operation.

9A. **Inspect rear door seals and tracks.**
   - Ensure that the door seals provide a tight closure.

10. **Check the license plate and mounting.**
   - The mounting of the license plate should be secure, current, and not damaged.

11. **Inspect side door(s), handles(s), hinges, lock(s), and tie backs.**
   - Check for damaged or missing hardware.

12. **Inspect operation of side door.**
   - Check side door(s) for ease of operation.

12A. **Inspect side door(s) seals and tracks.**
   - Ensure that the door seals provide a tight closure.

13. **Inspect all placards and mountings.**
   - Inspect for missing or damaged placards and mountings.

14. **Check all rivets and fasteners on bottom rail.**
   - Inspect lower rail for missing or loose fasteners.

15. **Record any physical damage to trailer.**

SECTION II

Undercarriage and Suspension

1. **Inspect upper coupler plate and kingpin.**
   - Check for damage to upper coupler and kingpin area. For more information refer to Chapter 13.

2. **Check cross-members and substructures.**
   - Inspect for bent, corroded, or loose cross-members. Inspect for broken welds and loose or missing fasteners.

3. **Inspect wiring harness(es) and air lines.**
   - Check for damage and security of wiring harnesses and air lines under the trailer. Look for any kinks or chaffing.

4. **Inspect trailer landing gear.**
   - Inspect legs, mounting brackets (wing plate), and braces for damage or broken welds. Check landing gear sand shoes for damage and security.

4A. **Test landing gear operation.**
   - Run landing gear up and down in both high and low range to ensure ease of operation.

5. **Inspect spring and hangers.**
   - Check springs for broken or shifted leaves as well as wear to the inside of the hanger bracket caused by rubbing springs.

6. **Torque U-bolts.**

   ■ Torque U-bolts to OEM specifications or general torque specifications listed on PMI sheet. For further information, see Chapter 11.

7. **Inspect torque rods, radius rods, and bushings.**

   ■ Check torque and radius rods for damage or worn bushings.

8. **Inspect spare tire carrier.**

   ■ Inspect spare tire carrier for mounting and cracked welds or damage.

9. **Inspect tandem slide and lock mechanism (if equipped).**

   ■ Check tandem slide for damage and locking bars for wear. Test to ensure that the release mechanism moves freely.

10. **Inspect inner wheel ends.**

    ■ With the backing plate(s) removed or inspection hole, check inner wheel ends for signs of a seal leaks.

11. **Inspect mud flaps and brackets.**

    ■ Check mud flaps and brackets for damage. See Fig-15-24 for a template of how a mud flap and bracket should be spaced out for drilling and mounting.

12. **Inspect pintle hook (if equipped).**

    ■ Check pintle hook for wear, damage, or loose or missing hardware. For more information see Chapter 13.

13. **Inspect glad hand and SAE J560 seven-pin connector (trailer rear, on doubles or triples units).**

    ■ Check the receptacle for proper operation and signs of damage or corrosion.

14. **Test air bags for leaks (air ride suspensions).**

    ■ Inspect air bags for wear, damage, or loose or missing hardware.

15. **Check height control valve and linkage (air ride suspensions).**

    ■ Inspect valve and linkages for looseness or damage.

SECTION III

Brakes

1. **Inspect brake linings and drums.**

   ■ Brake lining thickness must be the same on both of the brake shoes and on each side of the axle. Brake lining should not be less than 1/4 inch at the thinnest point. Brake drum replacement is required if the brake drum is cracked; the brake drum surface is heat checked, ground, or worn beyond the maximum diameter thickness limits; the mounting flange is cracked; the bolt holes are elongated; the drum is known to have been severely overheated; or the drum is out-of-round. For more information, see Chapter 9.

1A. **Record brake lining thickness (in inches).**

2. **Check the air brake system for air leaks.**

   ■ With air supplied to the glad hand or tractor attached, apply brakes. Check system for air leaks.

3. **Adjust all brakes (manual slack adjusters).**

   ■ For more information on adjustment procedures, see Chapter 9.

3A. **Check operation of automatic slack adjuster (if so equipped).**

   ■ Automatic slack adjusters are not a maintenance-free item. See Chapter 9 for additional information.

4. **Drain air reservoirs.**

   ■ Inspect for excess oil or moisture draining from the reservoirs.

5. **Inspect reservoir mounting and brake chambers.**

   ■ Inspect for damaged components.

SECTION IV

Hubs, Tires, and Wheels

HUBS

1. **Check wheel bearings.**

   ■ Spin the wheel on a hub while listening for noisy bearing. Bump side of tire and listen for popping sound from hub area.

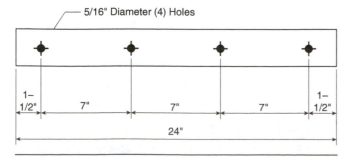

Figure 15-24 Mud flap mounting.

2. **Inspect hubcap for leaks.**
   - Inspect hubcap for leaks and damage.

3. **Check lubricant level.**
   - Make sure lube level is correct (at the full line on the hubcap).

TIRES

1. **Inspect tires for irregular wear.**
   - Tires with irregular wear patterns may indicate a variety of tire, wheel, rim, or alignment problems. For more information, see Chapter 8.

2. **Inspect tire sidewall for cuts or damage.**
   - Record any physical damage that may be found while inspecting the tire.

3. **Check dual mating with square.**
   - The matching of dual tires on any axle is important for safety as well as improving tire life (smaller tire will scrub). The maximum allowable diameter difference between a tire and its running mate is 1/4 inch for 9.00 cross-section (10d-22.5) and larger size tires. For more information, see Chapter 8.

4. **Check tread design match.**
   - Ensure that tires are appropriate for use on the same axle. Mismatched tires on the same axle can result in vibration complaints and other problems.

5. **Inspect tire valve caps.**
   - Check to make sure valve caps are present and tight. Metal heat-resistant sealing valve caps should always be used.

6. **Check and record tire tread depth.**
   - Federal Motor Vehicle Safety Standards require a minimum of 2/32-inch tread depth on all steer axle tires and a minimum of 1/32-inch on non-steer tires.

7. **Check and record air pressure.**
   - Record current air pressures and adjust if necessary. A tire is considered flat if the air pressure is 20 percent lower than the recommended or desired pressure, or if there is an air leak that can be heard or felt. Never re-inflate a flat tire. Remove it and have it inspected to determine the cause of the leak.

Shop Talk: Do not adjust air pressure when tires are hot; tires can take up to 4 hours to cool back down after driving has stopped.

WHEELS

1. **Inspect for loose wheel lugs.**
   - Replace and retorque any fastener that is missing, worn, or damaged. Retorque all wheel nuts to the proper torque. For disc wheels, the inner nuts on both sides of a failed fastener should also be replaced. For more information, refer to Chapter 8.

2. **Inspect wheels for cracks or damage.**
   - Inspect for signs of cracks or damage on any wheel component. Pay special attention to bolt hole mounting surfaces. Inspect both sides of the wheels and between duals. Replace any wheel that is cracked or has damaged components. You should also determine the cause of the damage before replacing the wheel/wheel component(s). For more information, refer to Chapter 8.

3. **Inspect for slipped wheels on spoke style.**
   - Check all metal surfaces thoroughly, including spacer bands and tire side of rims. Watch for excessive rust or corrosion buildup; cracks in metal; bent flanges or components; loose, missing, or damaged nuts, clamps; broken or stripped studs; and incorrectly matched rim parts. Check valve locator for slip damage and improper location. Correct any problems discovered. For more information, see Chapter 8.

4. **Check for lateral runout on spoke style.**
   - Use a tire runout gauge to check lateral runout. Lateral runout should be checked on a smooth surface along the tire's upper sidewall. For more information, see Chapter 8.

5. **Check hub odometer.**
   - Inspect hub odometer for security and damage.

SECTION V

Interior Inspection

1. **Inspect interior wall.**
   - Record any damage.

2. **Inspect roof bows and sheeting.**
   - Inspect for loose, damaged, or missing roof bows.

3. **Check interior light (if equipped).**
   - Make sure all interior lights are operable. Check wiring and lens(es).

4. **Check door(s) for proper sealing.**
   - Inspect for damaged doors, missing hardware, and security.

5. **Inspect trailer flooring.**
   - Record and flooring damage found. Inspect for any "soft spots" or cracked flooring. Pay close attention to the threshold area.

6. **Inspect all tie downs and logistics tracks.**
   - Check all tie downs and logistics tracks for security and damage.

SECTION VI

Lubrication

1. **Lubricate landing gear.**
   - Be sure to use the OEM-recommended lubricant.

2. **Lubricate the J560 seven-way receptacle with dielectric lubricant.**

3. **Lubricate all suspension fittings.**
   - Be sure to use the OEM-recommended lubricant.

4. **Lubricate doors.**
   - More information is available in this chapter.

4A. **Lubricate brake slack adjusters.**
   - Follow the slack adjuster manufacturer's recommendations.

5. **Service automatic/central lubrication system (if equipped).**

6. **Fill automatic/central lube reservoir (if equipped).**

SECTION VII

Extended Interval

1. **Check axle alignment.**
2. **Check tandem axle spacing.**
3. **Remove and inspect wheel bearings.**
   - See Chapter 8 for more information on wheel bearing adjustment procedures.

SECTION I EXTERIOR VISUAL INSPECTION

Sec I	Start at Front of Trailer		
Item		**A**	**C**
1	Check the registration holder and the annual inspection sticker.		
2	Inspect the SAE J560 seven-pin receptacle.		
3	Inspect gladhands.		
4	Check the operation of all trailer lights.		
5	Check all light lenses and reflectors.		
6	Inspect the trailer's wiring harness.		
7	Inspect rear underride guard.		
8	Inspect rear door(s), handle(s), hinges, lock(s) tie backs, and cables.		
9	Inspect operation of rear door(s).		
9A	Inspect rear door seals and tracks.		
10	Check the license plate and mounting.		
11	Inspect side door(s), handles(s), hinges, lock(s), and tie backs.		
12	Inspect operation of side door.		
12A	Inspect side door(s) seals and tracks.		
13	Inspect all placards and mountings.		
14	Check all rivets and fasteners on bottom rail.		
15	Record any physical damage to trailer.		

SECTION II UNDERCARRIAGE AND SUSPENSION

Sec II	Undercarriage and Suspension		
Item		**A**	**C**
1	Inspect upper coupler plate and kingpin.		
2	Check cross-members and substructures.		
3	Inspect wiring harness(es) and air lines.		
4	Inspect trailer landing gear.		

(continued)

Item		A	C
4A	Test landing gear operation.		
5	Inspect spring and hangers.		
6	Torque U-bolts.		
7	Inspect torque rods, radius rods, and bushings.		
8	Inspect spare tire carrier.		
9	Inspect tandem slide and lock mechanism (if equipped).		
10	Inspect inner wheel ends.		
11	Inspect mud flaps and brackets.		
12	Inspect pintle hook (if equipped).		
13	Inspect glad hand and SAE J560 seven-pin connector (rear).		
14	Test air bags for leaks (air ride suspensions).		
15	Check height control valve and linkage (air ride suspensions).		

SECTION III BRAKES

Sec III	Brakes		

Item		A	C
1	Inspect brake linings and drums.		
1A	Record brake lining thickness (in inches).		
2	Check the air brake system for air leaks.		
3	Adjust all brakes (manual slack adjusters).		
3A	Check operation of automatic slack adjuster (if so equipped).		
4	Drain air reservoirs.		
5	Inspect reservoir mounting and brake chambers.		

SECTION IV TIRES, WHEELS, AND HUBS

Sec IV	Tires, Wheels, and Hubs		

Item	Hubs	A	C
1	Check wheel bearings.		
2	Inspect hubcap for leaks.		
3	Check lubricant level.		

(continued)

Item		A	C
	Tires		
4	Inspect tires for irregular wear.		
5	Inspect tire sidewall for cuts or damage.		
6	Check dual mating with square.		
7	Check tread design match.		
8	Inspect tire valve caps.		
9	Check and record tire tread depth.		
10	Check and record air pressure.		
	Wheels		
11	Inspect for loose wheel lugs.		
12	Inspect wheels for cracks or damage.		
13	Inspect for slipped wheels on spoke style.		
14	Check for lateral runout on spoke style.		
15	Check hub odometer.		

SECTION V INTERIOR INSPECTION

Sec V	Interior Inspection		

Item		A	C
1	Inspect interior wall.		
2	Inspect roof bows and sheeting.		
3	Check interior light (if equipped).		
4	Check door(s) for proper sealing.		
5	Inspect trailer flooring.		
6	Inspect all tie downs and logistics tracks.		

SECTION VI LUBRICATION

Sec VI	Lubrication		

Item		A	C
1	Lubricate landing gear.		
2	Lubricate the J560 seven-way receptacle with dielectric lubricant.		
3	Lubricate all suspension fittings.		
4	Lubricate doors.		

(continued)

Sec VI	Lubrication		
Item		**A**	**C**
5	Lubricate brake slack adjusters.		
6	Service automatic/central lubrication system (if equipped).		
7	Fill automatic/central lube reservoir (if equipped).		

SECTION VII EXTENDED INTERVAL

Sec VI	Extended Interval		
Item		**A**	**C**
1	Check axle alignment.		
2	Check tandem axle spacing.		
3	Remove and inspect wheel bearings.		

Summary

- Maintenance programs are necessary to ensure the designed life expectancy of the vehicle and its components.

- Aluminium sheet and post vans are constructed so the posts act as the major sidewall structural members connecting the top and bottom side rails together.

- For unibody trailers, it is imperative that all rivets or fasteners connecting the top and bottom rails as well as all cross-member members be present and secure.

- Leaks in van trailers may be found by a light test, water test, smoke test, or ultrasonic test.

- FRP plywood is multiple layers of matt fibreglass cloth bonded to both sides of a plywood core, providing excellent strength and corrosion resistance.

- The two types of doors used on van-type trailers are hinged and roll-up.

- The trailer's floor system should be inspected on every PMI to ensure it can safely handle the loads it is intended to support and ensure its service life.

- New floorboards must be supported by four cross-members.

- Rear impact guard, or underride guard, should be inspected regularly for cracked welds, cracked or fractured vertical members, cuts and tears in any member, and for dimensional conformity.

- Poorly aligned trailer axles will cause the trailer to dog track and wear tires prematurely.

- The National Highway Safety Administration Regulations require that red and white alternating striped reflective tape, also known as conspicuity tape, be applied to the full trailer width at the rear of the trailer, near the van base.

Review Questions

1. In reference to aluminum sheet and post vans, the aluminum helps to tie the side rails together but its main function is to prevent the walls from "racking" as well as to protect the load from the elements.

   A. True   B. False

2. Multiple layers of matt fiberglass cloth bonded to both sides of a plywood core describes the construction of:

   A. an open top trailer.   C. an FRP trailer.

   B. a dump-style trailer.   D. a tank-style trailer.

3. Hinged doors have the disadvantage of a reduced opening size as well as reducing the interior trailer cube.

   A. True   B. False

4. What information will **not** be found on a trailer certification plate?

   A. Gross vehicle weight rating   C. Date of manufacture

   B. Date safety inspection is due   D. Manufacturer's name

5. It is important to repair any trailer roof damage immediately to protect cargo from the elements.

   A. True   B. False

6. A well-planned and executed PM program offers what results?

   A. Low maintenance costs   D. All of the above

   B. Better fuel economy   E. None of the above

   C. Fewer roadside breakdowns

7. To test a trailer for possible leakage, the best method is to use a smoke bomb and observe the trailer exterior for escaping smoke.

   A. True   B. False

8. Trailer flooring replacements must be made so that the new floor board is fastened to a minimum of _____ cross-members.

   A. two   C. four

   B. three   D. five

9. In reference to overhead doors, the cable is wound around the _____.

   A. counterbalance shaft   C. drum

   B. counterbalance spring   D. roller
      assembly

10. A properly adjusted overhead door should roll all the way up by itself once it is unlatched from the sill.

    A. True   B. False

11. Inspection of underride guards should be performed on all PMs. Look for:

    A. cracked welds.   D. conformity to all dimensional requirements.

    B. cracked or fractured vertical   E. all the above.
       members.

    C. cuts or tears in any member.

12. The underride guard must not be more than _____ inches above the ground.

    A. 22   C. 4

    B. 12   D. 3

13. Underride guards must be no more than _____ inches from the rear of the trailer.

    A. 22   C. 4

    B. 12   D. 3

14. The underride horizontal member must be at least _____ inches from the side of the trailer.

    A. 22   C. 4

    B. 12   D. 3

15. When an alignment of a trailer axle is performed, the _____ should be used for the centering reference point.

    A. axle extenders   C. kingpin

    B. wheel rims   D. axle center

Tank Trailer Service and Inspection

Objectives

Upon completion and review of this chapter, the student should be able to:

- List the general safety precautions that must be adhered to when working on a tank trailer.
- Explain the procedures that are performed during a 60-day PMI.
- List the various maintenance inspections and how often they must be performed on a tank trailer.
- List the items that must be inspected during an external visual inspection.
- List the items that must be inspected during an internal visual inspection.
- List the items that must be inspected during a pneumatic pressure test.
- Explain what items must be inspected when performing a thickness test.
- List the items that must be inspected during an upper coupler inspection.
- Explain what must be done during a hydrostatic pressure test.
- Describe the function of dry bulk tank trailers.
- List the items that are inspected during an external visual inspection.
- Describe the inspection procedure performed during an internal visual inspection and leakage test.

Key Terms

barrel

conspicuity tape

head

hydrostatic pressure test

Lunette eye

maximum allowable
   working pressure (MAWP)

outrigger

placard holders

pneumatics

ultrasonic thickness tester

INTRODUCTION

The purpose of this chapter is to go through a PMI on a tank trailer. For this example, we are using a bulk liquid tank trailer and will explain a general 60-day inspection. We also look at the different inspections that must be performed on a bulk liquid tank trailer that are unique to the tank trailer industry.

Figure 16-1 Worker entering a tank trailer. *(Courtesy of John Dixon)*

Figure 16-2 Items to be inspected before servicing a tank trailer. *(Courtesy of John Dixon)*

SYSTEM OVERVIEW

This chapter lists some of the procedures for both working around and entering a tank trailer. It also lists the steps to perform a 60-day PMI on a bulk liquid trailer. As well, we list the procedures for performing V, I, K, T, UC, and P inspections. Lastly we go through the procedures for a PMI on a dry bulk tank trailer, including an external visual inspection and an internal inspection and leakage test (see **Figure 16-1**).

SAFETY DURING MAINTENANCE

1. Follow the buddy system when doing internal tank work.
2. Conduct pressure testing in designated areas only, keeping people at a safe distance using "Tank under Pressure" signs.
3. Avoid having to go on top of the tank while it is under pressure.
4. Use protective gear and equipment appropriate for the tank. Watch for dried product residue atop corrosive tanks.
5. Follow degassing procedures thoroughly, and use explosion meters at frequent intervals.
6. Make sure manholes are open on tanks coming off wash racks into shop areas.
7. Use a two-step procedure to open manholes of tanks just tested or suspected of having any pressure in them.

Step 1: Open manhole latches slowly and for a short distance initially, until tank pressure is completely relieved.

Step 2: Open cover(s) fully.

SIXTY-DAY C INSPECTION

Tank trailers should be subjected to a C inspection every 60 days, and we will go through the general checks. The inspection for the running gear would be the same for all trucks and trailers:

- Before starting, make sure the tank trailer is safe to work on.
- Perform a circle check of the vehicle to inspect the general condition of the unit (see **Figure 16-2**).
- Inspect decals and the condition of the **conspicuity tape** (reflective tape).
- Inspect the **placard holders** for their numbers positioning and security to the trailer (see **Figure 16-3**).
- Apply the trailer service brakes and inspect for movement of all S-cams.
- With brakes applied, check for audible air leaks.

Figure 16-3 Placard holder. *(Courtesy of John Dixon)*

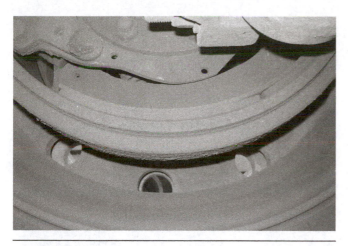

Figure 16-4 Brake assembly. *(Courtesy of John Dixon)*

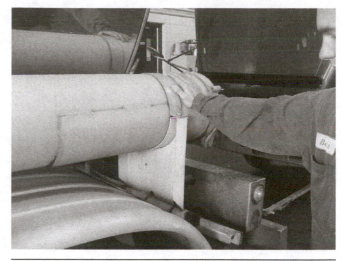

Figure 16-6 Inspecting fenders and mud flaps. *(Courtesy of John Dixon)*

- Measure and record the thickness of the brake shoes on all wheels (see **Figure 16-4**).
- Inspect the brake drums.
- Inspect S-cam bushings.
- Inspect slack adjusters' clevis pins (should spin in both the clevis and S-cam).
- Check brake adjustment.
- Measure and record tire tread depth.
- Measure and record tire pressure.
- Inspect tire matching, condition, and wear pattern of tires (see **Figure 16-5**).
- Check the condition of all brake hoses, air valves, maxi chambers, and air reservoir mounting.
- Check glad hand seals and replace as necessary.
- Inspect the subframes and cross-members for cracks and corrosion.
- Inspect the condition of all suspension components, including bushings in torque arms,

equalizers, and suspension beams. Inspect equalizers, hangers, beams, and axle seats for wear and cracks.
- Inspect leaf spring condition and U-bolts for tightness.
- Inspect air springs, shock absorbers, and height control system for wear and air leakage.
- Inspect fenders and compartment boxes for cracks and security (see **Figure 16-6**).
- Inspect mud flaps and brackets for mounting and damage.
- Jack up each axle and check for excessive wheel bearing end play.
- Check wheel end lubricant and inspect for any signs of contamination.
- Inspect for leaking wheel seal/brake hardware and condition.
- Check wheels for cracks, looseness, and that all mounting hardware is accounted for.
- Retorque all wheel mounting fasteners.
- Check landing gear operation if both high and low gear. Inspect mounting braces for cracks and structural integrity.
- Measure kingpin dimensions with kingpin gauge.
- Inspect upper coupler for cracks and dishing. Inspect kingpin and upper coupler mounting fasteners for tightness. This can be done by tapping the kingpin or mounting fasteners with a hammer. A loose kingpin or mounting fastener will make a dull thud sound, whereas it will resonate like a bell if it is tight.
- Inspect tank **barrel** (containment vessel) for cracks, dents, and corrosion, and inspect **outrigger** condition (see **Figure 16-7**).
- Inspect tank jacket for damage and looseness.

Figure 16-5 Dual wheel assembly. *(Courtesy of John Dixon)*

Figure 16-7 Tank barrel. *(Courtesy of John Dixon)*

Figure 16-9 Product line of a tank trailer. *(Courtesy of John Dixon)*

- Inspect ladder, catwalk, manhole gaskets, vents, caps, and gauges (see **Figure 16-8**).
- Inspect product line condition for leakage and security (see **Figure 16-9**).
- Inspect temperature gauge (see **Figure 16-10**).
- Inspect steam lines and air lines for leakage, security, and wear.
- Apply grease to all zerk fittings on the trailer. Fill auto greaser if so equipped.
- Inspect all lights and wiring for proper operation condition, and corrosion.
- Check ABS function and operation of ABS light if trailer is so equipped.
- Inspect hydraulic pump, lines, break-off, fusible plug, and oil contamination.
- Vacuum test external and internal valves (see **Photo Sequence 12**).

To test the internal and external valves, start by opening the internal valve and closing the external valve. Apply test fitting to the outlet pipe, turn vacuum pump on, and allow it to reach 20 inches of mercury (Hg). Turn off the vacuum pump and watch the gauge. If vacuum holds, the external valve is good; if not, repair external valve. Next, close internal valve and open external valve. Turn the vacuum pump on and allow it to reach 20 inches of vacuum. If vacuum holds, the internal valve is good; if not, make the necessary repairs.

- Inspect pintle hooks **(Lunette eye)** for wear and mount hardware tightness and condition of safety chains or cables (if equipped).

Figure 16-8 Pressure relief and vacuum valves. *(Courtesy of John Dixon)*

Figure 16-10 Temperature gauge. *(Courtesy of John Dixon)*

PHOTO SEQUENCE 12 # Vaccum Test External and Internal Valves

P12-1 Start by inspecting the condition of the external seal.

P12-2 Pump the manual hydraulic hand pump to open the internal valve.

P12-3 Close the external valve by turning the hand valve.

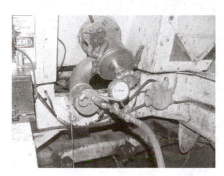

P12-4 Apply test fitting to the outlet pipe.

P12-5 Turn vacuum pump on. Shut off when 20 inches is reached. Turn off the vacuum pump and watch the gauge. If vacuum holds, the external valve is good.

P12-6 Turn the valve on the manual hand pump to allow the hydraulic pressure to bleed of allowing the internal valve to close.

P12-7 Open the external valve.

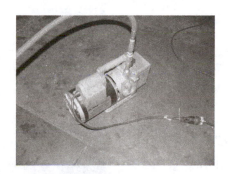

P12-8 Turn vacuum pump on. Shut off when 20 inches is reached. Turn off the vacuum pump and watch the gauge. If vacuum holds the external valve is good.

Figure 16-11 Tank trailer inspection symbols and dates. *(Courtesy of John Dixon)*

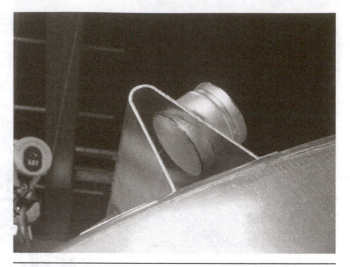

Figure 16-12 Inspect couplings and ports for damage. *(Courtesy of John Dixon)*

In addition to the items inspected at 60-day intervals, there are six other inspections that must be performed. These inspections and due dates are usually displayed in the form of a letter on the front **head** of the trailer (see **Figure 16-11**). The additional inspections are:

1. **V.** This stands for an external visual inspection and is performed yearly.
2. **I.** This stands for an internal visual inspection and is performed yearly.
3. **K.** This is a pneumatic pressure test, is performed at 80 percent of the tank's **maximum allowable working pressure (MAWP),** and is done yearly as well.
4. **T.** The "T" stands for a thickness test of the tank and is performed every 2 years.
5. **UC.** This is an inspection of the upper coupler requiring the upper coupler to be removed for a full visual inspection and must be performed every 2 years.
6. **P.** The "P" stands for a **hydrostatic pressure test** and is performed at 100 percent of the MAWP and is done at 5-year intervals.

EXTERNAL VISUAL INSPECTION

Items to be inspected annually during a "V" inspection:

- Inspect the tank jacket for damage and looseness.
- Inspect the tank's front and rear heads for any signs of damage such as dents, holes, and tears in the jacket.

- Inspect couplings and ports for any gouges or damage that can cause coupling to leak (see **Figure 16-12**).
- Inspect piping for any damage or signs of leakage (see **Figure 16-13**).
- Inspect internal valve, ensuring that it is there and intact.
- Inspect major appurtenances, which means inspect any specialty apparatus that is installed on the trailer.
- Inspect excess flow valves.
- Inspect gasket on opening head (if equipped) for condition (cracked, hard, dried out) and any signs of leakage.
- Inspect manhole gaskets for damage, signs of cracks, or dried out.

Figure 16-13 Inspect piping for any damage or signs of leakage. *(Courtesy of John Dixon)*

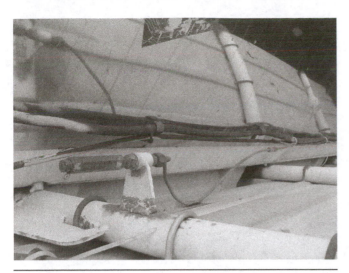

Figure 16-14 Remote closure device holds hydraulic pressure that keeps the spring-loaded internal valve open. *(Courtesy of John Dixon)*

Figure 16-15 Inspect flanges that hold piping together. *(Courtesy of John Dixon)*

- Inspect remote closure devices to make sure they are intact (see **Figure 16-14**). This remote closure device holds hydraulic pressure that keeps the spring-loaded internal valve open. It is located at the front of the trailer, while the internal product valve on this trailer is located at the rear head. In the case of an emergency, the remote closure device can be broken off with a hammer or other suitable object. This will allow the hydraulic pressure to bleed off, forcing the internal valve to close. The remote closure valve also incorporates a fusible plug in the end. If temperatures are elevated (as in a fire), the center of the fusible plug will melt out, also allowing the hydraulic pressure to bleed off and closing the internal valve.
- Inspect flanges that hold piping together for condition. Make sure bolts are secured and that gaskets are not leaking (see **Figure 16-15**).
- Inspect fusible plugs to make sure the solder is still intact and not pushing out the end (see **Figure 16-14**).
- Inspect lading seal provision and lading fasteners. The lading seal is a means of ensuring that the filler cap or manhole cannot be opened or removed after product is loaded, ensuring the product cannot be tampered with.
- Inspect for the presence of vacuum vents and that their size matches the tank specifications.
- Inspect that the void areas are not leaking and ensure the drain pipe at the bottom of the void is uncapped.
- Inspect overfill system (Skully system) if equipped (see **Figure 16-16**).

- Inspect for the presence of dust caps, either solid, or fusible caps. These caps prevent foreign material from contaminating the product.
- Inspect for the presence and legibility of an emergency shutoff decal.
- Inspect any bolted-on attachments such as hangers, bogie, flanges, etc.
- Inspect that overlay patches (sections where jacket connects to another section of jacket material) are secure.
- Inspect for the presence of DOT markings.
- Inspect for the presence of specification plate.
- Inspect for the presence of certification plate.
- Inspect self-closing stop valve.
- Inspect the condition of external valve.
- Inspect manifold valve.

Figure 16-16 Hook up for Skully system. *(Courtesy of John Dixon)*

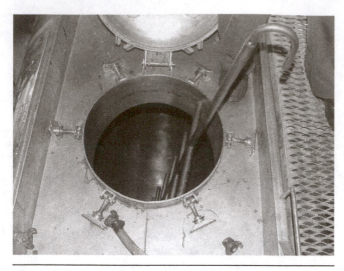

Figure 16-17 Manhole cover and latches. *(Courtesy of John Dixon)*

Figure 16-18 Inside of a bulk liquid tank trailer. *(Courtesy of John Dixon)*

- Inspect condition of manhole covers (see **Figure 16-17**).
- Inspect manhole hold-down hardware.
- Inspect frangible devices.
- Inspect the condition of vapor recovery system.
- Inspect the system that operates the internal valve (hydraulic pump, cable system, or pneumatic system).
- Inspect rupture disc (pressure relief device).
- Inspect vapor return lines.
- Inspect condition of air injection system (used to pressurize tank in order to offload product faster or to perform leakage tests).
- Inspect condition of spill dam drains. These are the tubes that run down either side of the tank from the manholes, preventing rainwater from entering the tank as well as draining any spillage.
- Inspect all hoses and lines.
- Inspect all attachment welds for cracks and corrosion.
- Inspect pressure relief vent.
- Upper coupler hammer test.

INTERNAL VISUAL INSPECTION

Items to be inspected annually during an "I" inspection:

- Inspect the entire inside surface of the tank for signs of corrosion (see **Figure 16-18**), abrasion, dents, and pitting. Pay special attention to the tank heads and area over top of the upper coupler (see **Figure 16-19**).
- Inspect condition of tank liner (if equipped).
- Inspect all gauging devices for vertical alignment and security.

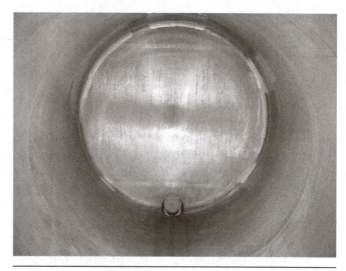

Figure 16-19 Internal head. *(Courtesy of John Dixon)*

- Inspect the areas around piping, sumps, and valves for foreign material that could prevent proper functioning of the valve (see **Figure 16-20**).
- Inspect areas of the heat panel for distortion, cracks, or pitting.
- Inspect sumps, ports, and manholes for cracks or damage.

LEAKAGE TEST

To be inspected annually during a "K" inspection:
This is a pneumatic test and is performed at 80 percent of the MAWP.

- Start by looking at the specification plate and confirming the MAWP.
- Using a regulator and required apparatus, including an accurate pressure gauge, pressurize

Figure 16-20 Internal valve. *(Courtesy of John Dixon)*

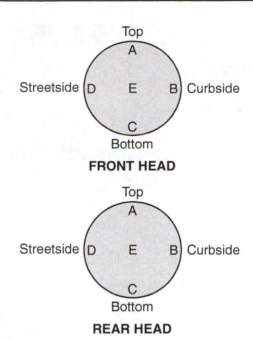

FRONT HEAD

REAR HEAD

Figure 16-22 Tank heads at five locations.

the compartment or each individual compartment to 80 percent of the MAWP.

- Once the correct pressure is achieved, disconnect or shut off the air supply and monitor the pressure gauge.
- Listen for any audible leaks and make note of them if found.
- The tank must hold 80 percent of the MAWP for a period of 10 minutes; if not, it must be repaired before being put back into service.

THICKNESS TEST

The thickness test is performed every 2 years during a "T" inspection.

This inspection is performed using an **ultrasonic thickness tester**, sometimes referred to as a sonic tester (see **Figure 16-21**). This test can be time

consuming, as the technician must test and record the material thickness from sometimes more than 100 different locations. Specific areas to be tested:

- The tank's shell
- Tank heads at five locations (see **Figure 16-22**).
- Test any area around piping that retains lading.
- Test areas of high stress (around the pick-up plate and boogie area).
- Test areas near openings.
- Areas around weld joints
- Areas around shell reinforcements (see **Figure 16-23**).
- Areas around appurtenance attachments
- Known thin areas and liquid level lines

UPPER COUPLER INSPECTION

The upper coupler must have an inspection every 2 years during a "UC" inspection (see **Figure 16-24**).

Figure 16-21 Ultrasonic thickness tester. *(Courtesy of John Dixon)*

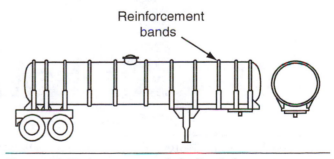

Figure 16-23 Areas around shell reinforcements.

Figure 16-24 Upper coupler assembly. *(Courtesy of John Dixon)*

During the inspection, the upper coupler must be removed from the trailer and the following inspected:

- Measure all kingpin dimensions using a suitable measuring instrument or use a Holland kingpin tester Part # TF-0110. Replace kingpin if minimum dimensions are not met (see **Figure 16-25**).
- Inspect for any cracks between mounting bolt holes.
- Measure for excessive bolster plate wear.
- Inspect the trailer cross-members and structures normally covered by the upper coupler assembly.
- Inspect for elongated bolt holes.
- Check coupler for corroded areas, abraded sections, dents, and bolster plate distortion.
- Inspect complete upper coupler for any weld defects.
- Replace all upper coupler mounting bolts and torque to specification.

PRESSURE TEST INSPECTION

This pressure test is usually a hydrostatic pressure test and is performed every 5 years. This test is symbolized by the letter "P":

- Before following procedures, verify that tank landing gear can support tank filled to capacity with water.
- Start by looking at the specification plate and confirming the MAWP.
- Next, the tank trailer is filled with water.
- Using a regulator and required apparatus, including an accurate pressure gauge, pressurize the compartment or compartments individually to 100 percent of the MAWP.
- Once the correct pressure is achieved, disconnect or shut off the air supply and monitor the pressure gauge.
- Inspect for any signs of water leakage and make note of any.
- The tank must hold 100 percent of the MAWP for a period of 10 minutes; if not, it must be repaired before being put back in service.

TANK ENTRY SAFETY PROCEDURES

1. Make sure tank has been steamed, washed, or degassed.
2. Notify foreman or night supervisor that you intend to enter tank.
3. Check out gas tester, oxygen tester, respirator, and safety harness (or wrist lines).
4. Test gas content with explosimeter. Tank must read non-explosive on meter.
5. Test oxygen content with oxygen tester. Tank must have between 19.5 and a maximum of 23 percent oxygen to prevent suffocation.
6. Put on safety harness (or wrist lines).

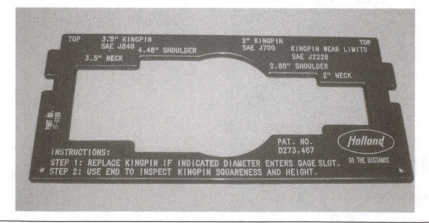

Figure 16-25 Kingpin tester. *(Courtesy of John Dixon)*

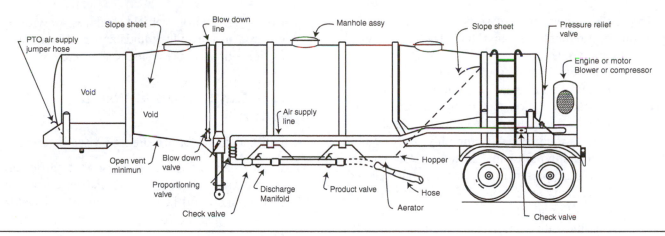

Figure 16-26 Components of a typical dry bulk tank trailer.

7. Put on respirator (fresh air or cartridge).

8. Make sure another employee is standing by on top of the tank to handle safety harness (or wrist lines).

9. Enter tank. Do not stay in tank for extended period of time. Leave immediately at first sign of dizziness, nausea, or labored breathing.

10. When use is completed, return all equipment to supervisor for cleaning and proper storage.

DRY BULK CARGO TANKS

There are two basic configurations for dry bulk tanks: tank type and hopper type.

The tank type incorporates a tank structure to stretch the tank out to get the fifth wheel and undercarriage under the ends and to meet bridge formulas. It uses slope sheets inside at each end of a hopper compartment to separate the tank-like ends from the vessel. The hopper type uses structures or frames attached to the vessel to accomplish the same purpose (see **Figure**).

Air pressure is used to unload most dry bulk tanks, so the term **pneumatics** is quite often applied to them. Some also have vacuum system for loading, with suction supplied by connecting the suction side of a blower to evacuate the vessel.

Bulk tanks are not designed or constructed to transport hazardous materials, although the variety of products they can handle is quite extensive. Examples of products shipped in these trailers are cement, salt, lime, fertilizers, plastics, gypsum detergent powders, ammonium nitrate, feed supplements, flour, sugar, and other edible products (see **Figure**).

General Maintenance for Dry Bulk Cargo Tankers

The most common operating problem for dry bulk tanks is discharge systems (manifold, hose) plugged with product. Each tank manufacturer provides instructions for unplugging by creating a pressure surge in the manifold and hose, or by reversing air flow and carrying the plugged material back into the vessel. This is done by opening and closing valves in the tank system, with appropriate blower operations per manufacturer's instructions.

Like all tanks loaded and or unloaded by air pressure, dry bulk tank systems are subject to repeated pressure applications. Consequently, leaks that develop will be detected during routine operations by sound of escaping air or blowing product dust. Leakage sources are usually the same as for liquid tanks: manhole assemblies and valves. Other common leak sources are banded sleeves between joining pipes from blowers, dome gaskets curled under, and vessel cracks.

CAUTION *When performing leak detection, never go on top of the tank while it is under pressure.*

Figure 16-27 Dry bulk tanker. *(Courtesy of John Dixon)*

Dry bulk tankers are not subject to the same inspection requirements as other tank trailers carrying hazardous materials.

At a minimum, all dry bulk tanks should be inspected externally and internally. The frequency of these services is a carrier option, but manufacturers recommend servicing and inspecting the various vehicular and tank components daily, weekly, monthly, every 5,000 miles, etc. Complying with manufacturer recommendations will support warranty claims, but detailed technical inspections at an annual frequency will prolong operating life and serve to detect flaws and concealed near failures that can result in serious accidents.

Dry Bulk Tank External Visual Inspection and Component Tests

Coverage of the following items is necessary in order to determine overall tank integrity and condition of operating systems:

1. Inspect all tank structures for corrosion or abrasion, dents, distortions, crack indications, defects in welds, and other defects apparent through visual inspection.
2. Inspect piping, valves, adapters, and hose assemblies and attachments for defects and indications of leakage.
3. Service blower, pump, or compressor as appropriate. Pay particular attention to lube oil leaks, drive belts, bearings, and lubrication points.
4. Check all components of the manhole assembly, with emphasis on gasket condition and correct functioning of all movable parts. Look for cracks and distortions, particularly in hold-down area of cover.
5. Perform complete systems function check on product handling system, pressurizing tank, and opening/closing of all valves, lines, etc.
6. Remove, clean, bench test, and recalibrate all spring-loaded pressure relief devices. Follow manufacturer's specifications when bench testing.
7. After ensuring that pressure relief valve is opening at correct pressure, reinstall on tank and pressurize tank until pressure relief device opens. Caution: this test should be performed outside with personnel clear of tank.
8. Replace all missing hardware and check tightness of all bolted components.
9. Perform inspection of upper coupler assembly, suspension system attachments, landing gear, ladder, catwalk, and platform.

10. Check all motor vehicle features previously covered in the bulk liquid tanker section, such as light, tires, brakes, suspension, etc.

Dry Bulk Tank External Visual Inspection

Activities to be performed during a visual inspection:

1. Inspect data plates for security, that entries are legible, that it is not painted, and for corrosion.
2. Inspect shell and heads for condition of welds, dents, gouges, corrosion, abrasions, distortion, or cracks. Circle area with chalk or a marking crayon and record findings.
3. Inspect frames and structures for rust, corrosion, and cracking. Record any findings.
4. Inspect upper coupler assembly for condition of plate, corrosion, deformation, and lubrication. Check bolt torque and measure all kingpin dimensions. Remove and service the upper coupler as needed.
5. Inspect landing gear frames and support bracing for corrosion and rust, as well as the condition of the welds. Check tightness of all bolts and check landing gear operation in both high and low gear. Inspect sand shoe security and lubricate the landing gear zerk fittings with chassis grease (see **Figure 16-28**).
6. Inspect void areas for sign of corrosion. Ensure fittings and drains are unplugged and operable. Inspect for the presence of product accumulation.
7. Check that placard holders are securely fastened and inspect the condition of the clips.

Figure 16-28 Landing gear and braces. *(Courtesy of John Dixon)*

8. Inspect condition of paint on the trailer, ensuring that it is up to company specifications.

9. Inspect bolted attachments such as piping brackets and supports, valve installations, dust cap retainers, and all tank-to-frame and/or undercarriage attachments.

10. Inspect hose tubes, troughs, or racks for physical condition, and condition of end doors and latches. Also inspect for any foreign items on the inside.

11. Inspect piping and all connecting valves and adapters; inspect for leaks at handles and levers. Lubricate as necessary.

12. Inspect discharge valve operation and condition of seats and discs. Lubricate as necessary (see **Figure 16-29**).

13. Inspect hose for condition of covers, reinforcement damage, condition of couplings, fittings, and other related hardware.

14. Inspect ladders, catwalk, and platform for proper security to the tank body. Check any mounting bolts for proper torque and any deformation of the structure.

15. Inspect that static grounding connections are present and tight.

16. Inspect the pressure gauge. Ensure that the needle is intact and resets to zero. Calibrate with a master pressure gauge if required. Make sure the gauge glass is clear, making viability of the gauge easy.

17. Inspect the manhole assemblies for each compartment for evidence of leakage, warpage, corrosion, and impact damage to the cover, weld collar, gasket seal surfaces, and all welds. Inspect for evidence of product accumulation and condition of latches, hinges, and all bolted connections, hold-downs, and safety catch. Lubricate as required.

18. Inspect pressure relief devices, verifying all vents are present and that venting is adequate for air supply marking on vents. Remove, clean, and bench test the pressure relief valve. Check all gaskets and O-rings involved in reinstallation of the valve.

19. Inspect any check valves for proper operation and that they are installed where they are supposed to be.

20. Inspect condition and presence of all safety labels; replace as necessary.

21. Inspect that proper tank markings are present and correct as per company policy.

Internal Visual Inspection and Leakage Test

Start by verifying a safe atmosphere with adequate oxygen level before entering the tank for any purpose.

The purpose of this service is to visually verify the condition of internal tank features, and then gain an overall structural integrity verification by pressure testing. Coverage of the following items is mandatory in order to accomplish this purpose:

1. Inspect all shell, head, and void areas that are accessible. Verify internal condition of suspect areas identified during external inspection.

2. Repair as necessary any defective condition that relates to pressure integrity of the tank.

3. Perform leakage test on each compartment at working pressure.

4. Record the results of the service in an inspection form.

CAUTION *All company tank entry safety procedures and OSHA regulations must be complied with.*

Use the following checklist as outlined to accomplish the service:

1. Review external inspection report and inspect interior of suspect areas recorded in the external inspection.

2. Inspect the structure for corrosion, abrasion, dents, pitting, or distortion at manhole area, around valve flanges and aerators. Inspect all structures for deformation.

Figure 16-29 Discharge valve. *(Courtesy of John Dixon)*

3. Check every inch of every weld in tank, making hand contact where possible. Pay special attention to welds in stress areas such as end frame supports above upper coupler, and undercarriage.

4. Check valve installation and security. Mack a visual inspection of valve surfaces. Check for foreign matter in valves and sumps.

5. Remove all equipment brought into tank. Then inspect again before leaving tank.

6. Conduct leakage test on vessel and associated valves and piping at working pressure, and hold for 5 minutes.

Summary

- Follow the buddy system when doing internal tank work.

- Tank trailers should be subjected to a C inspection every 60 days.

- Both the internal and external valves are leak tested with a vacuum pump and gauge.

- Use a two-step procedure to open manholes of tanks just tested or suspected of having any pressure in them.

- An external visual inspection is performed yearly.

- An internal visual inspection is performed yearly.

- A pneumatic pressure test is performed yearly at 80 percent of the tank's maximum allowable working pressure.

- A thickness test of the tank is performed every 2 years.

- An upper coupler inspection is performed every 2 years.

- A hydrostatic pressure test is performed at 5-year intervals.

- Oxygen content in tank must be between 19.5 and a maximum of 23 percent oxygen to prevent suffocation.

- Do not stay in tank for an extended period of time. Leave immediately at first sign of dizziness, nausea, or labored breathing.

- The two basic configurations for dry bulk tanks are tank type and hopper type.

- The most common operating problem for dry bulk tanks is discharge systems (manifold, hose) plugged with product.

- Dry bulk tanker leaks will be detected during routine operations by the sound of escaping air or blowing product dust.

- When performing leak detection, never go on top of the tank while it is under pressure.

Review Questions

1. Tank trailer personnel must have advanced technical knowledge of all tanker systems including:

   A. loading/unloading systems.

   B. product retention systems.

   C. safety relief devices.

   D. All of the above

   E. None of the above

2. When working inside a tank trailer:

   A. follow the buddy system.

   B. follow both company and OSHA procedures.

   C. test gas content with explosimeter. Tank must read non-explosive on meter.

   D. All of the above

   E. None of the above

3. It is recommended that personnel remain on top of the tank barrel during pressure testing to monitor tank pressure.

   A. True

   B. False

4. Oxygen content must be more than _____ percent before personnel enter a tank to prevent suffocation.

    A. 6   C. 16

    B. 10.5   D. 19.5

5. If a significant dent is found on a jacket of an insulated tanker, location and depth of dent should be entered in the inspection report.

    A. True   B. False

6. The tank liner is inspected during an internal visual inspection.

    A. True   B. False

7. What does Hg represent?

    A. Hazardous materials   C. Mercury

    B. Liquefied petroleum   D. Hazardous gas

8. What letter is assigned to a date on the inspection plate for an external visual inspection?

    A. I   C. V

    B. K   D. L

9. What letter is assigned to a date on the inspection plate for a hydrostatic pressure test?

    A. K   C. P

    B. V   D. UC

10. How much pressure must the tank hold when you perform a pneumatic pressure test?

    A. 50 percent of the tank's   C. 70 percent of the tank's MAWP
       MAWP
                                    D. 80 percent of the tank's MAWP
    B. 60 percent of the tank's
       MAWP

CHAPTER 17

Tractor PMI Class 7 and 8 Diesel Powered with Air Brakes

Objectives

Upon completion and review of this chapter, the student should be able to:

- The intent of this chapter it to put together a sample maintenance inspection sheet with explanations to expand the expectations of the PMI record sheet.

- The following are based upon the recommendations of The Maintenance Council (TMC) of the American Trucking Associations.

- The TMC is a great resource for any technician to access because all information is kept current, as rules and regulations within the industry do change regularly to improve the safety of commercial vehicles.

SECTION I

Cab Interior Inspection

1. **Inspect the ignition key and switch.**
   - Check the condition of the key and test the operation of the igniting switch. Both should be in good condition and working order.

2. **Inspect all warning lights and alarms.**
   - To perform this inspection, place the ignition switch in the start position. Check warning lights and/or alarms. All appropriate warning lights and alarms such as anti-lock braking systems (ABS), oil pressure, and seat belt warning lights and alarms that the vehicle is equipped with should operate initially when the engine is started and then go off. Consult vehicle operating and/or service manuals to determine the warning light or alarms with which the vehicle is equipped.

3. **Test starter operation.**
   - Engage the starter and listen for any unusual noises, starting difficulty, or starter drag.

4. **Check engine operation and record governor rpm.**
   - Once the engine oil pressure has stabilized, record governed engine speed.

5. **Inspect instruments and record oil pressure and electrical voltage.**
   - With the engine operating at high idle, record oil pressure and voltmeter readings.

6. **Test wiper and washer operation.**
   - Turn on the wiper and test for operation at varying speeds and ensure that the washers

operate. Top up washer fluid as required. Quality of the wiper blades, wiping quality, and the mounting and positioning of the wiper blade and wiper area should also be inspected.

7. **Test operation of electric and/or air horn.**
   - Both should operate properly and be clearly audible.

8. **Test hand throttle and shutdown cable operation.**
   - Ensure that the hand throttle valve operates smoothly and locks in place. The manual engine shutdown should operate smoothly and remain in the shutdown position when applied.

9. **Check A/C, heater, and defrost controls.**
   - Ensure that the A/C, heater, and defroster controls move easily. Check all speeds of the blower motor to ensure proper operation. For more information, see Chapter 10.

9A. **Check bunk climate controls.**
   - For the bunk/sleeper compartment, check the same as in 9.

10. **Test operation of all interior lighting.**
   - Check all interior lights, including bunk lighting, for operation and intensity.

11. **Test the operation of all switches and accessories.**
   - Test the operation of any accessories to ensure the operation properly (i.e., power mirrors, mirror heat, power windows, radio, etc.).

11A. **Check operation of fan clutch (optional).**

12. **Check clutch free travel and pedal pad.**
   - Depress the clutch pedal and determine the distance the pedal travels before the thrust load is felt and the clutch release bearing assembly engages. For more information, see Chapter 7.

13. **Test the operation of the clutch brake.**
   - With the engine idling, depress the clutch pedal to full travel. The transmission should shift with minimal effort and not gear clash. Clutch brake should engage approximately 0.5 in from the floor. For more information, see Chapter 7.

14. **Test the operation of the parking brake.**
   - With the spring parking brake set, the vehicle should not move when placed in gear and the clutch is slowly disengaged. See Chapter 9 for more information.

15. **Record air governor cutout setting (psi).**
   - At maximum air pressure (governor cutout). Air pressure should not exceed 135 psi for each of the air reservoirs. See Chapter 9 for more information.

16. **Test the operation of the air drier drain valve.**
   - Run the engine until the air system reaches governor cutout pressure and listen for the air drier to do its purge cycle. See Chapter 9 for more information.

17. **Inspect all safety equipment and decals.**
   - Check the condition of all onboard safety equipment (flares, triangles, fire extinguisher, and all required decals).

18. **Inspect the windshield and sun visor(s).**
   - Inspect the windshield's glass for cracks or discoloration. Sun visors should operate and adjust properly.

19. **Test the steering wheel for excessive play or binding.**
   - Rotate the steering wheel and determine amount of free play and check for binding. For more information, see Chapter 11.

SECTION II

Air Brake Pneumatic System Check

For more information on the following, see Chapter 9.

1. **Inspect the air system for leaks with the brakes released.**
2. **Inspect the air system for leaks with the brakes applied.**
3. **Test the operation of the single and double check valves.**
4. **Test the low air pressure warning devices.**
5. **Check governor cut-in pressure.**
6. **Test the spring brake inversion valve (if equipped).**
7. **Test tractor protection valve.**
8. **Check compressor air pressure build-up time.**

SECTION III

Cab and Body Walk Around Inspection Lights On

1. **Check seat operation and seat covering.**
   - Test the operation of the seat for ease of operation, adjustability, and security. Inspect seat cover for cuts, tears, and wear.

2. **Check floor mats/covering.**
   - Inspect the floor covering or mats for cleanliness, tears, or damage.

3. **Check seat belt and bunk restraints.**
   - Check the seat belts for latch and release of the locking assembly, proper adjustment, and cuts/severe abrasions, for both restraints.

4. **Check door locks, latches, and hinges.**
   - Test the operation of all door locks, latches, and hinges. Inspect for cracked or broken assemblies.

5. **Inspect the window operation.**
   - Ensure the operation of the windows, check window regulators for proper operation, and inspect door glass for cracks.

6. **Check steps and grab handles.**
   - Ensure all grab handles, steps, and mountings are secure and operational.

7. **Inspect wiper blades and arms.**
   - TMC suggests that the wiper blade should be positioned perpendicular to the glass surface in all wiping positions. Technicians should inspect the frame to ensure it has not become bent or distorted, and that no joint, including those between the frame claws and wiping element assembly, exhibits any excessive wear or looseness. In addition, arm and blade should not permit lateral motion of the blade relative to the arm.

8. **Inspect mirror mountings, brackets, and glass.**
   - Inspect the mounting and adjustment of all mirrors and brackets and inspect the glass for damage.

9. **Inspect headlamps, turn signals, fog, clearance, and brake lights.**

   - Note the operation and illumination of all lights. Inspect all lenses for damage.

10. **Check fuel tanks, mountings, lines, and caps.**
    - Examine the fuel tanks for leaks, shifting, or damage. Inspect fuel cap gaskets and tank mounting straps for cracks and security.

11. **Inspect air lines, holders, and gladhands.**
    - Check air lines, gladhands, and rubber gaskets for damage or wear. Inspect gladhand dummy connectors.

12. **Examine the condition of the SAE J 560 seven-pin connector, jumper cable, and holder.**
    - Check electrical connectors and the cable that connects the tractor to the trailer or converter dolly for damage or wear.

13. **Test the seven-pin connector and cable.**
    - Test the connector and jumper cable for proper connection and operation.

14. **Examine the mounting of the fifth wheel.**
    - Check fifth wheel mounts and welds for cracks and loose or missing hardware. For more information, see Chapter 13.

15. **Check the locking operation of the fifth wheel.**
    - Check fifth wheel locking and latching mechanism to ensure proper operation. For more information, see Chapter 13.

16. **Inspect the mud flaps and hangers.**
    - Inspect the mud flap and hangers for damage. For a template of how a mud flap and bracket should be spaced out for drilling and mounting, see Chapter 15.

17. **Inspect pintle hook (if equipped).**
    - Check the pintle hook for wear and proper locking operation. Inspect mounting hardware or welds.

18. **Inspect exhaust stack and mounting.**
    - Inspect exhaust stack for damage or looseness.

19. **Record any observed physical damage.**
    - Record any observed physical damage to the vehicle, such as dents, dings, scratches, etc.

SECTION IV

Tires and Wheels

FRONT

1. **Inspect for irregular wear patterns.**

   - Irregular wear patterns may indicate a variety of tire, wheel, rim, or alignment problems. To determine the probable cause of the irregular wear pattern encountered, see Figure 8.

2. **Inspect tire sidewall for cuts or damage.**

   - Record any physical damage that may be found while inspecting the tire.

3. **Inspect tire valve caps.**

   - Check to make sure valve caps are present and tight. Metal heat-resistant sealing valve caps should always be used.

4. **Check and record tire tread depth.**

   - Federal Motor Vehicle Safety Standards require a minimum of 2/32 inch tread depth on all steer axle tires and a minimum of 1/32 inch on non-steer tires.

5. **Check and record air pressure.**

   - Record current air pressures and adjust if necessary. A tire is considered flat if the air pressure is 20 percent lower than recommended or desired pressure or if there in an air leak that can be heard or felt. Never re-inflate a flat tire. Remove it and have it inspected to determine the cause of the leak.

Note: Do not adjust air pressure when tires are hot.

---

6. **Inspect for loose wheel lugs.**

   - Replace and retorque any fastener that is missing, worn, or damaged. Retorque all wheel nuts to the proper torque. For disc wheels, the inner nuts on both sides of failed fastener should also be replaced. For more information, refer to Chapter 8.

7. **Inspect wheels for cracks or damage.**

   - Inspect for signs of cracks or damage on any wheel component. Pay special attention to bolt hole mounting surfaces. Inspect both sides of the wheels and between duals.

Replace any wheel that is cracked or has damaged components. You should also determine the cause of the damage before replacing the wheel/wheel component(s). For more information, refer to Chapter 8.

7A. **Inspect for slipped wheels on spoke style.**

   - Check all metal surfaces thoroughly, including spacer bands and tire side of rims. Watch for excessive rust or corrosion buildup; cracks in metal; bent flanges or components; loose, missing, or damaged nuts, clamps; broken or stripped studs; and incorrectly matched rim parts. Check valve locator for slip damage and improper location. Correct any problems discovered. For more information, see Chapter 8.

REAR

8. **Inspect for irregular wear patterns.**

   - Irregular wear patterns may indicate a variety of tire, wheel, rim, or alignment problems. To determine the probable cause of the irregular wear pattern encountered, see Chapter 8.

9. **Inspect tire sidewall for cuts or damage.**

   - Record any physical damage that may be found while inspecting the tire.

10. **Check dual mating with square.**

    - The matching of dual tires on any axle is important for safety as well and improving tire life (smaller tire will scrub). The maximum allowable diameter difference between a tire and its running mate is 1/4 inch for 9.00 cross-section (10d-22.5) and larger size tires. For more information, see Chapter 8.

11. **Check tread design match.**

    - Ensure that tires are appropriate for use on the same axle. Mismatched tires on the same axle can result in vibration complaints and other problems.

12. **Inspect tire valve caps.**

    - Check to make sure valve caps are present and tight. Metal heat-resistant sealing valve caps should always be used.

13. **Check and record tire tread depth.**

    - Federal Motor Vehicle Safety Standards require a minimum of 2/32 inch tread depth on all steer axle tires and a minimum of 1/32 inch on non-steer tires.

14. **Check and record air pressure.**
   - Record current air pressures and adjust if necessary. A tire is considered flat if the air pressure is 20 percent lower than recommended or desired pressure or if there in an air leak that can be heard or felt. Never re-inflate a flat tire. Remove it and have it inspected to determine the cause of the leak.

Note: Do not adjust air pressure when tires are hot.

---

15. **Inspect for loose wheel lugs.**
   - Replace and retorque any fastener that is missing, worn, or damaged. Retorque all wheel nuts to the proper torque. For disc wheels, the inner nuts on both sides of the failed fastener should also be replaced. For more information, refer to Chapter 8.

16. **Inspect wheels for cracks or damage.**
   - Inspect for signs of cracks or damage on any wheel component. Pay special attention to bolt hole mounting surfaces. Inspect both sides of the wheels and between duals. Replace any wheel that is cracked or has damaged components. You should also determine the cause of the damage before replacing the wheel/wheel component(s). For more information, refer to Chapter 8.

17. **Inspect for slipped wheels on spoke style.**
   - Check all metal surfaces thoroughly, including spacer bands and tire side of rims. Watch for excessive rust or corrosion buildup; cracks in metal; bent flanges or components; loose, missing, or damaged nuts and clamps; broken or stripped studs; and incorrectly matched rim parts. Check valve locator for slip damage and improper location. Correct any problems discovered. For more information, see Chapter 8.

SECTION V
Engine and Electrical

1. **Examine the mounting of the radiator and core.**
   - Inspect the radiator mounting for security and cracks. The radiator core should be clean and free of bugs and debris.

2. **Examine condenser mounting and core.**
   - Inspect condenser for mounting security and cracks. The condenser core should be clean and free of bugs and debris.

3. **Examine engine fan assembly and shroud.**
   - Check for excessive play in fan shaft. Check fan blades and shroud for cracks. For more information, refer to Chapter 6.

4. **Examine vibration damper.**
   - Check vibration damper for cracks, bulges, shifting, or damage. For more information, refer to Chapter 6.

5. **Pressure test cooling system.**
   - Test the system only at the rated pressure of the cooling system and do not exceed it. If the system will not hold pressure, start by looking under the vehicle for any coolant dripping on the ground. Then check radiator caps, hoses, water pump, radiator, heater, and external engine gaskets. Also check for internal engine coolant leaks from head gaskets, oil coolers, and cracked heads and blocks. When checking for internal leaks, follow engine manufacturer's recommendations. For more information, refer to Chapter 5.

6. **Examine coolant clamps and hoses.**
   - Any cracked, swollen, dried out, or non-flexible hoses should be replaced immediately. Squeeze hoses to identify softness and hairline cracks. Hoses with spring inserts should be inspected for both proper location in the hose and damage to the spring. Hose clamps should be secure and as close to the end of the pipe as possible to prevent tube erosion. Clamps should be tightened firmly and replaced if they show any signs of wear. Hose clamps should be replaced at OEM-recommended intervals. For more information on the cooling system, refer to Chapter 5.

7. **Inspect coolant recovery system.**
   - If the cooling system uses a coolant recovery system, inspect to ensure no coolant leaks out of the system or air leaks into the system. If

replacing coolant, remember to flush the recovery system as well as to remove accumulated sediment. For more information on cooling systems, refer to Chapter 5.

8. **Check coolant protection level.**

   ■ Test the freezing point to ensure the coolant is at the correct strength. Optical refractometers are much more accurate than hydrometers for this test. For more information on cooling system testing, refer to Chapter 5.

8A. **Test supplemental coolant additives.**

   ■ Test the coolant chemistry and adjust as required. For more information on cooling system testing, refer to Chapter 5.

9. **Perform a pressure test on the radiator cap.**

   ■ Test the pressure and vacuum relief valves for proper sealing. For more information on cooling system testing, refer to Chapter 5

10. **Examine A/C compressor, mounting, and refrigerant lines.**

   ■ Check the mounting of the compressor for security, condition of mounting bracket, and refrigerant lines for condition and security. For more information, refer to Chapter 10.

11. **Examine alternator mounting and wiring.**

   ■ Check the mounting of the alternator for security and the routing of the wiring harness for the charging system. Make sure it is not rubbing on anything. For more information, refer to Chapter 6.

12. **Inspect drive belt for proper tension and condition.**

   ■ Check engine drive belt for loss of tension and unacceptable condition. For more information, refer to Chapter 6.

13. **Check power steering fluid level and hoses.**

   ■ Check the fluid level in the power steering reservoir and replenish as required. Use only the OEM-recommended power steering fluid. Inspect power steering hoses for physical condition and leaks.

14. **Examine the fuel pump governor and seals.**

   ■ Check the fuel system and governor seals for signs of tampering or leaks.

15. **Examine the fuel pump throttle linkage.**

   ■ Inspect throttle linkage for seal, binding, and smoothness of operation. Ensure that

there are two operational throttle return springs.

16. **Service fuel water separator (if equipped).**

   ■ Follow OEM recommendations. For more information, refer to Chapter 6.

17. **Replace fuel filter.**

   ■ Follow OEM recommendations. For more information, refer to Chapter 6.

18. **Examine engine for external oil leaks.**

   ■ Check all surfaces and gasket areas for signs of leaks.

19. **Examine air induction system.**

   ■ For more information, refer to Chapter 6.

20. **Examine engine exhaust system.**

   ■ Check exhaust manifold and piping for overall condition, broken fasteners, and exhaust leaks. For more information, refer to Chapter 6

21. **Inspect turbocharger.**

   ■ Visually inspect turbocharger for signs of oil, exhaust, or intake leaks.

22. **Examine the wiring harness in the engine compartment.**

   ■ Inspect the engine wiring harness for security and signs of chafing or damage.

23. **Examine battery box(es) cover(s) and mounting.**

   ■ Check the mounting of the battery box for security and latching mechanism for cover. Look for cracks and general cleanliness.

24. **Examine battery hold-down, cables, and connections.**

   ■ The battery terminals must be securely fastened to the battery and kept clean. Both the terminal and the connector should be wire brushed to a tarnish-free condition before assembling the cables to the battery. A baking soda solution can be used to wash away any buildup of corrosion that has occurred. For more information on battery maintenance, see Chapter 12.

25. **Test and record the battery state-of-charge.**

   ■ There are two basic ways of checking the battery's state-of-charge. The first is to measure the specific gravity of the electrolyte. The other is to measure the open-circuit voltage of the battery. The specific gravity check is preferable, but in the case of maintenance-free batteries, an open-circuit voltage check can be substituted. For more information on battery testing, refer Chapter 12.

SECTION VI
Chassis and Undercarriage

1. **Examine steering box mounting.**
   - With someone in the cab rocking the steering wheel, look for movement between the steering box and the frame section it is mounted to.

2. **Examine steering shaft and linkage.**
   - Check steering shaft U-joints, pinch bolts, splines, and Pitnam arm to steering sector shaft. With the front end on safety stands, check linkage and drag line for wear. For more information on steering system maintenance, refer to Chapter 11.

3. **Check kingpins for wear.**
   - With a pry bar in and up/down motion, check for excessive play between kingpin and spindle, and thrust bearing clearance. For more information, refer to Chapter 11.

4. **Check front wheel bearings.**
   - Bump side of tire while listening for a popping sound from the hub area. Rotate the wheel on the hub and listen for noisy bearings.

5. **Check lubricant level in front hubs.**
   - Make sure lube level is correct (at the full line on the hubcap).

6. **Inspect hubcap for leaks.**
   - Inspect hubcap for leaks and damage.

7. **Examine front brake chambers and air lines.**
   - Ensure front brake chambers are securely mounted and not damaged. Inspect brake hoses for chafing or cuts.

8. **Examine front brake lining condition.**
   - The thickness of the front brake lining must be same on both shoes of the brake and as well as on each side of the axle. Brake lining should not be less than 0.25 inch at the thinnest point. For more information, refer to Chapter 9.

8A. **Record the thickness of the front brake linings.**
   - For more information, refer to Chapter 9.

9. **Examine condition of front brake drums.**
   - Brake drum replacement is required if the brake drum is cracked; the brake drum surface is heat checked, ground, or worn beyond the maximum diameter thickness limits; the mounting flange is cracked; the bolt holes are elongated; the drum is known to have been severely overheated; or the drum is out-of-round. For more information, refer to Chapter 9.

10. **Adjust front brakes (manual slack adjusters).**
   - For more information on adjustment procedures, see Chapter 9.

10A. **Check operation of automatic slack adjusters (if so equipped).**
   - Automatic slack adjusters are not a maintenance-free item. See Chapter 9 for additional information.

11. **Examine front springs hangers and shackles.**
   - Check springs for broken or shifted leaves as well as wear to the inside of the hanger bracket caused by rubbing springs.

12. **Inspect engine mounts.**
   - Inspect engine mounts for cracks and engine mounting pads for deterioration.

13. **Examine engine starter motor and electrical connections.**
   - Inspect the starter motor for security and tightness and condition of electrical connections.

14. **Inspect clutch linkage.**
   - Check linkage for looseness or binding. For more information on clutch linkage, refer to Chapter 7.

15. **Examine transmission for fluid leaks.**
   - Inspect transmission case and seals for cracks and leaks.

16. **Inspect transmission breather.**
   - Clean dirt from around the transmission breather vent. Spin the vent cap, and also make sure the vent cap will move up and down slightly. If not, replace the transmission vent.

17. **Examine transmission mounts.**
- Check for cracks at mounts or worn mounting pads.

18. **Check oil level in transmission.**
- Check the transmission oil level and adjust accordingly. For more information on transmission servicing, refer to Chapter 7.

19. **Examine driveline, U-joints, and slip yokes.**
- Examine U-joints for looseness and proper phasing. For more information on transmission servicing, refer to Chapter 7.

20. **Examine rear axle housing(s).**
- Check the rear axle(s) housing for cracks and fluid leaks.

21. **Check axle breather(s).**
- Clean dirt from around the axle breather vent. Spin the vent cap, and also make sure the vent cap will move up and down slightly. If not, replace the axle breather.

22. **Examine the rear hubs for oil leaks.**
- Check for lubricant leaks on both the inner and outer sides of the hubs.

23. **Check rear axle(s) oil level.**
- Check rear axle oil level and adjust accordingly. For more information, refer to Chapter 7.

24. **Examine rear spring and suspension.**
- Check springs for broken or shifted leaves as well as wear to the inside of the hanger bracket caused by rubbing springs.

25. **Examine the rear brake chambers and air lines.**
- Ensure rear brake chambers are securely mounted and not damaged. Inspect brake hoses for chafing or cuts.

26. **Inspect rear lining thickness.**
- The thickness of the front brake lining must be same on both shoes of the brake and as well as on each side of the axle. Brake lining should not be less than 0.25 inch at the thinnest point. For more information, refer to Chapter 9.

26A. **Record rear brake lining thickness (in inches).**

27. **Inspect rear brake drum(s) condition.**
- Brake drum replacement is required if the brake drum is cracked; the brake drum surface is heat checked, ground, or worn beyond the maximum diameter thickness limits; the mounting flange is cracked; the bolt holes are elongated; the drum is known to have been severely overheated; or the drum is out-of-round. For more information, refer to Chapter 9.

28. **Adjust rear brakes (manual slacks).**
- For more information on adjustment procedures, see Chapter 9.

28A. **Check operation of rear automatic slack adjusters (if equipped).**
- Automatic slack adjusters are not a maintenance-free item. See Chapter 9 for additional information.

29. **Check rear wheel bearings.**
- Bump side of tire while listening for a popping sound from the hub area. Rotate the wheel on the hub and listen for noisy bearings.

SECTION VII
Lubrication and Engine Oil Service

1. **Lubricate all grease fittings.**
- With a clean rag, wipe grease fitting before lubricating with OEM-recommended chassis grease.

2. **Lubricate door and hood hinges, latches, and door strikers.**
- Wipe any excessive lubricant from these surfaces.

3. **Lubricate all cables and linkages.**
- Wipe any excessive lubricant from these surfaces.

4. **Change engine oil and filters.**
- Follow federal guidelines for deposal of used oil and filters. For more information on engine oil change, refer to Chapter 6.

5. **Obtain an oil sample (optional).**
- An analysis of used engine oil can be an extremely useful tool in a preventive maintenance program. It can determine the overall condition of lubricants and assist in monitoring engine component conditions.

6. **Service water filter (optional).**
- Follow OEM-recommended procedures. For more information, refer to Chapter 6.

7. **Test run engine to check for oil, water, or fuel leaks.**
   - After starting the engine, inspect oil and fuel filters as well as oil drain plug for signs of leakage.

8. **Recheck engine oil level.**
   - Check engine oil after starting and adjust as necessary.

SECTION VIII

Road Test

1. **Check gear shift operation.**
   - The transmission should shift smoothly. Listen for any unusual noises.

2. **Check the operation of all dash gauges.**
   - Monitor the speedometer, tachometer, and other gauges for proper operation.

3. **Check the steering wheel for play or bind.**
   - Steering should not exhibit excessive play or binding. For more information, refer to Chapter 11 as well as OEM literature.

4. **Test operation of engine electronic controls.**
   - Vehicle and engine should not exceed OEM setting for speed controls.

5. **Check road speed governor.**
   - Check for proper operation of road speed governor and report signs of tampering.

6. **Test cruise control operation.**
   - The cruise control should engage and disengage with all designed methods of control. Ensure the vehicle does not surge when in cruise control mode.

7. **Examine ease of road handling.**
   - Check how well the vehicle tracks down the road in a straight line. Make note of any unusual noises or other problems.

8. **Test service brakes.**
   - Make an application and release of the service brakes. Feel for any brake drag or pulling to one side.

9. **Check for excessive exhaust smoke.**
   - Determine if exhaust smoke is excessive or exceeds state or local limits.

SECTION IX

PMI "C" Supplemental Items

1. **Change power steering fluid and filter.**
   - Power steering fluid and filter (if equipped) should be replaced at OEM-recommended intervals.

2. **Change drive axle(s) lubricant and filter (nonsynthetic).**
   - Rear end fluid and filter (if equipped) should be replaced at OEM-recommended intervals.

2A. **Draw a drive axle oil sample(s) (synthetic).**
   - If the manufacturer has not established a synthetic oil change interval, an oil analysis should be performed.

3. **Change transmission lubricant (nonsynthetic).**
   - Change transmission oil at OEM-recommended interval using only their approved lubricants.

3A. **Draw a transmission oil sample (synthetic).**
   - If the manufacturer has not established a synthetic oil change interval, an oil analysis should be performed.

4. **Remove and inspect the front wheel bearings.**
   - Refer to Chapter 8.

5. **Remove and inspect rear wheel bearings.**
   - Refer to Chapter 8.

6. **Complete an alternator output test.**
   - Refer to Chapter 12 for more information.

7. **Complete a starter draw test.**
   - Refer to Chapter 12 for more information.

8. **Adjust the engine valves.**
   - Always follow the OEM recommendations and specifications when adjusting valves.

9. **Adjust fuel injectors.**
   - Always follow the OEM recommendations and specifications when adjusting fuel injectors.

10. **Check front axle toe-in.**
   - Perform alignment procedures.

11. **Check tandem axle alignment and spacing.**
   - Perform alignment procedures for rear axle.

12. **Service air drier**
   - Follow OEM recommendation for servicing air driers.

SECTION I

Cab Interior Inspection

Sec I	Cab Interior Inspection	A	C
Item			
1	Inspect the ignition key and switch.		
2	Inspect all warning lights and alarms.		
3	Test starter operation.		
4	Check engine operation and record governor rpm.		
5	Inspect instruments and record oil pressure and electrical voltage.		
6	Test wiper and washer operation.		
7	Test operation of electric and/or air horn.		
8	Test hand throttle and shutdown cable operation.		
9	Check A/C, heater, and defrost controls.		
9A	Check bunk climate controls.		
10	Test operation of all interior lighting.		
11	Test the operation of all switches and accessories.		
11A	Check operation of fan clutch (optional).		
12	Check clutch free travel and pedal pad.		
13	Test the operation of the clutch brake.		
14	Test the operation of the parking brake.		
15	Record air governor cutout setting (psi).		
16	Test the operation of the air drier drain valve.		
17	Inspect all safety equipment and decals.		
18	Inspect the windshield and sun visor(s).		
19	Test the steering wheel for excessive play or binding.		

SECTION II

Air Brake Pneumatic System Check

Sec II	Air Brake Pneumatic System Check	A	C
Item			
1	Inspect the air system for leaks with the brakes released.		
2	Inspect the air system for leaks with the brakes applied.		
3	Test the operation of the single and double check valves.		
4	Test the low air pressure warning devices.		
5	Check governor cut-in pressure.		
6	Test the spring brake inversion valve (if equipped).		
7	Test tractor protection valve.		
8	Check compressor air pressure build-up time.		

SECTION III

Cab and Body Walk Around Inspection Lights On

Sec III	Cab and Body Walk Around Inspection Lights On	A	C
Item			
1	Check seat operation and seat covering.		
2	Check floor mats/covering.		
3	Check seat belt and bunk restraints.		
4	Check door locks, latches, and hinges.		
5	Inspect the window operation.		
6	Check steps and grab handles.		

(continued)

Item		A	C
7	Inspect wiper blades and arms.		
8	Inspect mirror mountings, brackets, and glass.		
9	Inspect headlamps, turn signals, fog, clearance, and brake lights.		
10	Check fuel tanks, mountings, lines, and caps.		
11	Inspect air lines, holders, and gladhands.		
12	Examine the condition of the SAE J 560 seven-pin connector, jumper cable, and holder.		
13	Test the seven-pin connector and cable.		
14	Examine the mounting of the fifth wheel.		
15	Check the locking operation of the fifth wheel.		
16	Inspect the mud flaps and hangers.		
17	Inspect pintle hook (if equipped).		
18	Inspect exhaust stack and mounting.		
19	Record any observed physical damage.		

SECTION IV

Tires and Wheels

Sec IV	Tires and Wheels		
Item		**A**	**C**
	Front		
1	Inspect for irregular wear patterns.		
2	Inspect tire sidewall for cuts or damage.		
3	Inspect tire valve caps.		
4	Check and record tire tread depth.		
5	Check and record air pressure.		
6	Inspect for loose wheel lugs.		
7	Inspect wheels for cracks or damage.		
7A	Inspect for slipped wheels on spoke style.		

Item		A	C
	Rear		
8	Inspect for irregular wear patterns.		
9	Inspect tire sidewall for cuts or damage.		
10	Check dual mating with square.		
11	Check tread design match.		
12	Inspect tire valve caps.		
13	Check and record tire tread depth.		
14	Check and record air pressure.		
15	Inspect for loose wheel lugs.		
16	Inspect wheels for cracks or damage.		
17	Inspect for slipped wheels on spoke style.		

SECTION V

Engine and Electrical

Sec V	Engine and Electrical		
Item		**A**	**C**
1	Examine the mounting of the radiator and core.		
2	Examine condenser mounting and core.		
3	Examine engine fan assembly and shroud.		
4	Examine vibration damper.		
5	Pressure test cooling system.		
6	Examine coolant clamps and hoses.		
7	Inspect coolant recovery system.		
8	Check coolant protection level.		
8A	Test supplemental coolant additives.		
9	Perform a pressure test on the radiator cap.		
10	Examine A/C compressor, mounting, and refrigerant lines.		
11	Examine alternator mounting and wiring.		
12	Inspect drive belt for proper tension and condition.		

(continued)

(continued)

Sec V	Engine and Electrical (*continued*)		
Item		**A**	**C**
13	Check power steering fluid level and hoses.		
14	Examine the fuel pump governor and seals.		
15	Examine the fuel pump throttle linkage.		
16	Service fuel water separator (if equipped).		
17	Replace fuel filter.		
18	Examine engine for external oil leaks.		
19	Examine air induction system.		
20	Examine engine exhaust system.		
21	Inspect turbocharger.		
22	Examine the wiring harness in the engine compartment.		
23	Examine battery box(es) cover(s) and mounting.		
24	Examine battery hold-down, cables, and connections.		
25	Test and record the battery state-of-charge.		

SECTION VI

Chassis and Undercarriage

Sec VI	Chassis and Undercarriage		
Item		**A**	**C**
1	Examine steering box mounting.		
2	Examine steering shaft and linkage.		
3	Check kingpins for wear.		
4	Check front wheel bearings.		
5	Check lubricant level in front hubs.		
6	Inspect hubcap for leaks.		
7	Examine front brake chambers and air lines.		
8	Examine front brake lining condition.		
8A	Record the thickness of the front brake linings.		

Item		A	C
9	Examine condition of front brake drums.		
10	Adjust front brakes (manual slack adjusters).		
10A	Check operation of automatic slack adjusters (if equipped).		
11	Examine front springs hangers and shackles.		
12	Inspect engine mounts.		
13	Examine engine starter motor and electrical connections.		
14	Inspect clutch linkage.		
15	Examine transmission for fluid leaks.		
16	Inspect transmission breather.		
17	Examine transmission mounts.		
18	Check oil level in transmission.		
19	Examine driveline, U-joints, and slip yokes.		
20	Examine rear axle housing(s).		
21	Check axle breather(s).		
22	Examine the rear hubs for oil leaks.		
23	Check rear axle(s) oil level.		
24	Examine rear spring and suspension.		
25	Examine the rear brake chambers and air lines.		
26	Inspect rear lining thickness.		
26A	Record rear brake lining thickness (in inches).		
27	Inspect rear brake drum(s) condition.		
28	Adjust rear brakes (manual slacks).		
28A	Check operation of rear automatic slack adjusters (if equipped).		
29	Check rear wheel bearings.		

(continued)

SECTION VII

Lubrication and Engine Oil Service

Sec VII Item	Lubrication and Engine Oil Service	A	C
1	Lubricate all grease fittings.		
2	Lubricate doors, hood hinges, latches, and door strikers.		
3	Lubricate all cables and linkages.		
4	Change engine oil and filters.		
5	Obtain an oil sample (optional).		
6	Service water filter (optional).		
7	Test run engine to check for oil, water, or fuel leaks.		
8	Recheck engine oil level.		

SECTION VIII

Road Test

Sec VIII Item	Road Test	A	C
1	Check gear shift operation.		
2	Check the operation of all dash gauges.		
3	Check the steering wheel for play or bind.		
4	Test operation of engine electronic controls.		
5	Check road speed governor.		
6	Test cruise control operation.		
7	Examine ease of road handling.		
8	Test service brakes.		
9	Check for excessive exhaust smoke.		

SECTION IX

PMI "C" Supplemental Items

Sec IX Item	PMI "C" Supplemental Items	A	C
1	Change power steering fluid and filter.		
2	Change drive axle(s) lubricant and filter (nonsynthetic).		
2A	Draw a drive axle oil sample(s) (synthetic).		
3	Change transmission lubricant (nonsynthetic).		
3A	Draw a transmission oil sample (synthetic).		
4	Remove and inspect the front wheel bearings.		
5	Remove and inspect rear wheel bearings.		
6	Complete an alternator output test.		
7	Complete a starter draw test.		
8	Adjust the engine valves.		
9	Adjust fuel injectors.		
10	Check front axle toe-in.		
11	Check tandem axle alignment and spacing.		
12	Service air drier.		

Glossary

Ackerman arm Steering component that connects a tire rod end to the steering knuckle.

adjustable pliers Tools with a multipositional slip joint that permits a wide spread of jaw dimensions.

adjustable wrench Wrench with one fixed and one movable jaw, which can span a wide range of dimensions. The position of the movable jaw is determined by rotating a helical adjusting worm.

aerobic Requiring the presence of air.

air spring suspensions Any of a number of different types of pneumatic suspension.

American Petroleum Institute Main U.S. trade association for the oil and natural gas industry.

ampere-hour rating Amount of current that a fully charged battery can feed through a circuit before the cell voltage drops to 1.75V. In a typical 6 cell, 12-V battery, this would be equal a battery voltage of 10.5V.

anaerobic In the absence of air.

anti-lock braking system (ABS) Electronically managed hydraulic or pneumatic air brake system that senses wheel speeds and manages application pressures to prevent wheel lock-up.

applied stroke Movement of the slack adjuster while a brake application is being made.

aqueous Solution in which a component(s) is dissolved in water.

articulation Movement permitted between two components connected by a pivot. For example, a tractor/semitrailer articulates at the fifth wheel connection.

axial runout Wobble from an axial plane of a rotor, shaft, or wheel. For example, in a wheel, axial runout is the wobble as viewed from the front or rear of the vehicle.

ball joint Spherical bearings that connect the control arms to the steering knuckles. More specifically, a ball joint is a steel bearing stud and socket enclosed in a steel casing. The bearing stud is tapered and threaded and fits into a tapered hole in the steering knuckle.

barn style Type of trailer rear or side door assembly.

barrel tank vessel including shell and heads.

BCI (Battery Council International) Trade organization that brings together the leading lead-acid battery manufacturers in North America and other major players from around the world. Externally, BCI provides information and resources on the industry to numerous outside organizations and researchers.

bias ply Tire construction geometry in which the plies are laid either diagonally or at an angle of 35 degrees.

bleeding the fuel system Act of removing air from a fuel system. Bleeding the fuel system is necessary when the fuel tank as been allowed to run dry. The injection must be brimmed and the air bleed (removed) from the fuel system.

bolster plate Term for the support pad on a trailer upper coupler, also known as a upper coupler pad. Bolster plates are normally rigid but in certain specialty applications, such as trailer dumps, may be designed to oscillate.

box-end wrench Hex or double-hex wrench designed to completely enclose a bolt or nut and evenly apply torque to it.

brake fade Vehicle braking system fade, or brake fade, is the reduction in stopping power that can occur after repeated application of the brakes, especially in high-load or high-speed conditions.

brazier head Type of buck rivet.

breather Device that allows a circuit to be vented in to the atmosphere. It is used on reservoirs, axles, and so on.

bucking bar Tool used to produce the bucked head on a buck rivet.

camber Position of a wheel assembly when viewed head-on from the front of the vehicle. A wheel that leans outward at the top (away from the vehicle) has positive camber and one that leans inward at the top has negative camber.

canister Filter element that is positioned in a reusable housing or case.

cartridge Filter and filter housing built into one spin-on unit.

caster Forward or rearward tilt of the kingpin centerline to vertical when viewed from the side. Zero caster indicates that the kingpin is located at true vertical. Positive caster occurs when the kingpin tilts rearward (leans behind) and negative caster occurs when the kingpin tilts forward.

center support bearings Hanger bearing used to suspend the first section of driveshaft in a double driveshaft assembly. It is used on truck drivelines when the distance between the transmission and drive carrier exceeds 70 inches or so. The center support bearing consists of a sealed bearing assembly, supported by a rubber insulator, mounted in a clamp assembly.

centrifuge Piece of equipment, generally driven by a motor, that puts an object in rotation around a fixed axis, applying a force perpendicular to the axis. The centrifuge works using the sedimentation principle, where the centripetal acceleration is used to separate substances of greater and lesser density.

chisels Impact tool, hardened steel bar dressed with a cutting edge and driven by a hammer.

clutch Device used to break torque flow between an engine and transmission, and to apply/release application force in other units.

clutch break Circular disc with friction surfaces on both sides mounted on the transmission input shaft. Its purpose is to slow or stop the input shaft during initial shifts to prevent gear clash.

coalesced In fuel, when small droplets of water come together to combine into larger droplets.

cold cranking amps Amperes that a battery can sustain for 30 seconds at $0°F$ ($-17.8°C$).

combination wrench Wrench that has an open end on one side and box end on the other, with both ends of the same nominal dimension.

commercial vehicle operator's registration (CVOR) CVO Registration creates a Commercial Vehicle Operator Record. Once an operator record is set up, licensing and compliance data (accidents, convictions, inspections) are stored on the record. A record may also be created for a non-CVO registrant should a conviction, accident, or inspection event be received for the operator.

Commercial Vehicle Safety Alliance (CVSA) Nonprofit organization that defines out-of-service (OOS) criteria used by vehicle safety enforcement officers across North America (United States, Mexico, and Canada) to standardize truck safety standards. Nonprofit organization: publishes OOS criteria and inspection forms online and on DVDs that can be obtained from http://www.cvsa.org.

compensating fifth wheel Fifth wheel designed to articulate fore and aft, and side-to-side. It is used to limit trailer torque effect transfer: the Holland Kompensator would be an example.

compressor pump down Service procedure used to remove refrigerant from the compressor crankcase so that service work may be preformed on the compressor. For example, replacing compressor oil.

ConMet PreSet Preset hub assembly that eliminates the need for the technician to adjust wheel bearings. A ConMet hub is installed, then torqued to 300 ft lbs.

conspicuity tape Nighttime reflective tape required on trucks, trailer, school buses, and highway equipment to maximize their conspicuity on the highway.

convoluted Style of suspension air spring.

coolant A coolant is a fluid that flows through a device in order to prevent it from overheating and transferring the heat produced to other devices that utilize or dissipate it.

corrosive Substance capable of chemically attacking another.

coupler Something that links two components or circuits, such as the fitting used in hydraulic circuits. Another example of couplers would be gladhands, or a style of connecting a tractor to a trailer or a trailer to a trailer. In most cases, couplers have a male and female component.

dampening To buffer or insulate.

deadline To take a truck or trailer out of service (OOS) because of a safety or running defect. The CVSA defines most truck chassis OOS standards.

deburr To remove sharp or rough edges from a cut.

deep cycle The rapid, continuous, full-discharging and recharging of batteries.

Department of Transportation (DOT) Federal government agency that oversees all matters related to highway and transportation safety. Oversees NHTSA and can be accessed online at http://www.dot.gov.

diagonal cutting pliers *See* sidecutters.

differential carrier Drive axle assembly that houses the differential gearing.

distilled water Utilizing a water distiller, water is heated to its boiling point, thus producing steam and thereby separating impurities from the water.

dog track Misalignment condition in which the rear wheels of a truck run on a different track line than the front steering wheels. This creates a thrustline irregularity. More often seen on tractor-semi combinations in which the trailer travels at an angle to the tractor—usually caused by bogie to kingpin misalignment.

drag link Connecting linkage between the steering gear Pitman arm and the steering control arm, it is sometimes adjustable.

drawbars Mating part to the pintle hook or coupler in the coupling system.

driver's inspection report Pre-trip and post-trip inspections that are mandatory tasks of the vehicle operator.

driveshaft Assembly of one or more propeller shafts (hollow tubes) connected by universal joints (U-joints) to drive-and-driven driveline components. In a typical tandem drive highway tractor, driveshafts are used to transmit torque from the transmission and forward differential carrier, and between the tow differential carriers.

dry positive filtration In regards to air filters, all of the air entering the intake system must pass through the filtering media.

elastic A term describing reversible deformations of materials.

electrolyte Electrically conductive substance in a voltaic cell that interacts between electrodes (anode and cathode). The electrolytic media in a standard lead-acid battery is a solution of 34 percent sulfuric acid and 66 percent pure (distilled) water.

elevating fifth wheel Style of fifth wheel usually found on shunt tractors.

emulsified Dispersion of one liquid into another, such as water, in the from of fine droplets into diesel fuel.

Environmental Protection Agency (EPA) Federal regulating body the sets and monitors noxious emission standards among other functions.

ethylene glycol (EG) Type of antifreeze solution.

extended life coolant (ELC) Type of antifreeze solution.

fast charging Rate of charging a battery using high amperage for approximately 2 hours.

fifth wheels Lower coupler assembly used on most highway tractors and lead trailers. It typically consists of a pivot plate and locking jaws that engage to the kingpin upper coupler on a trailer.

file Hand tool used to remove small amounts of metal for shaping, smoothing, or sharpening parts.

flammable liquids Capable of burning.

flux Substance used in the joining of metals when heat is applied to promote the fusion of metals.

footprint Critical contact interface between a tire and the road surface it is in contact with.

free state Water in fuel that appears in large globules and because of its greater weight than diesel will readily collect in puddles at the bottom of fuel tanks or storage containers.

freestroke Brake slack adjuster travel.

FRP (fiberglass reinforced plywood) Description of a type of van trailer or truck box wall construction.

full field test Method of testing the actual output of the alternator.

fully oscillating fifth wheel A fifth wheel type with fore, aft, and side-to-side oscillation.

gasification Process that converts carbonaceous materials, such as coal, petroleum, or biomass, into carbon monoxide and hydrogen by letting the raw material react with a controlled amount of oxygen and/or steam at high temperatures.

gasses To give off gas.

gladhand Pneumatic couplers used to connect tractor air and brake systems with those on trailers. When coupled, a pair of gladhands have the appearance of a handshake.

glazing To wear to a mirror finish.

governor cut-in pressure Pressure at which the air compress starts to pump air into the reservoirs.

governor cut-out pressure Pressure at which the air compressor stops compressing air.

gross axle weight rating (GAWR) Total weight of a fully equipped vehicle and its load. GVW defines much of the specified equipment on a commercial vehicle, including engine, transmission, brakes, axles, suspension, frame, and so on.

hacksaw Bowed frame saw with a replaceable blade used for cutting metals.

hanger bearings Center support bearings used to suspend the first section of driveshaft in a double driveshaft assembly. It used on truck drivelines when the distance between the transmission and drive carrier exceeds 70 inches or so. Hanger bearings consist of a sealed bearing assembly, supported by a rubber insulator, mounted in a clamp assembly.

harmonic balancer (vibration damper) Component used to reduce the amplitude of vibration to the flywheel's mass.

hazardous materials Substances that can have a harmful effect on a person's health.

head Vertical height of a column of liquid above a given point. It is expressed in linear units (feet). Head can be used to express gauge pressure. Head pressure is equal to the height of a liquid in a column, multiplied by the density of the liquid.

heating ventilation and air conditioning (HVAC) Climate control system for a truck or bus.

height control valve Simple lever-actuated valve used on an air suspension system to automatically maintain chassis ride height at a preset dimension. It provides the suspension with its adaptive capability. Also known as a leveling valve.

Heli-Coil™ Insert used to repair blind internal threads.

hex Term used to describe a six-point fastener head or a socket or wrench used to hold or apply torque to the fastener. Any bolt head with six flats in a hex head. A hex head bolt can be gripped on 6 angular planes by a hex wrench and 12 angular planes by a double-hex wrench.

high hitch Situation where the locks of the fifth wheel close and lock only around the lower portion of the kingpin.

hinged doors *See* barn style.

hub-piloted Mounting style for disc wheels.

Huck fastener Commonly utilized production line fastener that ensures accurate clamping pressure. It uses air-over-hydraulic pressure to establish stud clamping force before locking with a crimp type fastener. The Huck fastener, is often robotically installed on an assembly line. It is less commonly used in service shops because of the bulk and awkwardness of the installation tooling, and so is replaced with nuts and bolts.

hydrometer Device used to measure the specific gravity of a coolant used in determining is freeze protection. Specific hydrometers may also be used to test a batteries state of charge.

hydrostatic pressure test Type of leakage test preformed on tank trailers.

hygroscopic Readily absorbing and retaining moisture.

impact sockets Heavier wall, softer steel sockets designed for use with impact wrenches.

impact wrench Another term for impact gun. The impact wrench is any of various different types of hammer wrenches. It is usually pneumatically driven, though electric versions exist. Impact wrenches drive impact sockets. The common size impact wrench used by truck technicians uses a 1/2-inch drive lug.

inboard Style of braking system in which brake drums require the removal the entire hub assembly in order to remove the bolts securing the brake drum to the wheel assembly.

interaxle differential Differential gearing used between the drive axles on a tandem-drive, highway tractor, it divides torque between the two drive axles. It is also known as a power divider.

interior cube Capacity of a tuck or van or volume they able to hold.

International Organization for Standardization International standard-setting body composed of representatives from various national standards organizations. Founded on 23 February 1947, the organization promulgates worldwide proprietary, industrial, and commercial standards.

jounce Most compressed condition of a spring. Literally, the spring's bounce.

jump starting Practice of using a second vehicle or a generator to start a vehicle with dead batteries.

kingpin (1) The linkage pin used in a steering knuckle. Which permits the assembly to pivot, or the tempered steel, coupling pin mounted on a trailer upper coupler that engages with fifth wheel jaws; (2) the lugged pin built into a trailer upper coupler assembly, which engages to jaws in a fifth wheel assembly, permitting articulation at the fifth wheel. The kingpin is forged from middle alloy steels and tempered.

kingpin setting Distance that a trailer kingpin is positioned from the front edge of the bolster plate.

leaf spring Any of a number of types of steel springs consisting of single or clamped multiple plates used in truck and trailer suspensions.

liftgate Hydraulically powered elevating structure on the rear of some trucks and trailers used to load or offload products from the vehicle.

loaded Condition of air compressor when it is pumping air. When the air compressor is unloaded it is not pumping air.

load test (battery) A common battery capacity test performed by connection a carbon pile load tester across the battery or battery bank terminal. Also known as a capacity test.

locking pliers Better known as Vice-Grips™.

lockstrap Removable locking tong that holds the clutch ring in position. It has to be removed, and the clutch ring rotated to make a clutch adjustment.

low side pump down Term used in a refrigeration system when the receiver tank outlet valve had been front seated and all refrigerant in the system is now trapped between the compressor outlet and the receiver tank. Anything downstream of the receiver can be serviced when the unit is in this state.

Lunette eye A steel eye mounted on the drawbar of a trailer or dolly designed to couple with a pulling vehicle having a pintle hook.

Magnehelic gauge A tool used to measure both positive and negative low air or gas pressures.

Material Safety Data Sheet (MSDS) Data information sheet, mandated by WHIMIS, that must be displayed on any known hazardous substance.

maximum allowable working pressure (MAWP) Pressure rating for tank trailers that cannot be exceeded.

National Highway Traffic Safety Administration (NHTSA) Agency of the executive branch of the U.S. government, and part of the Department of Transportation. It describes its mission as "Save lives, prevent injuries, reduce vehicle-related crashes."

needlenose pliers Pliers with slim tapered jaws used for electrical work and grasping small objects.

neutralize Make (something) ineffective.

oil bath type A style of engine air filter.

open circuit voltage test What is measured when a battery is not delivering or receiving power. For a fully charged 12-volt battery, the reading should be 12.66 volts, or 2.11 volts per cell.

open-end wrench Wrench with parallel jaws in the shape of a Y. It contacts a hex nut on two opposing flats only at a given moment.

original equipment manufacturer (OEM) Typically, a company that uses a component made by a second company in its own product, or sells the product of the second company under its own brand.

oscillate Act of oscillating. This word is used to describe the natural dynamics of frame rails, the vibrations of drivelines, wheel dynamics, and electrical wave pulses.

outboard Brake system style where the brake drums are easily replaced without removing the hubs.

out of service (OOS) To withdraw a truck or trailer from operation because of a safety or running defect. Also known as deadline.

outrigger Structural, load-carrying members attached to and extending outward from the main longitudinal frame members of a trailer.

parts requisition Soft (electronic) or hard (on paper) order for parts from within a company or for purchase from outside.

pencil Inspection the driver or the technician responsible for the inspections checks of the item on the PM checklist without actually verifying the condition.

permeation To penetrate through the pores.

Phillips screwdriver Screwdriver tip with four radial prongs in the shape of a cross, designed to engage with a similarly shaped recess in a screw.

pintle hooks Standard coupling mechanism used in truck full-trailers. The function of a pintle hook is to engage to the drawbar eye, which itself is bolted to the tongue or drawbar assembly of the trailer.

Pitman arm Steering linkage arm that connects a steering gear to a drag link. A Pitman arm is usually splined to the steering gear sector shaft. In a typical steering gear, the Pitman arm converts the rotary motion of the sector shaft into linear motion at the drag link, plus adds leverage.

placard holders Sign holder designed to hold warning signs affixed to the body of trucks and trailers. They are used to identify dangerous cargo.

pliers Gripping tool with a pair of jaws mounted on a pivot. There are numerous types of pliers, including lineman, needlenose, and slip-joint.

PM form Preventive maintenance form in the form of a checklist that is followed by a technician so that none of the areas to be serviced or inspected are overlooked.

pneumatics In truck technology, this term is used mainly to describe chassis air under pressure, as in pneumatic brakes or pneumatic suspension.

Posi-Drive™ screwdriver A screwdriver similar in appearance to a Phillips, but with a blunter tip—used a lot in robotic assembly.

positive filtration Describe a filter that operates by forcing all the fluid to be filtered through the filtering medium.

post Vertical structure that connects the upper and lower rails together in a unibody style trailer. Wall panels are also attached to the posts.

power take off (PTO) Any of a number of different means of gear-driving auxiliary apparatus, such as hydraulic pumps, compressors, and cement mixer cylinders, using engine power. A PTO can be mounted to the engine, transmission (most common), transfer case, or elsewhere on the vehicle drivetrain.

preload Predetermined force applied to a component such as a bearing on assembly, usually with the objective of preventing unwanted (end) play in operation.

pre-trip inspection Safety inspection of the vehicle that must be performed by the driver of the vehicle every 24 hours.

preventative maintenance Service (PMS) Act of maintaining a vehicle and repairing small problems before they become catastrophic.

primary reservoir Air reservoir that stores air pressure for use on the primary braking system. This would be the drive axles in a tractor.

propylene glycol (PG) Type of antifreeze solution.

pull-type clutch Any clutch that pulls the release bearing toward the transmission to release it.

punches Impact tool used to drive out pins, rivets, or shafts, or to scribe/mark components for identification.

push-type clutch Any clutch that is disengaged by forcing the release bearing toward the clutch plate where it acts on release fingers.

radial tires Tire with circumferential belts in which the plies run at right angles to the beads.

ratchet Hand tool that drives a socket wrench.

reactive Something that alters its physical or chemical status after contact with another substance.

rebound Reactive response of a spring—after being compressed it kicks back. Suspensions attempt to limit jounce and rebound cycles by using dampening devices such as shock absorbers.

refractometer Device used to measure the refractive index of a solution such as battery electrolyte or antifreeze solution.

release bearing Usually refers to the clutch release mechanism actuated by clutch pedal movement.

repair order Form usually written by a service manager to instruct technicians of the work to be preformed on a vehicle.

reserve capacity rating Amount of time a vehicle can be driven with its headlights on in the event of a total charging system failure. The current output will vary with the type of vehicle electrical system but the critical low voltage valve is 10.5V.

Resource Conservation and Recovery Act (RCRA) Unites States' primary law governing the disposal of solid and hazardous waste.

reversible sleeve Style of suspension air spring.

Right-to-Know Law Hazard communication federal legislation administered by OSHA. It requires any company that produces or uses hazardous chemicals or substances must inform its employees and customers of the potential danger. Designed to ensure transparency of handling hazards in the workplace.

rivet Any of a number of different types of one-off use fasteners used to clamp plate For example, a bucked rivet has an oval crown head on one side and after insertion of a hole through both plates. The head on the other side is formed by a pneumatic impact against the bucking bar.

rivet set Tool used in a rivet gun to drive rivets against a bucking bar.

roll-up door Sectioned or slatted van door on runner rails that is opened vertically, assisted by spring force.

sacrificial anode Metallic anode used in cathodic protection where it is intended to be dissolved to protect other metallic components.

sand shoes Bottom component of a trailer landing gear leg. It is designed to spread the weight out over a larger area.

screw extractor Hand tool used to remove a bolt or stud that has broken below the surface.

scrubbing Tire terminology for a tire being dragged sideway on the road.

scuff Tire terminology fot skidding over the road.

secondary reservoir Air reservoir stores for front brakes, suspension, and all accessories.

semi-absorbed In regards to diesel fuel, semi-absorbed water is usually water in solution with alcohol—a direct result of the methyl hydrate (type of alcohol added to fuel tanks as deicer or in fuel conditioner) added to fuel tanks to prevent winter freeze-up.

semi-oscillating fifth wheel Standard that articulates on pivot pins only in a fore/aft direction.

semi-trailer Any trailer that has part of its weight supported by the towing vehicle. Most highway tractors haul semi-trailers, coupled by fifth wheels that support a percentage of the trailer weight depending on how the axles are arranged.

sidecutters Pliers with cutting edges on offset jaws, used to strip/cut electrical wiring, steel wire, and cotter pins.

slip splines (slip joint) The splined section of a driveshaft or steering column. A slip joint is designed to accommodate variations in shaft length but has no axial play.

slow charging Rate at which a battery is charged.

socket wrenches An enclosed cylindrical wrench forged with a hex opening, designed to engage with a bolt or nut on one end, and a square drive lug recess on the other. Socket wrenches are designed with a full range of standard and metric nominal sizes and are further classified by the size of drive lug required to turn them—1/4, 3/8, 1/2, 3/4, 1-inch drive sizes are typical.

solvents Substances that dissolve other substances.

specific gravity (SG) The weight of a liquid or solid versus that of the same volume of water. A hydrometer is used to make SG measurements.

spider The axle hub mounting to which truck foundations brakes are mounted. Also a slang term for a U-joint cross or trunnion.

spontaneous combustion Substance with a relatively low ignition temperature begins to release heat. This may occur in several ways, for example, through oxidation, fermentation.

spring pack Multileaf, steel spring assembly with plates clamped together with a center-bolt to form an elliptical shape.

steering control arm Forged link that connects a drag link to the steering knuckle, Also known as a steering arm and a control arm.

steering gear Steering control gearing of a chassis steering system. Steering input is from a driver-controlled steering column, output is through a sector shaft. The steering gear is also known as a steering box.

steering knuckle Pivot point of the steering system, which allows the wheels to turn

stud-piloted Style of disc wheel mounting.

stud removers Tool consisting of hardened, knurled eccentric jaws that grip the shank of a stud. It uses both jaws to remove and install studs.

supplemental coolant additive (SCA) Prevents scaling and corrosion when mixed with coolant in a truck's cooling system.

supplemental restraint system (SRS) Usually referred to as an air bag.

supply reservoir (wet tank) In a vehicle air break circuit, the supply tank is the first reservoir to receive air from the air compressor. The supply tank feeds the primary, secondary, and auxiliary air tanks in the circuit.

swinger doors *See* barn style.

Technical and Maintenance Council (TMC) Division of the American Trucking Association (ATS) responsible for setting maintenance and equipment standards in the U.S. trucking industry. Formerly (before 2005) known as The Maintenance Council.

thread chaser Tool used for cleaning or restoring existing threads using a tap or die.

throat Any kind of inlet passage such as the entry guide channel to the jaws of a fifth wheel.

throw-out bearing Another term for a release bearing, the linkage actuated disengagement mechanism used in both push-and pull-type clutches.

tire square Tire tool used for checking dual diameter matching on the vehicle.

toe Steering geometry dimension that specifies the distance between the extreme front and extreme rear of a pair of tires on either end of the axle.

toe-in When the extreme front of a pair of tires on an axle points inward toward the vehicle.

toe-out When the extreme front of a pair of tires on an axle points outward, away from the vehicle.

torque-to-yield Fasteners that intentionally torqued just barely into a yield condition, although not quite enough to create the classic Coke bottle shape of a necked-out bolt.

towed vehicle weight (TVW) Main factor for selecting a coupling device, pintle hook, or drawbar. The component should be specked out to the greatest TVW expected.

toxic Poisonous.

TransSynd Brand of synthetic transmission oil.

trunnions In U-joints, the cross or spider assembly with bearing races machined on the ends. Also, the suspension sub-frame assembly used on some tandem drive highway tractors.

two-stage filtering Used when a primary and secondary filter are used in series.

U-joint Assembly consisting of a trunnion cross and bearings that connect a pair of yokes for purposes of transmitting drive torque on an angled plane.

ultrasonic thickness tester Tool used to locate leaks or weak points in a tank vessel.

unibody Strong yet light method of manufacturing van trailers.

Unitized Hub System (UHS) A hub assembly that is factory filled with synthetic grease, eliminating the need for bearing adjustment and seal replacement when removed from the axle.

unsprung weight Weight of a vehicle not supported by the suspension. Ideally, this weight is kept as low as possible because unsprung weight forces react through the suspension.

upper coupler Kingpin deck plate on a trailer that rides on the fifth wheel in a tractor or lead trailer. The upper coupler is built into the sub-frame of the trailer and the kingpin may be welded or bolted (riveted in much older trailers) into the upper coupler assembly.

vibration damper *See* harmonic balancer.

Vise Grips™ Locking pliers manufactured with various different types of jaw design. It is a very commonly used multipurpose tool.

wear compensator Component that automatically adjusts for wear of the clutch friction surface each time the clutch is actuated.

wet tank Commonly used slang term for supply tank. *See* supply reservoir.

Workplace Hazardous Material Information Systems (WHMIS) National system designed to ensure that all employers obtain the information that they need to inform and train their employees properly about hazardous materials used in the workplace.

wrench Tool for turning or holding fastener heads (nuts and bolts). The width of the jaws or span determines its nominal size.

yield point Expressed in psi, it is the highest stress a material can withstand without permanent deformation. In steels, yield strength is typically about 10 percent less than tensile strength.

yoke Any of a number of different fork-shaped components. For example, shift yokes (on shift rails) fit on to dog clutches in a transmission to move them in and out of engagement; a pair of driveshaft yokes can be used when coupled to each other using a U-joint as an intermediary.

zerk-type standard greased nipple Its orifice is sealed with a spring-loaded ball that unseats when pressurized grease is applied to it.

zipper Term used to describe the sound that can be made by a truck tire when the side wall fails under pressure. It sounds like the sound a zipper makes a fraction of a second before it explodes.

Index

Note: Page numbers referencing figures are italicized and followed by an "*f*." Page numbers referencing tables are italicized and followed by a "*t*."